Progress in Mathematics
Vol. 25

Edited by
J. Coates and
S. Helgason

Birkhäuser
Boston · Basel · Stuttgart

Phillip A. Griffiths

Exterior Differential Systems and the Calculus of Variations

1983

Birkhäuser
Boston • Basel • Stuttgart

Author:

Phillip A. Griffiths
Department of Mathematics
Harvard University
Cambridge, MA 02138

Library of Congress Cataloging in Publication Data

Griffiths, Phillip.
Exterior differential systems and the calculus of variations.

(Progress in mathematics ; v. 25)
Includes index.
1. Calculus of variations. 2. Exterior differential systems. I. Title. II. Series: Progress in mathematics (Cambridge, Mass.) ; v. 25.
QA316.G84 1982 515'.64 82-17878
ISBN 3-7643-3103-8

CIP-Kurztitelaufnahme der Deutschen Bibliothek

Griffiths, Phillip A.:
Exterior differential systems and the calculus of variations / Phillip A. Griffiths. - Boston ; Basel ; Stuttgart : Birkhauser, 1982.
(Progress in mathematics ; Vol. 25)
ISBN 3-7643-3103-8

NE: GT

ISBN 3-7643-3103-8
Printed in USA

To the memory of my mother

Jeanette Field Griffiths

TABLE OF CONTENTS

LIST OF COMMONLY USED NOTATIONS

(Note: The references for the undefined terms used below may be found in the index.)

$A^*(X)$	Exterior algebra of smooth differential forms on a manifold X
$\{\Sigma\}$	Algebraic ideal in $A^*(X)$ generated by a set Σ of forms on X
(I,ω)	Exterior differential system with independence condition
$\mathcal{V}(I,\omega)$	Set of integral manifolds of (I,ω)
$T_N(\mathcal{V}(I,\omega))$	Tangent space to $\mathcal{V}(I,\omega)$ at N
$(I,\omega;\varphi)$	Variational Problem (cf. Chapter I, Sec. a)
$\Phi:\mathcal{V}(I,\omega)\to\mathbb{R}$	Functional on $\mathcal{V}(I,\omega)$
$\delta\Phi:T_N(I,\omega)\to\mathbb{R}$	Differential of Φ
$\mathcal{V}(I,\omega;[A,B]))$	Subset of $\mathcal{V}(I,\omega)$ given by endpoint conditions
$T_N(\mathcal{V}(I,\omega;[A,B]))$	Tangent space to $\mathcal{V}(I,\omega;[A,B])$
$\equiv \bmod I$	Congruence modulo an ideal $I\subset A^*(X)$
$\overline{\overline{\equiv}}$	Congruence modulo the image of $I\wedge I\to A^*(X)$ (cf. (II.b.4))
$\mathbb{P}E$	Projectivization of a vector space E
θ_N	Restriction of $\theta\in A^*(X)$ to a submanifold $N\subset X$
$d\theta=\Theta$	Exterior derivative of a differential form; little θ is frequently denoted by capital Θ
$\mathcal{F}(\cdot)$	Frame manifold
$L_v\varphi$	Lie derivative of a form φ along a vector field v
Y	Momentum space associated to $(I,\omega;\varphi)$
Q	Reduced momentum space associated to $(I,\omega;\varphi)$

INTRODUCTION

This monograph is a revised and expanded version of lecture notes from a class given at Harvard University, Nankai University, and the Graduate School of the Academia Sinica during the academic year 1981-82. The objective was to present the formalism, together with numerous illustrative examples, of the calculus of variations for functionals whose domain of definition consists of integral manifolds of an exterior differential system. This includes as a special case the Lagrange problem of analyzing classical functionals with arbitrary (i.e., non-holonomic as well as holonomic) constraints. A secondary objective was to illustrate in practice some aspects of the theory of exterior differential systems. In fact, even though the calculus of variations is a venerable subject about which it is hard to say something new,(1) we feel that utilizing techniques from exterior differential systems such as Cauchy characteristics, the derived flag, and prolongation allows a systematic treatment of the subject in greater generality than customary and sheds new light on even the classical Lagrange problem.

As indicated by the table of contents the text is divided into four chapters, with most of the general theory being presented in the first and last. We break somewhat with current tradition in that an unusually large amount of space is devoted to examples. Perhaps even more of a break (or is it a regression?) is the special concern given to the explicit integration of the Euler-Lagrange equations, Jacobi equations, Hamilton-Jacobi equations, etc. in these examples—in a word we want to get out *formulas*. Much of the middle two chapters are devoted to methods for doing this; again the theory of exterior differential systems provides an effective computational tool.(2)

For reasons of space, and even moreso because the several variable theory is incomplete at several crucial points, the discussion is restricted to the case of one independent variable; i.e., we consider functionals defined on integral *curves* of an exterior differential system.

We will now describe an example that may help motivate developing the theory in such generality. Let $\gamma \subset \mathbb{E}^n$ be a smooth curve given parametrically by its position vector $x(s) \in \mathbb{E}^n$ viewed as a function of arclength. It is well-known that in general γ has curvatures $\kappa_1(s), \ldots, \kappa_{n-1}(s)$ that are Euclidean invariants and that uniquely determine γ up to a rigid motion (when $n = 3$ these are the usual curvature and torsion). We consider a functional

$$\Phi(\gamma) = \int_\gamma L(\kappa_1(s), \ldots, \kappa_{n-1}(s))\, ds \tag{1}$$

and ask standard questions such as i) find the Euler-Lagrange equations and explicitly integrate them if possible; ii) find the Jacobi equations and information on conjugate points; and iii) if $L = L(\kappa_1, \ldots, \kappa_r)$ depends only on the first r curvatures and if the matrix $\|\partial^2 L/\partial\kappa_i \partial\kappa_j\|_{1 \leqq i, j \leqq r} > 0$, then show that a solution to the Euler-Lagrange equations having no conjugate points is a local minimum for (1). It is clear that this problem may be set up in coordinates as a classical higher order variational problem, and it is equally clear that in this formulation the resulting computations will be quite lengthy. Alternatively, we may consider the Frénet frame associated to γ as a curve N in the group $E(n)$ of Euclidean motions. Then N is an integral manifold of a left invariant exterior differential system (I, ω) on $E(n)$, and (1) may be viewed as an invariant functional defined on *any* integral manifold of (I, ω). Once the general formalism of the calculus of variations is in place for functionals defined only on integral manifolds of differential systems, we may hope that in examples such as this the theory should provide an effective computational tool. For instance, it is known that the classical theory of rigid body motion extends to Lagrangians defined by left-invariant metrics on any Lie group (theory of Kirilov-Kostant-Souriau; cf. [50] and [61]), and it is reasonable to try to further extend this theory to invariant functionals defined only on integral manifolds of invariant exterior differential systems and apply the result to the study of (1). This will be done in Chapter III.[3]

We shall now describe in more detail some of the contents of this monograph, where we refer to the text for explanation of notations and undefined terms (there is an index at the end).

Chapter 0 is preliminary and is intended only for reference. (It is suggested that the reader begin with Chapter I.) In it are first collected some terminology and notations from standard manifold theory. Next there is a very brief description of the language of jet manifolds and of moving frames. The former provides a useful formalism for introducing derivatives as new variables (cf. [31], [38], [43], and [62]). The latter is especially relevant due to the fact that a general curve in many homogeneous spaces G/H have a "Frénet frame"; i.e., a canonical lifting to G (cf. [34], [44]), and consequently the aforementioned analysis of the functional (1) may be expected to reflect rather general phenomena. Finally, in Chapter 0 we record some of the definitions and elementary facts from the theory of exterior differential systems. Again this is only meant to establish language; the more substantial aspects of the theory are introduced as needed during the text.(4)

In Chapter I we explain the basic setup and derive the main equations of the theory, the Euler-Lagrange equations. Assume given an exterior differential system (I,ω) on a manifold X and denote by $V(I,\omega)$ the set of integral manifolds $N\subset X$ of (I,ω). For an example in addition to the Frenet liftings mentioned above, we consider the Lagrange problem:(5) Let $J^1(\mathbb{R},\mathbb{R}^m)$ denote the space of 1-jets of maps from $\mathbb{R}$ to $\mathbb{R}^m$. On $J^1(\mathbb{R},\mathbb{R}^m)$ we have a natural coordinate system $(x;y^1,\ldots,y^m;\dot{y}^1,\ldots,\dot{y}^m)$ and canonical differential ideal I_0 generated by the Pfaffian forms

$$\theta^\alpha = dy^\alpha - \dot{y}^\alpha\, dx \qquad \alpha = 1,\ldots,m\,. \qquad (6)$$

Setting $\omega = dx$, $V(I_0,\omega)$ consists of 1-jets $x\to(x,y(x),\frac{dy(x)}{dx})$ of parametrized curves in $\mathbb{R}^m$. Let $X\subset J^1(\mathbb{R},\mathbb{R}^m)$ be a submanifold and let (I,ω) be the restriction of (I_0,ω) to X. We may think of X as defined by equations

$$g^\rho(x,y,\dot{y}) = 0\,, \qquad (2)$$

and then $V(I,\omega)$ consists of 1-jets of parametrized curves that satisfy the constraints

$$g^\rho\left(x,y(x),\frac{dy(x)}{dx}\right) = 0\,.$$

A special case is when the constraints (2) are of the form

$$g^\rho_\alpha(y)\dot{y}^\alpha = 0\,.$$

Then they correspond to the sub-bundle $W^* = \text{span}\{g^\rho_\alpha(y)dy^\alpha\}$ of the co-tangent bundle of $\mathbb{R}^m$ (or dually to a sub-bundle of the tangent bundle; i.e., a distribution). (Note: In general on a manifold M the differential ideal generated by the sections of a sub-bundle $W^* \subset T^*(M)$ will be called a *Pfaffian differential system*. In this text essentially all differential ideals will be of this type. However, they will usually be defined on manifolds lying over the one of interest.) Another special case of (2) is given by the canonical embeddings

$$J^k(\mathbb{R},\mathbb{R}^\ell) \subset J^1(\mathbb{R},\mathbb{R}^{k(\ell)})$$

of higher jet-manifolds into 1-jets.[7]

Returning to the general situation, on X we assume given a differential form φ and consider the *functional*

$$\Phi(N) = \int_N \varphi \quad , \qquad N \in V(I,\omega) \ . \tag{3}$$

Eventually we will restrict Φ to N's satisfying suitable boundary or endpoint conditions, but this is a somewhat subtle matter involving the structure theory of (I,ω). In particular, at first glance it appears to involve the derived flag of I, which roughly speaking tells how many derivatives are implicit in I. Endpoint conditions will be discussed in Chapter IV; in Chapter I we simply finesse the matter and argue formally.

By the *variational problem* $(I,\omega;\varphi)$ will be meant the analysis of the functional (3). For $X \subset J^1(\mathbb{R},\mathbb{R}^m)$ given by (2) above, if we take

$$\varphi = L(x,y,\dot{y})\,dx$$

then $(I,\omega;\varphi)$ is a classical variational problem with constraints (Lagrange problem). Another example is given by the functional (1). In general, understanding a variational problem $(I,\omega;\varphi)$ clearly will involve at least some of the structure of (I,ω) and how $d\varphi$ relates to this structure.

The first order of business is to derive the *Euler-Lagrange equations* expressing the condition that $N \in V(I,\omega)$ be an *extremal* of $(I,\omega;\varphi)$; i.e., the "differential" of Φ should vanish at N, written

$$\delta\Phi(N) = 0 \quad . \tag{4}$$

For the Lagrange problem this is to some extent accomplished in the classic treatise [13] and is discussed in many other sources (e.g. [5] and [40]). The traditional method is to use Lagrange multipliers. In general, the obvious difficulty in deriving the Euler-Lagrange equations is that only certain normal vector fields to $N \subset X$ represent infinitesimal variations of N *as an integral manifold of* (I,ω), (e.g., see page 344 of [13]). In particular there may be no compactly supported such infinitesimal variations of $N \in V(I,\omega)$, and consequently at first glance it would seem that already at this stage we must worry about endpoint conditions, thereby dragging in the structure theory of (I,ω) and running the considerable risk of becoming bogged down at the very outset.

This difficulty may be avoided by proceeding indirectly. If we simply assume the existence *in complete generality* of Euler-Lagrange equations having certain functoriality properties, then heuristic reasoning leads to a unique and remarkably symmetrical system of equations that "should" be the equations (4).[(8)] Rather than try to justify the heuristic reasoning at this point, we simply *define* these to be the Euler-Lagrange equations associated to the variational problem $(I,\omega;\varphi)$ and proceed to investigate these equations in their own right.

In somewhat more detail, after setting up the variational problem in Chapter I, Section a), in Chapter I, Section b) the "tangent space" $T_N(V(I,\omega))$ is defined to be the kernel of a certain linear differential operator on normal vector fields (cf. (I.b.16); in doing this the reference [38] has been helpful). In order to better understand $T_N(V(I,\omega))$, we turn in Chapter I, Section c) to some of the structure theory of Pfaffian differential systems. In particular, *Cauchy characteristics*, the *derived flag*, the important concepts of a *Pfaffian system in good form* and its basic invariant the *Cartan integer* s_1, and finally the *prolongation* of an exterior differential system are discussed. In this setting we establish an "infinitesimal Cartan-Kähler theorem," which states that a general $v \in T_N(V(I,\omega))$ depends on s_1 arbitrary functions of one variable (plus a certain number of constants).

In Chapter I, Section d) the Euler-Lagrange equations are defined (cf. (I.d.14)), and a number of examples are computed to show that in classical cases they give the right answer. We also analyze the Euler-Lagrange equations associated to the functional

$$\Phi(\gamma) = \frac{1}{2} \int_{\gamma} \kappa^2 \, ds \tag{5}$$

where γ is a curve on a surface S and κ is its geodesic curvature. When S has constant Gaussian curvature it is found that these equations may be explicitly integrated by elliptic functions whose modulus depends on the curvature of S and on an "energy level."

In Chapter I, Section e) the basic step in this presentation of the theory is taken by writing the Euler-Lagrange equations as a Pfaffian differential system (J,ω) on an associated manifold Y that we call the *momentum space* (cf. Theorem (I.e.9)). (Note: Although we give the construction of (J,ω) on Y explicitly, from the viewpoint of the general theory of exterior differential systems it may be explained very simply: The Euler-Lagrange equations (I.d.14) contain mysterious "functions λ_α to be determined." We adjoin the λ_α as new variables, write the resulting equations as a differential system, and then (J,ω) is simply the involutive prolongation of this system.) For unconstrained and non-degenerate classical variational problems, Y is the usual momentum space $\mathbb{R} \times T^*(M)$ where $X = \mathbb{R} \times T(M)$, but in general even the dimension of Y will depend on the numerical invariants of $(I,\omega;\varphi)$, especially the Cartan integer. We call (J,ω) the *Euler-Lagrange system* and note the remarkable fact that, despite the apparent generality of the variational problem $(I,\omega;\varphi)$, (J,ω) is a very simple standard Pfaffian system: On Y there is canonically given a 1-form ψ_Y with exterior derivative $\Psi_Y = d\psi_Y$, and J is the Cauchy characteristic system (i.e., $\Psi_Y^\perp$) of Ψ_Y. In particular this leads naturally to the definition of a *non-degenerate variational problem* $(I,\omega;\varphi)$ to be one where $\dim Y = 2m+1$ (for some m) and where

$$\psi_Y \wedge (\Psi_Y)^m \neq 0 \quad . \tag{6}$$

Thus far, all "natural" examples have turned out to be non-degenerate in this sense. By the theorem of Pfaff-Darboux the Euler-Lagrange system of a non-degenerate variational problem has a standard local normal form; frequently, this normal form is even global.

Next, also in Chapter I, Section e), we associate to a variational problem $(I,\omega;\varphi)$ satisfying a mild internal structural condition (one that is satisfied in almost all our examples, and is also satisfied on any prolongation) a quadratic form $\mathcal{Q}$. If I is the Pfaffian differential ideal generated by a sub-bundle $W^* \subset T^*(X)$ and if the sub-bundle $W_1^* \subset W^*$ generates the 1st derived system, then intrinsically $\mathcal{Q}$ is a quadratic form on the rank-s_1 bundle $(W^*/W_1^*)^*$. The variational problem is defined to be *strongly non-degenerate* in case $\mathcal{Q}$ is pointwise non-degenerate, and it is shown that "strong non-degeneracy ⇒ non-degeneracy." Strongly non-degenerate problems turn out to have the important property that only the 1st derived system and not the whole derived flag intervenes in their basic structural properties.

Before continuing this introduction we should like to emphasize our feeling that the numerous examples scattered throughout the text are of equal importance to the general theory (they may even be more important). Moreover, these examples as well as the general theory show that exterior differential systems constitute a computationally effective and theoretically natural setting for the calculus of variations. This latter philosophy is by no means original (9), and in this regard we should like to point out the sources [16], [31], and [38] as being especially helpful to us. Although by and large they deal with unconstrained problems (however, see [40]), they contain intrinsic formulations, using jet bundles and differential forms, of the classical theory and our debt to them is apparent.

As previously mentioned, one of the main concerns of this monograph is in variational problems $(I,\omega;\varphi)$ whose Euler-Lagrange differential systems (J,ω) are explicitly solvable in the old-fashioned sense of being "suitably" integrable by quadratures. Here a most useful tool is (a suitable generalization of) Emmy Noether's theorem [55], which associates a 1st integral of (J,ω) to each infinitesimal symmetry of $(I,\omega;\varphi)$. In Chapter II, Section b) Noether's theorem is combined with the general formalism to show that several natural differential-geometric variational problems are quasi-integrable by quadratures. One of these is the functional, motivated by physical considerations,

$$\Phi(\gamma) = \frac{1}{2}\int_\gamma \kappa^2\, ds \qquad (\kappa = \text{curvature}) \tag{7}$$

defined for curves $\gamma \subset \mathbb{E}^3$. A partial result here is due to Radon [57] and is discussed in Blaschke [3]. Using the formalism developed thus far it is essentially known *a priori* that the Euler-Lagrange equations associated to (7) are quasi-integrable by quadratures, and it is a simple matter to carry out the integration. The same is also true of the variational problem associated to the functional (7) with the integral constraint

$$\int_\gamma ds = \ell = \text{constant}$$

(this one turns out to be equivalent to the unconstrained problem (7) on a surface of constant curvature), and of the *Delauney problem*

$$\begin{cases} \Phi(\gamma) = \int_\gamma ds \\ \text{with the constraint } \kappa = \text{constant} \end{cases} \tag{8}$$

which was much discussed classically (a treatment may be found in [14]). The Euler-Lagrange equations associated to (7) and (8) have "phase portraits" given respectively by elliptic and rational algebraic curves.

These examples begin to make clear the general point that once one accepts the basic construction

$$(I,\omega;\varphi) \text{ on } X \rightsquigarrow (J,\omega) \text{ on } Y ,$$

the computation of examples, 1st integrals, and later on Jacobi equations, has an algorithmic character. Carrying out this algorithm in practice has as its essential step the computation of the structure equations of the differential system (I,ω) and the relation of $d\varphi$ to these equations. Once this is done the determination of (J,ω), 1st integrals, Jacobi equations, etc. is reduced to formal algebraic manipulations that seem to always have the same flavor.

As stated earlier, in this monograph we have restricted attention to the case of one independent variable. Now it seems likely that even more interesting problems will arise in higher dimensions, and in this regard we should like to call attention to the functional[(10)]

$$\Phi(M) = \frac{1}{k} \int_M \|II\|^k \, dA , \tag{9}$$

defined on submanifolds $M^n \subset \mathbb{E}^{n+r}$, where $\|II\|$ is the length of the

2nd fundamental form. The case $n=1$, $k=2$ is the functional (7). In general we might think of (9) as standing in the same relation to (7) as does the minimal surface functional

$$V(M) = \int_M dA \tag{10}$$

to geodesics. To reduce higher dimensional problems to one independent variable it is natural to look at surfaces of revolution. In the minimal surface case it is well-known that the Euler-Lagrange equations associated to (10) are integrable by quadratures (catenary). Using Noether's theorem we are almost, but not completely, able to integrate the very interesting Euler-Lagrange equations associated to (9) in the case $n=k=2$ of a surface of revolution (cf. the end of Chapter II, Section a)).

As suggested at the beginning of the Introduction, one of the goals of this text is to treat variational problems for functionals (3) where I is an invariant exterior differential system and φ is an invariant form on a Lie group. Of course a principle motivation is the aforementioned fact that general curves γ in many homogeneous spaces G/H have canonical liftings to integral manifolds of an invariant differential system I on G. Now it is a well-known and beautiful fact that a left invariant positive definite quadratic Lagrangian system on any Lie group G has associated *Euler equations* that describe the motion along integral curves of the Euler-Lagrange equations of the momentum vector λ in the dual $\mathfrak{g}^*$ of the Lie algebra $\mathfrak{g}$ of G. In particular λ moves on a coadjoint orbit (Kostant-Souriau; loc. cit., and an Appendix to [2]). In Chapter III, Section a) these results are generalized to the setting of an invariant variational problem $(I,\omega;\varphi)$, and then the resulting Euler equations and coadjoint orbit description are used to integrate several problems, including the Euler-Lagrange equations associated to the functional (7) defined on curves in a Riemannian manifold of constant curvature.(11)

With regard to the Lagrange problem on Lie groups we should like to call attention to the papers [8], [9], which are related to the discussion in Chapter III, Section a).

Thus far the Euler-Lagrange equations have only been arrived at by heuristic reasoning; in particular, they have not yet been shown to yield critical values of the functional (3). In fact, without specifying endpoint conditions this doesn't even make sense. Examination of special cases leads in Chapter IV, Section a) to a natural class of variational problems $(I,\omega;\varphi)$, that are said to be *well-posed*, whose endpoint conditions are given by the leaves of a canonical foliation on the momentum space Y. The situation may be summarized by the diagram

$$\begin{array}{ccc} & Y & \\ {\scriptstyle\tilde{\omega}}\swarrow & & \searrow{\scriptstyle\pi} \\ Q & & X \end{array} \qquad (11)$$

where Q is the quotient (assumed to exist) by the endpoint foliation. We call Q the *reduced momentum space*, and the understanding of the geometry of the basic diagram (11) turns out to be the key to the deeper aspects of the theory.[12]

For well-posed variational problems it is shown in Chapter IV, Section a) that the solutions to the Euler-Lagrange equations do in fact give critical points of the restriction of the functional (3) to subsets $V(I,\omega;[A,B]) \subset V(I,\omega)$ consisting of integral manifolds satisfying endpoint conditions. Following this the endpoint conditions are interpreted in a number of examples,[13] and it is proved that a strongly non-degenerate variational problem is well-posed.

In Chapter IV, Section b) the important concepts of *Jacobi vector fields* and *conjugate points* are defined for well-posed variational problems. (Actually the definition of the Jacobi equations makes sense in general.) The point here is to work on the momentum space Y and not down on X. Not only is this natural theoretically, but as shown by examples the definition "upstairs" leads to effective computation of some examples.[14]

In the classical calculus of variations there is an intimate and very beautiful connection between the Jacobi equations and the 2nd variation. In the general setting when one tries to compute the 2nd variation down on X this causes considerable difficulty, but when lifted to Y the situation becomes quite simple and elegant. Again, as in the classical unconstrained case it is possible to define the *index form* as a quadratic form on the space of Jacobi vector fields and establish a simple connection between the index and 2nd variation.

Following this it is proved in Chapter IV, Section c) that, if $(I,\omega;\varphi)$ is a strongly non-degenerate variational problem whose quadratic form $\mathcal{Q}$ is positive definite, then for sufficiently close endpoint conditions a solution to the Euler-Lagrange equations yields a local minimum for the functional (3). This is an extension of a standard classical result, but it is not just a direct generalization since the point turns out to be to show that in the diagram (11) there is a (unique) exterior differential system (G,ω) on the reduced momentum space Q such that

$$\tilde{\omega}^*G = \pi^*I_1$$

where $I_1 \subset I$ is the 1st derived system (recall that $\mathcal{Q}$ is a quadratic form on $(W^*/W_1^*)^*$). However, once one insists on setting up the calculus of variations in the general framework of exterior differential systems and introducing only those concepts intrinsic to this theory, results such as the one just mentioned become quite natural and the proofs not more difficult than in the classical case.

Next, in Chapter IV, Section d) the analogues of the classical concept of a *field* (sometimes called geodesic field) and the *Hamilton-Jacobi equation* are defined. Some examples are computed, and then these concepts are used, as in the classical case, to show that if N is a solution to the Euler-Lagrange equations of a positive-definite strongly non-degenerate variational problem, and if moreover N contains no pair of conjugate points, then N gives a local minimum for the functional (3). Again the essential ingredient beyond the classical case is the relation of the 1st derived system of I to the diagram (11).

In summary, it would appear that the concept of a *positive strongly non-degenerate variational problem* is a good notion that includes the classical cases together with the Lagrange problem in sufficient generality to be useful, while at the same time allowing an extension of the main points of the classical theory.

Finally, in Chapter IV, Section e) we specifically discuss the classical Lagrange problem. It is shown that, with a minor modification of the previous endpoint conditions, this problem fits into the general theory. It is interesting that our theory does *not* reduce to the classical method of Lagrange multipliers; the exact relation together with several examples are also discussed.

There is an appendix devoted to several tangential and recreational topics. First there is a discussion of how to set up and derive the Euler-Lagrange equations for variational problems with integral constraints. Again, once one gets into the swing of writing the Euler-Lagrange equations as a differential system on an auxiliary manifold, the whole proceudre has an algorithmic character. This is illustrated by a couple of examples.

Next, since this a monograph on the calculus of variations we would be seriously remiss to say nothing about the classical problems in the subject (brachistochrone, minimal surface of revolution, etc.). Certainly it is difficult to improve any on the excellent classical texts, so instead we illustrate our general approach by setting up the classical problems intrinsically, in effect viewing them through the structure equations of a moving frame.

Concerning prerequisites, the reader should have a good foundation in general manifold theory and the rudiments of Lie group theory (e.g. [6] and [64]). Moreover, although not logically necessary it certainly will help to be familiar with the classical calculus of variations (e.g., [29]). Otherwise, the main prerequisite is the omnipresent one of having had sufficient experience with mathematics of this general sort (so-called mathematical maturity).

It is a pleasure to thank the many students and colleagues with whom this material was discussed. A preliminary version of the manuscript was reproduced by Nankai University and the suggestions made by the mathematicians there have been very helpful. Discussions with Robert Bryant on both the general theory of exterior differential systems and on the calculus of variations have been very helpful. I would also like to express appreciation to Rennie Mirollo for an extremely careful and constructive job of proofreading. Special thanks go to Renate D'Arcangelo for a marvelous job of typing an unbelievably messy manuscript. Finally, I would like to express gratitude to the Guggenheim Foundation for financial support during the preparation of this manuscript.

FOOTNOTES FOR THE INTRODUCTION

(1) Our main sources for the classical theory are the beautiful books [13], [29], and (for a treatment using differential forms) [16]. In addition we refer to the excellent texts [1], [2], [66] for a discussion of the seminal relation between classical mechanics and the calculus of variations. For a modern and intrinsic approach, using differential forms, to the classical theory we have found [31], [38], and [43] very helpful (cf. also [23] and [26]). We have not attempted a complete bibliography, but have just listed our main sources as well as certain specific references that serve to supplement (frequently by examples) the material in this text.

(2) Here, what we are mainly doing is to use moving frames to analyze the Euler-Lagrange equations in various examples, especially those in which there is a symmetry group operating. Our main sources for moving frames have been [18], [34], [35], and [44]. In addition we have drawn numerous examples from differential geometry, especially Riemannian geometry and surface theory. Here as references we suggest [6], [15], [22], [28], [53]), [63], and [64]. Again we emphasize that this is just a sample of the possible references; further sources may be found in bibliographies of our references.

(3) In this regard we should like to point out that control theory provides an excellent motivation for studying invariant Lagrange problems on homogeneous spaces. Here there is considerable common ground with the material in Chapter III, and as a few references (through which other sources may be found) we mention [7], [9], [10] and [43].

(4) For this the main classical references are [17] and [47]. Exterior differential systems are discussed [11], [12], [23], [32], [33], [42], [52], [60], [62]. Our notations and terminology are those of [11] and its (hopefully soon to appear successor) [12]. Actually, aside from relatively elementary topics such as the theorem on Cauchy characteristics (§2 of [11]) and the theorem of Pfaff-Darboux (§3 of [11]), both of which are explained below, this text is self-contained as far as the theory of exterior differential systems goes.

The deeper aspects of the theory of exterior differential systems, such as prolongation and involution and the theory of characteristic varieties, are required in the case of the calculus of variations for several independent variables. It is partly for this reason that we have restricted to the case of one independent variable.

(5) Cf. Chapter 18 of [13] for a guide to the classical theory and [43] for an exposition of and references to more recent developments.

(6) In this discussion we may replace $\mathbb{R}^m$ by any manifold M.

(7) This is just considering a higher order O.D.E. system as a (much larger) 1st order system by introducing the intermediate derivatives as new variables. One of the traditional difficulties in doing the Lagrange problem in complete generality is intrinsically recognizing what order problem the given 1st order one may be a "disguised version" of. We shall find that this has a rather natural formulation in the general setting of exterior differential systems.

(8) An amusing by-product is an intrinsic interpretation of the "functions λ_α to be determined": If I is the differential ideal generated by a sub-bundle $W^* \subset T^*(X)$, then a crucial ingredient in the theory is the restriction to W^* of the standard canonical 1-form on $T^*(X)$; this 1-form is just λ.

(9) For example, Chern mentioned in a conversation that some time ago he worked out the classical Euler-Lagrange equations in this setting.

(10) In this regard we should like to point out that the extremals of (7) arise from the physical problem of determining the shape assumed by a piece of wire or thin hacksaw blade (cf. [25]). Similarly, the extremals of a functional closely related to (9) turn up in the motion of a vibrating plate discussed in [29].

(11) In particular, there were several computations in [57] that seemed to us unpredictable and somewhat mysterious; these are now explained by Theorem (III.a.25).

(12) For classical 1st order variational problems, $Q = X$ and the diagram (11) collapses to the fibering $T^*(X) \to X$. In general we shall find that this fibering has so to speak two aspects, one of which is given by π and the other by $\tilde{\omega}$.

(13) In all cases they give the correct endpoint conditions.

(14) In [13] the Jacobi equations were given in canonical coordinates, which amounts to "working upstairs."

0. PRELIMINARIES

a) Notations from Manifold Theory ([6], [38], and [64]).

We shall denote manifolds by M, N, X, Y,... . Typical local coordinates on M are $(x^1,\ldots,x^n)$; sometimes we simply write (x^i). The tangent space to M at the point p is $T_p(M)$, and the tangent bundle is $T(M) = \cup_{p\in M} T_p(M)$. We sometimes write points in $T(M)$ as (p,v) where $p\in M$ and $v\in T_p(M)$. A local coordinate system (x^i) on M induces a local coordinate system $(x^1,\ldots,x^n;\dot{x}^1,\ldots,\dot{x}^n) = (p,v)$ on $T(M)$ where p has coordinates $(x^1,\ldots,x^n)$ and $v = \dot{x}^i\ \partial/\partial x^i$. *Here, and throughout, we use the summation convention.*

The cotangent space and cotangent bundle will be denoted by $T_p^*(M)$ and $T^*(M) = \cup_{p\in M} T_p^*(M)$. Points of $T^*(M)$ will be denoted by (p,ω) where $p\in M$ and $\omega\in T_p^*(M)$. A local coordinate system (x^i) on M induces a local coordinate system $(x^1,\ldots,x^n;\xi_1,\ldots,\xi_n) = (p,\omega)$ on $T^*(M)$ where p has coordinates (x^i) and $\omega = \xi_i\ dx^i$. Associated to the cotangent bundle are its exterior powers $\Lambda^q T^*(M) = \cup_{p\in M}\Lambda^q T_p^*(M)$.

The differential at $p\in M$ of a smooth mapping $f:M\to X$ is denoted by $f_*:T_p(M)\to T_{f(p)}(X)$ with dual $f^*:T^*_{f(p)}(X)\to T_p^*(M)$.

If $E\to M$ is a vector bundle then $C^\infty(M,E)$ denotes the space of C^∞ cross-sections. For example, $C^\infty(X,T(X))$ are the vector fields on X. If $M\subset X$ is a submanifold then $C^\infty(M,T(X))$ denotes the space of tangent vectors to X *defined along* M

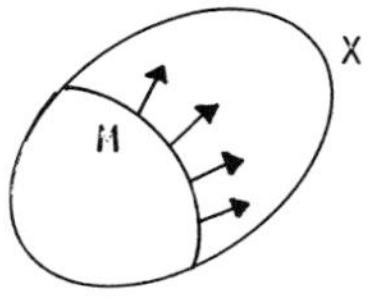

The *normal bundle* to M in X is $E = \cup_{p\in M} E_p$ where $E_p = T_p(X)/T_p(M)$. If $v\in C^\infty(M,T(X))$ then $[v]$ denotes the corresponding normal vector field.

An exception to these notations is that $A^q(X)$ (and not $C^\infty(X,\Lambda^q T^*(X))$) will denote the C^∞ q-forms on X. A map $f:M\to X$ induces the usual pullback map $f^*:A^q(X)\to A^q(M)$ on forms. For $M\subset X$ a submanifold we sometimes denote by ω_M the restriction to M of a form ω on X. Otherwise, when no confusion should arise we will omit the restriction map.

For a manifold M it is well-known that there is a *canonical 1-form* θ *on* $T^*(M)$ defined in $T^*_{(p,\omega)}(T^*(M))$ by

$$\langle\theta,v\rangle = \langle\omega,\pi_* v\rangle$$

where $\pi:T^*(M)\to M$ is the projection. In the above local coordinates, $\theta=\xi_i dx^i$. The exterior derivative $d\theta = d\xi_i\wedge dx^i$ is the *standard symplectic form on* $T^*(M)$.

For a vector field $v\in C^\infty(M,T(M))$ we will denote by

$$\exp(tv):\ M\to M$$

the local 1-parameter group it generates by flowing along the integral curves of v. The *Lie derivative*

$$L_v:\ A^q(M)\to A^q(M)$$

is defined as usual by

$$L_v(\varphi) = \frac{d}{dt}(\exp(tv)^*\varphi)_{t=0} \quad .$$

The contraction of a vector field v by a differential form φ is denoted by

$$v \lrcorner \varphi \ .$$

On innumerable occasions we will use the *H. Cartan formula*

$$L_v(\varphi) = v\lrcorner d\varphi + d(v\lrcorner\varphi) \quad . \qquad (0.a.1)$$

The bracket of vector fields v,w will be denoted as usual by [v,w]. We shall also use the relation

$$L_{[v,w]} = [L_v,L_w] \qquad (0.a.2)$$

where the right hand side is the commutator.

If G is a Lie group then its Lie algebra $\mathfrak{g}$ consists of the *left* invariant vector fields. The dual space $\mathfrak{g}^*$ of left invariant 1-forms will sometimes be called the *Maurer-Cartan forms*. If $\{\omega^i\}$ is a basis for $\mathfrak{g}^*$ then we have the *Maurer-Cartan equation*

$$d\omega^i = -\frac{1}{2} c^i_{jk}\omega^j \wedge \omega^k, \quad c^i_{jk} + c^i_{kj} = 0 \quad . \qquad (0.a.3)$$

The reason for the minus sign is that if $\{v_i\}$ is the dual basis of $\mathfrak{g}$ then

$$[v_i, v_j] = c^k_{ij} v_k \quad ,$$

so that the c^i_{jk} are the structure constants for $\mathfrak{g}$.

If $\pi: M \to N$ is a surjective C^∞ map between manifolds whose differential π_* has everywhere maximal rank (i.e., π is a submersion), then the kernel of π_* defines the sub-bundle $V_\pi(M) \subset T(M)$ of *vertical tangent vectors*. A differential form $\varphi \in A^k(M)$ is *horizontal* if

$$v \lrcorner \varphi = 0$$

for all $v \in C^\infty(M, V_\pi(M))$. This is equivalent to

$$\varphi(p) \in \text{image}\left\{\pi^*: \Lambda^k T^*_{\pi(p)}(N) \to \Lambda^k T^*_p(M)\right\}$$

for each point $p \in M$. Assuming that the fibres of π are connected, it is well-known and easy to prove that the condition

$$\varphi = \pi^* \eta$$

for a (necessarily unique) form $\eta \in A^k(N)$ is that both φ and $d\varphi$ be horizontal.

b) The Language of Jet Manifolds (cf. [31], [38], [62] and the references cited there).

We shall find it convenient to use the formalism of the jet manifolds $J^k(\mathbb{R}, M)$. Recall that two maps

$$f, \tilde{f}: \mathbb{R} \to M$$

have the same k-jet at $x \in \mathbb{R}$ if $f(x) = \tilde{f}(x)$ and if in one, and hence in any, local coordinate system around $f(x)$ the Taylor's series expansions of f and $\tilde{f}$ agree up *through* order k. The equivalence class determined by f will be denoted by $(j^k f)(x)$. The totality of all k-jets forms the *jet manifold* $J^k(\mathbb{R},M)$ and, as we shall now illustrate, the coordinate x on $\mathbb{R}$ and any local coordinate system $(y^1,..,y^m) = (y^\alpha)$ on M induces a coordinate system on $J^k(\mathbb{R},M)$.

A 1-jet consists of a point $x \in \mathbb{R}$ together with a point $(p,v) \in T(M)$. We may think of it as a linear map

$$T_x(\mathbb{R}) \to T_p(M)$$

sending d/dx to v. Thus

$$J^1(\mathbb{R},M) \cong \mathbb{R} \times T(M). \tag{0.b.1}$$

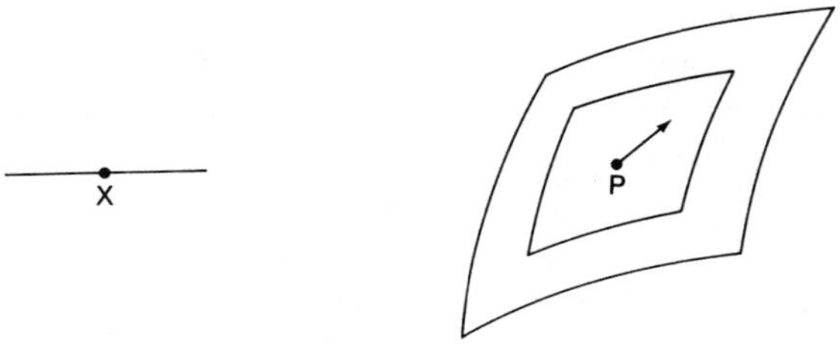

Local coordinates (y^α) on M give what we shall call *standard local coordinates*

$$(x;y^1,...,y^m;\dot{y}^1,...,\dot{y}^m) = (x;y^\alpha;\dot{y}^\alpha)$$

on $J^1(\mathbb{R},M)$ where

$$\begin{cases} p = (y^1,..,y^m) \in M \\ v = \dot{y}^\alpha \,\partial/\partial y^\alpha \in T_p(M) \end{cases} .$$

A 2-jet consists of points $x \in \mathbb{R}$, $p \in M$ together with a 2nd order element of an arc passing through p.

Local coordinates on $J^2(\mathbb{R},M)$ are

$$(x;y^1,\dots,y^m;\dot y^1,\dots,\dot y^m;\ddot y^1,\dots,\ddot y^m) = (x;y^\alpha;\dot y^\alpha;\ddot y^\alpha)$$

where the 2nd order Taylor's series has α^{th} component given by

$$y^\alpha + (t-x)\dot y^\alpha + \frac{1}{2}(t-x)^2\ddot y^\alpha \quad .$$

It is clear how to generalize these considerations to $J^k(\mathbb{R},M)$.

To each such jet manifold there is associated a canonical sub-bundle

$$W^* \subset T^*(J^k(\mathbb{R},M)) \quad . \tag{0.b.2}$$

We describe this first when $k=1$. For this we consider the projection

$$\begin{cases} \pi\colon J^1(\mathbb{R},M) \to \mathbb{R}\times M \\ \pi(x,p,v) = (x,p) \quad . \end{cases}$$

There is a canonical tangent vector (cf. (0.b.1))

$$\frac{d}{dx} + v \in T_{(x,p)}(\mathbb{R}\times M) \qquad\qquad (\cong \mathbb{R}^{m+1})$$

and the annihilator

$$\left(\frac{d}{dx} + v\right)^{\perp} \subset T^*_{(x,p)}(\mathbb{R}\times M)$$

is a hyperplane $(\cong \mathbb{R}^m)$. We set

$$W^*_{(x,p,v)} = \pi^*\left(\left(\frac{d}{dx} + v\right)^{\perp}\right) \quad .$$

In the above local coordinate system

$$W^* = \mathrm{span}\{\theta^\alpha,\dots,\theta^m\}$$

where

$$\theta^\alpha = dy^\alpha - \dot y^\alpha\, dx \quad .$$

When $k=2$ we give $W^* \subset T^*(J^2(\mathbb{R},M))$ locally by

$$W^* = \mathrm{span}\{\theta^\alpha,\dots,\theta^m;\dot\theta^\alpha,\dots,\dot\theta^m\} = \mathrm{span}\{\theta^\alpha;\dot\theta^\alpha\}$$

where

$$\begin{cases} \theta^\alpha = dy^\alpha - \dot{y}^\alpha dx \\ \dot{\theta}^\alpha = d\dot{y}^\alpha - \ddot{y}^\alpha dx \end{cases} .$$

As in the previous case, W^* may be defined in a coordinate free manner. One convenient way to do this is to give an embedding

$$J^2(\mathbb{R},M) \subset J^1(\mathbb{R},J^1(\mathbb{R},M)) , \tag{0.b.3}$$

and then $W^* \subset T^*(J^2(\mathbb{R},M))$ will be induced from the canonical subbundle of $T^*(J^1(\mathbb{R},J^1(\mathbb{R},M)))$. To give (0.b.3) we remark that the differential of a map $f:\mathbb{R}\to M$ gives $f_*:\mathbb{R}\to T(M)$ where by definition $f_*(x) = f_*(d/dx)$, and if $(j^k f)(x) = (j^k\tilde{f})(x)$ then $(j^{k-1}f_*)(x) = (j^{k-1}\tilde{f}_*)(x)$. Using the identification (0.b.1) this map

$$(j^2 f)(x) \to (j^1 f_*)(x)$$

induces the inclusion (0.b.3). Local coordinates (y^α) on M induce local coordinates $(x;y^\alpha;\dot{y}^\alpha)$ on $J^1(\mathbb{R},M)$, and these induce local coordinates

$$(t;(x;y^\alpha;\dot{y}^\alpha);(\dot{x};\dot{y}^\alpha;(\dot{\dot{y}}^\alpha)))$$

on $J^1(\mathbb{R},J^1(\mathbb{R},M))$. The inclusion (0.b.3) is given in local coordinates by

$$(x;y^\alpha;\dot{y}^\alpha;\ddot{y}^\alpha) \to (x;(x;y^\alpha;\dot{y}^\alpha);(1;\dot{y}^\alpha;\ddot{y}^\alpha)) \quad .$$

It is clear that similar considerations will apply to all $J^k(\mathbb{R},M)$. We shall frequently just use $J^k(\mathbb{R},\mathbb{R}^m)$ so as to have global coordinates available. However, all constructions will be intrinsic and therefore valid where $\mathbb{R}^m$ is replaced by any manifold M.

c) Frame Manifolds (cf. [34], [35], and [44]).

Euclidean N-space will be denoted by $\mathbb{E}^N$. It is an affine vector space (no preferred origin) having a translation invariant inner product. By a *frame* $(x;e_1,\ldots,e_N)$ we mean a position vector $x \in \mathbb{E}^N$ and orthonormal basis $e_1,\ldots,e_N \in T_x(\mathbb{E}^N)$. The set of all frames forms a manifold $F(\mathbb{E}^N)$ which, upon choice of a reference frame (i.e., an isomorphism $\mathbb{E}^N \cong \mathbb{R}^N$), may be identified with the group $E(N)$ of Euclidean motions:

the position vector x gives the translation part of the Euclidean motion, and the rotation is that which takes the coordinate frame in $\mathbb{R}^N$ to $e_1,..,e_N$.

If we fix an origin $0 \in \mathbb{E}^N$ then the set of all frames in $T_0(\mathbb{E}^N)$ will be denoted by $F_0(\mathbb{E}^N)$. Clearly a choice of reference frame gives an identification of $F_0(\mathbb{E}^N)$ with the orthogonal group $O(N)$.

By $\mathbb{A}^N$ we will mean affine N-space having a translation invariant volume form Ω. By an *affine frame* we mean $(x;e_1,..,e_N)$ where $x \in \mathbb{A}^N$ and $e_1,..,e_N \in T_x(\mathbb{A}^N)$ give a basis with $\Omega(e_1,..,e_n) = 1$. Frequently we will simply write this as $e_1 \wedge .. \wedge e_N = 1$. Upon choice of a reference frame we may identify the manifold $F(\mathbb{A}^N)$ of affine frames with the group $A(N)$ of affine linear transformations $T:\mathbb{R}^N \to \mathbb{R}^N$ given by $T(x) = Ax + b$ where $b \in \mathbb{R}^N$ and $A \in SL_N = \{n \times n$ matrices A with $\det A = +1\}$.

Let F denote either of the frame manifolds $F(\mathbb{E}^N)$ or $F(\mathbb{A}^N)$. The position vector gives a map

$$x\colon F \to V$$

where V is either $\mathbb{E}^N$ or $\mathbb{A}^N$. At a point $(x;e_1,..,e_N) = (x;e_i)$ its differential is

$$dx\colon T_{(x;e_i)}(F) \to T_x(V)$$

(we write dx instead of x_*). In terms of the basis e_i of $T_x(V)$ we have

$$dx = \omega^i e_i \qquad (0.c.1)$$

where the $\omega^i \in T^*_{(x;e_i)}(F)$. Similarly, using the unique up to translation identification $T_x(V) \cong V$ we have

$$e_i\colon F \to V$$

with differential

$$de_i\colon T_{(x;e_i)}(F) \to T_x(V) \quad ,$$

and as before we write

$$de_i = \omega_i^j e_j \qquad (0.c.2)$$

where $\omega_i^j \in T^*_{(x;e_i)}(F)$. (<u>Note</u>: Here it may be convenient to choose an isomorphism $V \cong \mathbb{R}^N$ and consider x and the e_i as $\mathbb{R}^N$-valued functions on F. Then their differentials are $\mathbb{R}^N$-valued 1-forms on F, and as such may be expressed in terms of the basis e_i. The 1-form coefficients ω^i, ω_i^j are then *independent* of the particular identification $V \cong \mathbb{R}^N$.)

The exterior derivatives of (0.c.1) and (0.c.2) give respectively

$$\begin{cases} d\omega^i = \omega^j \wedge \omega_j^i \\ d\omega_i^j = \omega_i^k \wedge \omega_k^j \end{cases} .$$

We collect these together as the *structure equations of a moving frame:*

$$\left.\begin{array}{lll} \text{(i)} & dx = \omega^i e_i \\ \text{(ii)} & de_i = \omega_i^j e_j \\ \text{(iii)} & d\omega^i = \omega^j \wedge \omega_j^i \\ \text{(iv)} & d\omega_i^j = \omega_i^k \wedge \omega_k^j \end{array}\right\} \qquad (0.c.3)$$

For $F(\mathbb{E}^N)$ the additional relation

$$\text{(v)} \qquad \omega_i^j + \omega_j^i = 0 ,$$

holds, and on $F(\mathbb{A}^N)$ we have (recall that we use summation convention)

$$\text{(vi)} \qquad \omega_i^i = 0 .$$

If we choose a reference frame and identify F with $E(N)$ or $A(N)$, then the ω^i and ω_i^j give left invariant 1-forms on the appropriate group. They actually give a basis if we take the relations (v), (vi) into account (e.g., for $F(\mathbb{E}^N)$ the ω_i^j for $i < j$ give a basis). Equations (iii), (iv) are just the Maurer-Cartan equations (0.a.3).

For $F_0(\mathbb{E}^N)$ the structure equations (0.c.3) reduce to

$$\left.\begin{array}{l} de_i = \omega_i^j e_j \\ \omega_i^j + \omega_j^i = 0 \\ d\omega_i^j = \omega_i^k \wedge \omega_k^j \end{array}\right\} \qquad (0.c.4)$$

d) Differential Ideals ([11], [12], [17], [23], [38], [47], and [60]).

By a *differential ideal* we shall mean a graded ideal $I = \oplus_{q\geq 0} I^q$ in the exterior algebra $A^*(X) = \oplus_{q\geq 0} A^q(X)$ of C^∞ forms on a manifold X with the property that $dI \subset I$. Thus if

$$\varphi = \varphi^0 + \varphi^1 + \cdots + \varphi^n \in I \qquad (\varphi^k \in A^k(X))$$

then

$$\begin{cases} \varphi^k \in I \\ \varphi \wedge \eta \in I & \text{for all} \quad \eta \in A^*(X) \\ d\varphi \in I \end{cases}$$

For any subset $\Sigma \subset A^*(X)$ we denote by $\{\Sigma\}$ the *algebraic ideal* generated by Σ. In practice, Σ will frequently be a finite set $\theta^1, \ldots, \theta^s$ of forms and we write

$$\{\Sigma\} = \{\theta^\alpha\} .$$

If $d\Sigma$ denotes the set of forms $d\varphi$ for $\varphi \in \Sigma$, then the *differential ideal generated by* Σ is $\{\Sigma, d\Sigma\}$. If Σ is the finite set $\theta^1, \ldots, \theta^s$ this is denoted by $\{\theta^\alpha, d\theta^\alpha\}$. In case the θ^α are 1-forms we will say that $I = \{\theta^\alpha, d\theta^\alpha\}$ is a *Pfaffian differential ideal.*

By an *integral manifold of a differential ideal* I on X we mean a manifold N and maximal rank mapping

$$f : N \to X \tag{0.d.1}$$

such that $f^*(I) = (0)$; i.e.,

$$f^*\varphi = 0 \qquad \text{for all} \quad \varphi \in I . \tag{0.d.2}$$

Later on, for notational simplicity we shall drop the f and think of integral manifolds as submanifolds $N \subset X$ satisfying

$$\varphi_N = 0 \qquad \text{for all} \quad \varphi \in I .$$

If $I = \{\Sigma, d\Sigma\}$ then since

$$\begin{cases} f^*(\varphi \wedge \eta) = f^*\varphi \wedge f^*\eta \\ f^*(d\varphi) = df^*(\varphi) \end{cases} ,$$

the condition (0.d.2) is equivalent to

$$\begin{cases} f^*\varphi = 0 \\ f^*d\varphi = 0 \qquad \text{for } \varphi \in \Sigma . \end{cases} \tag{0.d.3}$$

By an n-dimensional *integral element of a differential ideal* we shall mean the pair (p,E) consisting of a point $p \in X$ and an n-plane $E \subset T_p(X)$ such that

$$\varphi(p)|E = 0 \qquad \text{for all } \varphi \in I .$$

If $G_n(X) \to X$ denotes the Grassmann bundle whose fibre over $p \in X$ is

$$\begin{aligned} G_n(X)_p &= G_n(T_p(X)) \\ &= \left\{\begin{matrix}\text{Grassman manifold of} \\ \text{n-planes } E \subset T_p(X)\end{matrix}\right\} , \end{aligned}$$

then the set of n-dimensional integral elements of I forms naturally a subset

$$V_n(I) \subset G_n(X) .$$

An integral manifold is given by a maximal rank mapping (0.d.1) such that

$$f_*(T_p(N)) \in V_n(I)$$

for all $p \in N$ (where $\dim N = n$).

A Pfaffian differential ideal $\{\theta^\alpha, d\theta^\alpha\}$ is said to be *completely integrable* in case

$$d\theta^\alpha \in \{\theta^\alpha\} .$$

We frequently write this as

$$d\theta^\alpha \equiv 0 \bmod \{\theta^\alpha\} .$$

Suppose that $I = \{\theta^\alpha, d\theta^\alpha\}$ is a Pfaffian differential ideal such that for each $p \in X$ the subspaces

$$W_p^* = \operatorname{span}\{\theta^1(p), \ldots, \theta^s(p)\} \subset T_p^*(X)$$

have constant dimension s. Then $W^* = \cup_{p\in X} W_p^*$ gives a sub-bundle which completely determines I. We sometimes abuse notation and write

W^* instead of I: *for a sub-bundle* $W^* \subset T^*(X)$ *the Pfaffian system* W^* *means the differential ideal generated by* $C^\infty(X,W^*)$. For a sub-bundle $W^* \subset T^*(X)$ the annihilator $W^{*\perp} = S \subset T(X)$ is a *distribution* or sub-bundle of the tangent bundle. It is well-known that the Pfaffian system W^* is completely integrable if, and only if, S is an *involutive distribution* in the sense that

$$v,w \in C^\infty(X,S) \Rightarrow [v,w] \in C^\infty(X,S) .$$

In this case the *Frobenius theorem* says that there exist local coordinates $(u^1,..,u^s;v^1,..,v^{n-s})$ on X such that

$$\begin{cases} W^* = \mathrm{span}\{du^1,..,du^s\} \\ S = \mathrm{span}\{\partial/\partial v^1,..,\partial/\partial v^{n-s}\} . \end{cases}$$

As usual the level sets $u^\alpha = \text{constant}$ are called the *leaves of the foliation* determined by W^*. We shall make extensive use of the Frobenius theorem.

We want to give an important example, but before doing this we need a short digression on symplectic linear algebra (cf. [2]). Let V be a vector space and $\Omega \in \Lambda^2 V^*$ an alternating bilinear form on V. We define the *rank* ρ of Ω by

$$\begin{cases} \Omega^\rho \neq 0 \\ \Omega^{\rho+1} = 0 \end{cases}$$

where $\Omega^k = \underbrace{\Omega \wedge .. \wedge \Omega}_{k\text{-times}}$. If $v_1,..,v_n \in V$ is a basis and we set

$$\Omega_{ij} = \Omega(v_i,v_j) = -\Omega_{ji} ,$$

then the skew-symmetric matrix $\|\Omega_{ij}\|$ has rank 2ρ. It is well-known that we may choose the basis $v_1,..,v_n$ with dual basis $\omega^1,..,\omega^n \in V^*$ such that

$$\Omega = \omega^1 \wedge \omega^2 + .. + \omega^{2\rho-1} \wedge \omega^{2\rho} . \tag{0.d.4}$$

We define

$$\begin{cases} C(\Omega) = \{v \lrcorner \Omega : v \in V\} \subset V^* \\ \Omega^\perp = \{v : v \lrcorner \Omega = 0\} \subset V \end{cases} .$$

Then using the normal form (0.d.4)

$$\begin{cases} C(\Omega) = \mathrm{span}\{\omega^1, \ldots, \omega^{2\rho}\} \\ \Omega^\perp = \mathrm{span}\{v_{2\rho+1}, \ldots, v_n\} \end{cases} . \tag{0.d.5}$$

A special case is when $\dim V = 2\rho + 1$; then $\Omega^\perp \subset V$ is a line called the *characteristic direction* associated to Ω.

We may now discuss the

(0.d.6) Example. Let Ψ be a 2-form on X and define the *Cartan system* $C(\Psi)$ to be the Pfaffian differential ideal generated by the set of 1-forms

$$\{v \lrcorner \Psi \quad \text{where} \quad v \in C^\infty(X, T(X))\} \subset A^1(X) .$$

If we assume that the 2-forms $\Psi(p) \in \Lambda^2 T^*_p(X)$ have constant rank ρ, then $C(\Psi)$ is given by a sub-bundle $W^* \subset T^*(X)$ of rank 2ρ and the distribution $W^{*\perp} = S \subset T^*(X)$ is a sub-bundle of rank $n-2\rho$. Since

$$\langle v \lrcorner \Psi, w\rangle = \langle \Psi, v \wedge w\rangle = -\langle w \lrcorner \Psi, v\rangle$$

it follows that for the space of sections of S we have

$$C^\infty(X,S) = \{v \in C^\infty(X, T(X)) : v \lrcorner \Psi = 0\} .$$

It is well-known that

(0.d.7) *If* $d\Psi = 0$, *then* $C(\Psi)$ *is completely integrable.*

In fact a much stronger statement holds (cf. [2], [64]):

(0.d.8) THEOREM (Darboux). *If* Ψ *is a closed 2-form of constant rank* ρ, *then locally there exist functions* $u_1, \ldots, u_\rho, v^1, \ldots, v^\rho$ *such that*

$$\Psi = du_1 \wedge dv^1 + \cdots + du_\rho \wedge dv^\rho .$$

In this case the Cartan system

$$C(\Psi) = \{du_1,..,du_\rho;dv^1,..,dv^\rho\}$$

is obviously completely integrable.

Finally, we consider a 1-form ψ with exterior derivative $d\psi = \Psi$. We assume that in each cotangent space

$$\begin{cases} \psi(p) \wedge \Psi(p)^\rho \neq 0 \\ \psi(p) \wedge \Psi(p)^{\rho+1} = 0 \end{cases}.$$

Then we have the following (cf. footnote(26) to Chapter I).

(0.d.9) THEOREM (Pfaff-Darboux [11], [17]). *Locally there exist functions* $x,u_1,..,u_\rho,v^1,..,v^\rho$ *such that*

$$\psi = -H(x,u,v)dx + u_1dv^1 + \cdots + u_\rho dv^\rho . \quad (26)$$

It may be shown that the Pfaff theorem implies the Darboux theorem. In this text we shall make essential use of both results.

Somewhat surprisingly our main construction (the Euler-Lagrange differential system (J,ω) defined in Chapter I, Section e)) will be the Cartan system $C(\Psi)$ of a 2-form $\Psi = d\psi$ *where the rank of* Ψ *is definitely not constant.* In order to understand this one must of course have the constant rank case well in hand.

e) Exterior Differential Systems ([11], [12], [17], [38], [47], and [60]).

Definitions: i) An *exterior differential system* (I,ω) on a manifold X is given by a differential ideal $I \subset A^*(X)$ together with an n-form ω. We call n the *number of independent variables* in (I,ω).

ii) An *integral element* of (I,ω) is given by the pair (p,E) consisting of a point $p \in X$ and an n-plane $E \subset T_p(X)$ satisfying the two conditions

$$\begin{cases} \varphi|E = 0 & \text{for all } \varphi \in I \\ \omega|E \neq 0 \end{cases}.$$

We denote by $V(I,\omega) \subset G_n(X)$ the set of integral elements of (I,ω).

iii) An *integral manifold* of (I,ω) is given by a smooth mapping

$$f:N\to X$$

where N is a *connected* manifold of dimension n and where $f_*(T_p(N))\in V(I,\omega)$ for all $p\in N$.

We shall frequently drop the f and think of an integral manifold as given by a submanifold $N\subset X$ satisfying

$$\begin{cases} \varphi_N = 0 & \text{for all } \varphi\in I \\ \omega_N \neq 0 \end{cases}$$

The second of these is called the *independence condition*. The set of all integral manifolds of (I,ω) is denoted by $V(I,\omega)$.

In this monograph we shall almost exclusively be concerned with the case when I is a *Pfaffian differential system* generated by a sub-bundle $W^*\subset T^*(X)$ and the number of independent variables $n=1$. Since the independence form ω is only well defined modulo I, what we will be concerned with then is the data of vector bundles

$$W^*\subset L^*\subset T^*(X) \tag{0.e.1}$$

where

i) W^* generates a Pfaffian system I; and

ii) L^*/W^* is a line bundle and $\omega\in C^\infty(X,L^*)$ induces a non-zero section of L^*/W^*.

There are several aspects of the general theory of exterior differential systems that we shall need to bring into our discussion; among them are i) the associated system and Cauchy characteristics, ii) the derived flag, iii) prolongation, iv) the Cartan integer, and v) involution (this enters much more in the case of $n>1$ independent variables). In fact, doing the calculus of variations in a general setting provides a good opportunity to see how these concepts arise naturally and are useful. Rather than explain them here, we shall deal with them as the need arises (cf. especially Chapter I, Section c)).

Here we shall simply mention the universal example and a few of its special cases.

(0.e.2) <u>Example.</u> On $X = J^1(\mathbb{R}, M)$ we let I be the Pfaffian system given by the sub-bundle (0.b.2) above, and we let $L^* = \pi^*(T(\mathbb{R} \times M))$. This gives the data (0.e.1), which we shall call the *canonical exterior differential system* (I, ω).

Using standard local coordinates $(x; y^\alpha; \dot{y}^\alpha)$ on X as described in Chapter 0, Section b) above, the 1-forms $\theta^\alpha = dy^\alpha - \dot{y}^\alpha dx$ $(\alpha = 1, \ldots, m)$ give a local basis for W^* and $\omega = dx$ gives a local basis for L^*/W^*. We therefore write (I, ω) locally as

$$\begin{cases} \text{(i)} & \theta^\alpha = dy^\alpha - \dot{y}^\alpha dx = 0 \\ \text{(ii)} & \omega = dx \neq 0 \; . \end{cases} \tag{0.e.3}$$

These equations have the following meaning:

(i) is a set of differential equations to be satisfied by solutions or integral manifolds, and

(ii) is a transversality condition on solutions.

An integral manifold of (0.e.3) is given by a smooth mapping

$$f : N \to X$$

where $N = \{a \leqq x \leqq b\}$ (we shall always include the endpoints) and the conditions

$$f^*\theta^\alpha = 0, \qquad f^*\omega \neq 0$$

are satisfied. Using the 2nd condition we may take x as a local coordinate on N and locally give f parametrically by

$$x \to (x; y^\alpha(x); \dot{y}^\alpha(x)) \quad . \tag{0.e.4}$$

Then $f^*\theta^\alpha = 0$ gives

$$dy^\alpha(x) - \dot{y}^\alpha(x)\,dx = 0 \; ;$$

i.e.,

$$\dot{y}^\alpha(x) = \frac{dy^\alpha(x)}{dx} \quad .$$

Thus (0.e.4) is

$$x \to \left(x; y^\alpha(x); \frac{dy^\alpha(x)}{dx}\right) \quad .$$

It follows that the integral manifolds of this differential system are given by 1-*jets of parametrized immersed curves* in M.

We shall generally not be so fussy and simply say that: the integrals of (I,ω) are the 1-jets of curves in M.

Given a differential system (I',ω') on a manifold X' we may restrict everything to a submanifold $X\subset X'$ to obtain another differential system (I,ω). This may be thought of as "imposing constraints" on the original system.

(0.e.5) Example. Let $X\subset J^1(\mathbb{R},M)$ be a submanifold and denote by (I,ω) the restriction to X of the canonical system (0.e.3) on $J^1(\mathbb{R},M)$. Locally X is given by equations

$$g^\rho(x;y^\alpha;\dot{y}^\alpha) = 0 \ ,$$

and so the integral manifolds $N\subset X$ of (I,ω) are 1-jets of curves $x\to y^\alpha(x)$ in M satisfying the O.D.E.'s

$$g^\rho\left(x;y^\alpha(x);\ \frac{dy^\alpha(x)}{dx}\right) = 0 \ .$$

For instance suppose that $X=J^2(\mathbb{R},M)$ embedded in $J^1(\mathbb{R},J^1(\mathbb{R},M))$ as in Section b). Then (I,ω) on $J^2(\mathbb{R},M)$ is locally given by

$$\begin{cases} \theta^\alpha = dy^\alpha - \dot{y}^\alpha dx = 0 \\ \dot{\theta}^\alpha = d\dot{y}^\alpha - \ddot{y}^\alpha dx = 0 \\ \omega = dx \neq 0 \end{cases} \tag{0.e.6}$$

Integral manifolds of this differential system are locally given by 2-jets

$$x\to\left(x;y^\alpha(x);\ \frac{dy^\alpha(x)}{dx}\ ;\ \frac{d^2y^\alpha(x)}{dx^2}\right)$$

of curves in M.

During the course of this text we shall give several other kinds of differential systems arising from the theory of moving frames as well as from certain physical considerations. However, they will all fall into the class of what are called Pfaffian systems in good form, as defined in Chapter I, Section c) for the case of one independent variable and in [12] (also implicitly in [47]) in general.

I. EULER-LAGRANGE EQUATIONS FOR DIFFERENTIAL SYSTEMS WITH ONE INDEPENDENT VARIABLE

a) Setting Up the Problem; Classical Examples.

On a manifold X we consider a Pfaffian differential system (I,ω) whose independence condition is a 1-form ω. We give (I,ω) locally by Pfaffian equations

$$\theta^1 = \cdots = \theta^s = 0, \quad \omega \neq 0 , \tag{I.a.1}$$

and recall (cf. (0.e.1)) that the notation has the following meaning: The cotangent bundle of X has a 2-step filtration by sub-bundles

$$W^* \subset L^* \subset T^*(X) , \qquad \text{rank } L^*/W^* = 1$$

and $\theta^1,..,\theta^s,\omega$ are locally given 1-forms such that

$$\begin{cases} W^* = \text{span}\{\theta^1,..,\theta^s\} \\ L^* = \text{span}\{\theta^1,..,\theta^s,\omega\} \end{cases} .$$

For the time being we will give integral manifolds of (I,ω) by maps

$$f: N \to X \tag{I.a.2}$$

where N is a connected 1-dimensional manifold and f is a smooth map, and where the equations (I.a.1) mean that

$$f^*\theta^1 = .. = f^*\theta^s = 0, \quad f^*\omega \neq 0 . \tag{I.a.3}$$

Later on we shall drop the f and simply think of submanifolds $N \subset X$ such that $\theta^1_N = .. = \theta^s_N = 0$, $\omega_N \neq 0$. We recall our notations φ_N for the restriction of a differential form φ to N and $V(I,\omega)$ for the set of integral manifolds of (I,ω).

We now assume given a 1-form φ on X, and for each integral manifold (I.a.2) of (I,ω) we set

$$\Phi(N,f) = \int_N f^*\varphi \tag{I.a.4}$$

Of course, we agree to consider only integral manifolds for which this integral exists; i.e., N may be non-compact and the improper integral (I.a.4) should converge. We may view the assignment

$$(N,f) \to \int_N f^*\varphi$$

as a *functional* (perhaps not everywhere defined)

$$\Phi : V(I,\omega) \to \mathbb{R} \quad ,$$

and we consider the following

PROBLEM: *Determine the variational equations of the functional* Φ.[1]

Of course, as it stands this is too loosely posed since we must worry about such matters as smoothness and endpoint conditions, but the intent should be clear: Determine a Pfaffian system (J,ω) on a manifold Y whose integral manifolds are in a natural one-to-one correspondence with the integral manifolds of (I,ω) that satisfy the Euler-Lagrange equations corresponding to (I.a.4).

NOTATION: We will denote by $(I,\omega;\varphi)$ the variational problem associated to the functional (I.a.4).

(I.a.5) Example. Let M be a manifold with associated 1-jet manifold $X = J^1(\mathbb{R},M)$, and let L be a function on $J^1(\mathbb{R},M)$ (we shall sometimes call L a *Lagrangian*). Letting x be the coordinate on $\mathbb{R}$ we set

$$\varphi = L\,dx$$

and take for (I,ω) the canonical Pfaffian system given in example (0.e.2).

Definition. $(I,\omega;\varphi)$ will be called a *classical variational problem*.

A coordinate system $y = (y^1,\ldots,y^m)$ on M induces a coordinate system $(x,y,\dot{y}) = (x;y^1,\ldots,y^m;\dot{y}^1,\ldots,\dot{y}^m)$ on $J^1(\mathbb{R},M)$ and (I,ω) is given in the form (I.a.1) by

$$\theta^\alpha = dy^\alpha - \dot{y}^\alpha dx = 0, \qquad \omega = dx \neq 0 \quad .$$

Integral manifolds $N \subset J^1(\mathbb{R},M)$ of (I,ω) are locally the 1-jets

$$x \to \left(x, y(x), \frac{dy(x)}{dx}\right)$$

of curves $y: \mathbb{R} \to M$ (here we are dropping the f in (I.a.2)). If $L = L(x,y,\dot{y})$ then the functional under consideration is

$$\Phi(N) = \int L\left(x, y(x), \frac{dy(x)}{dx}\right) dx \quad . \tag{I.a.6}$$

Of course, here we should specify both the endpoint data and class of competing curves; i.e., C^∞, C^1, piecewise smooth, etc. In this case there are obvious natural choices and the situation is carefully discussed in classical texts, such as [29]. In general these questions are more subtle and will be deferred until Chapter IV.

(I.a.7) <u>Special Case of (I.a.5)</u>. We identify $J^1(\mathbb{R},M)$ with $\mathbb{R} \times T(M)$ and assume given a function T on $T(M)$ that induces a positive definite quadratic form in each fibre $T_y(M) \subset T(M)$. In the above local coordinates, $T = T(y,\dot{y})$ where

$$T(y,\dot{y}) = \frac{1}{2} g_{\alpha\beta}(y)\dot{y}^\alpha \dot{y}^\beta \tag{I.a.8}$$

(throughout we use summation convention). Here $g(y) = \|g_{\alpha\beta}(y)\|$ is a C^∞ positive symmetric matrix. Clearly T is just one half the square length function associated to the *Riemannian metric*

$$ds^2 = g_{\alpha\beta}(y)\, dy^\alpha dy^\beta \tag{I.a.9}$$

on M. For $U = U(y)$ a function on M, Lagrangians of the form

$$L = T - U$$

are said to give *mechanical systems* on the manifold with T being the *kinetic energy* and U the *potential energy*. If we think of an integral manifold as coming from a path $\gamma \subset M$, then (I.a.6) is called the *action functional* associated to γ. The statement that the motion of a mechanical system occurs so as to minimize this integral is called the *principle of least action* ([1], [2], and [29]). Historically the calculus of variations for mechanical systems provided a major impetus for the development of the theory.(2)

(I.a.10) Further Special Case of (I.a.5). A *Riemannian manifold* (M,ds^2) is given by a manifold M together with a Riemannian metric (I.a.9). We set

$$L = \sqrt{2T} \qquad \text{(positive square root);}$$

in local coordinates

$$\begin{aligned} L(y,\dot{y}) &= \sqrt{g_{\alpha\beta}(y)\dot{y}^{\alpha}\dot{y}^{\beta}} \\ &= \|\dot{y}\| \end{aligned}$$

where $\|\dot{y}\|$ is the length of the tangent vector $\dot{y} = \dot{y}^{\alpha}\,\partial/\partial y^{\alpha} \in T_y(M)$. For a curve $\gamma \subset M$ the functional (I.a.6) is

$$\begin{aligned} \ell(\gamma) &= \int \left\| \frac{dy(x)}{dx} \right\| dx \\ &= \text{length of } \gamma \,. \end{aligned}$$

As usual, curves that locally minimize length are called *geodesics* (cf. especially [49], [53]).

(I.a.11) Example (Continuation of Example (I.a.5)). We let $X \subset J^1(\mathbb{R},M)$ be a submanifold such that $dx \neq 0$ on $T(X)$, and we denote again by (I,ω) the restriction to X of the canonical system on $J^1(\mathbb{R},M)$. In terms of standard local coordinates $(x,y,\dot{y})$ on $J^1(\mathbb{R},M)$ we may think of X as given by equations

$$g^1(x,y,\dot{y}) = \cdots = g^r(x,y,\dot{y}) = 0 \,. \tag{I.a.12}$$

Let L be a function on X (in practice L will be the restriction of a function on $J^1(\mathbb{R},M)$) and set

$$\varphi = L\omega \,.$$

Definition. $(I,\omega;\varphi)$ will be called a *classical variational problem with constraints*.

This simply means that the domain of the functional (I.a.4) consists of curves

$$x \to \left(x, y(x), \frac{dy(x)}{dx}\right)$$

where

$$g^{\rho}\left(x, y(x), \frac{dy(x)}{dx}\right) = 0 \qquad \rho = 1,\ldots,r \,.$$

When the constraints are equivalent to ones of the form $h^{\rho}(x,y) = c^{\rho}$ = constant they are said to be *holonomic*. The further special case where $h^\rho = h^\rho(y)$ are functions on M means that $V(I,\omega)$ consists of 1-jets of curves lying in the submanifold $\{h^1 = c^1, .., h^r = c^r\}$ of M.

In general the constraints (I.a.12) are said to be *non-holonomic*. The description of the integral manifolds of (I,ω) is more complicated and involves the structure theory of the differential ideal I. A classical example of non-holonomic constraints will be given in (II.a.40) below and a general discussion appears in Chapter IV, Section e).

(I.a.13) Example. Let M be a manifold with associated 2-jet bundle $J^2(\mathbb{R},M)$, and let L be a function on $J^2(\mathbb{R},M)$ (we sometimes call L a *2nd order Lagrangian*). Letting x be the coordinate on $\mathbb{R}$ we set

$$\varphi = L\,dx$$

and take for (I,ω) the canonical Pfaffian system given in example (0.e.5) (cf. (0.e.6)).

Definition. $(I,\omega;\varphi)$ will be called a *classical 2nd order variational problem.*

A coordinate system $y = (y^1, .., y^m)$ on M induces a coordinate system $(x,y,\dot{y},\ddot{y}) = (x; y^1, .., y^m; \dot{y}^1, .., \dot{y}^m; \ddot{y}^1, .., \ddot{y}^m)$ on $J^2(\mathbb{R},M)$ and the canonical Pfaffian system (I,ω) is given locally in the form (I.a.1) by

$$\begin{cases} \theta^\alpha = dy^\alpha - \dot{y}^\alpha dx = 0 \\ \dot{\theta}^\alpha = d\dot{y}^\alpha - \ddot{y}^\alpha dx = 0 \\ \omega = dx \neq 0 \end{cases} \tag{I.a.14}$$

Integral manifolds $N \subset J^2(\mathbb{R},M)$ of (I,ω) are locally 2-jets

$$x \to \left(x, y(x), \frac{dy(x)}{dx}, \frac{d^2y(x)}{dx^2}\right)$$

of curves $y : \mathbb{R} \to M$. The functional under consideration is of the form

$$\Phi(N) = \int L\left(x, y(x), \frac{dy(x)}{dx}, \frac{d^2y(x)}{dx^2}\right) dx .$$

Remarks: i) There is a natural embedding

$$J^2(\mathbb{R},M) \subset J^1(\mathbb{R}, J^1(\mathbb{R},M))$$

which we discussed in Chapter 0, Section b). Using this inclusion a classical 2nd order variational problem may be considered as a classical variational problem with constraints.

ii) Of course this example (I.a.13) extends to jet bundles $J^k(\mathbb{R},M)$ of any order k, which we shall call a *classical* k^{th} *order variational problem.*

iii) Finally, by considering submanifolds $X \subset J^k(\mathbb{R},M)$, as was done in example (I.a.11) when $k=1$, we may consider *classical* k^{th} *order variational problems with constraints.*

(I.a.15) Special Case of (I.a.13). We let $X \subset J^2(\mathbb{R},\mathbb{R}^2)$ be the open set given by

$$X = \{(x,y,\dot{y},\ddot{y}) : \dot{y} \wedge \ddot{y} \neq 0\} ,$$

and on X we set

$$\begin{cases} L = L(\dot{y},\ddot{y}) = |\dot{y} \wedge \ddot{y}| \\ \varphi = L\, dx \end{cases}$$

Integral manifolds $N \subset X$ of the canonical 2-jet system (I.a.14) are given by curves $y : \mathbb{R} \to \mathbb{R}^2$ satisfying

$$\frac{dy(x)}{dx} \wedge \frac{d^2y(x)}{dx^2} \neq 0 \quad .$$

The functional

$$\Phi(N) = \int L\left(\frac{dy(x)}{dx} , \frac{d^2y(x)}{dx}\right) dx \tag{I.a.16}$$

is the *affine arc length* of the corresponding curve $\gamma \subset \mathbb{R}^2$.(3) The variational problem $(I,\omega;\varphi)$ is invariant under the special affine linear group A(2) consisting of area preserving linear maps

$$\begin{cases} T(y) = Ay + b \\ \det A = \pm 1 \end{cases}$$

of $\mathbb{R}^2$. Curves that locally minimize (I.a.16) are called *affine geodesics.*

Clearly this example generalizes to $\mathbb{R}^m$ with

$$\Phi(N) = \int \left| \frac{dy(x)}{dx} \wedge \ldots \wedge \frac{d^my(x)}{dx^m} \right| dx \quad .$$

Remark. In discussing geodesics it is convenient to allow piecewise smooth curves. The analogue for affine geodesics in $\mathbb{R}^m$ would be given by maps $y_i : [a_i, b_i] \to \mathbb{R}^m$ such that

$$y_i(b_i) = y_{i+1}(a_{i+1}), \quad \frac{dy_i}{dx}(b_i) = -\frac{dy_{i+1}}{dx}(a_i), \ldots, \frac{d^{m-1}y_i}{dx^{m-1}}(a_i)$$

$$= (-1)^{m-1} \frac{d^{m-1}y_{i+1}}{dx^{m-1}}(a_{i+1}) \quad .$$

Once we discuss endpoint conditions in Chapter IV it will be natural to extend a functional Φ from $V(I,\omega)$ to a larger class of broken or piecewise smooth integral manifolds whose exact definition will of course depend on the structure theory of $(I,\omega;\varphi)$.

It is obvious that the variety of higher order classical variational problems with constraints is endless. One feature of phrasing these variational problems in the setting of exterior differential systems is that it unifies and isolates the essential features of the examples.

b) Variational Equations for Integral Manifolds of Differential Systems.(4)

On a manifold X we consider a Pfaffian system (I,ω) given locally by (I.a.1). We will derive the variational equations for an integral manifold (I.a.2) of (I,ω); i.e., roughly speaking we shall compute the quantity (to be precisely defined in Chapter I, Section c))

$$T_{(N,f)}(V(I,\omega)) = \textit{tangent space to } V(I,\omega) \textit{ at } (N,f) \; .$$

For this, and also for extensive later use, we shall formulate the basic general computation (I.b.5) (cf. [31], [38]).

Let N and X be manifolds and

$$f\colon N \to X \tag{I.b.1}$$

a smooth mapping. Setting $[0,\varepsilon] = \{t\colon 0 \leq t \leq \varepsilon\}$ a *variation* of (I.b.1) is given by a smooth mapping

$$F\colon N \times [0,\varepsilon] \to X \tag{I.b.2}$$

such that if we let

$$f_t\colon N \to X \qquad t \in [0,\varepsilon]$$

be the restriction of F to $N \times \{t\} \cong N$, then $f_0 = f$. The associated *infinitesimal variation*

$$v \in C^\infty(N, f^*T(X))$$

is defined by

$$v(x) = F_*(\partial/\partial t) \qquad (I.b.3)$$

where

$$F_*: T_{(x,0)}(N \times [0,\varepsilon]) \to T_{f(x)}(N)$$

is the differential of F at $(x,0)$.

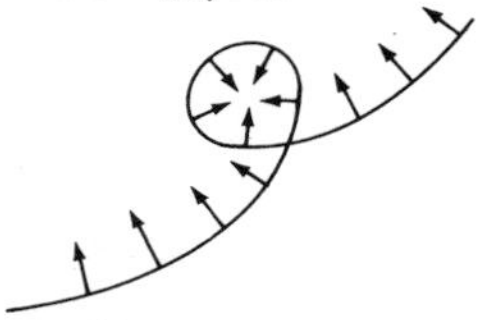

Remark. In all cases of concern to us, f will be an immersion and v induces a section

$$[v] \in C^\infty(N,E)$$

of the *normal bundle*

$$E = f^*(T(X)/f_*T(N))$$

to $f(N)$ in X. Roughly speaking, two variations F, $\tilde{F}$ of (I.b.1) whose associated infinitesimal variations v, $\tilde{v}$ satisfy $[v] = [\tilde{v}]$ will have the property that

$$f_t = \tilde{f}_t \circ g_t \mod t^2 \qquad (I.b.4)$$

where g_t are diffeomorphisms of N (in fact, $v - \tilde{v} = w$ is a vector field on N and in 1st approximation $g_t = \exp t w$). Since we are primarily interested in the images $f_t(N) \subset X$ and (I.b.4) means that $f_t(N) = \tilde{f}_t(N)$ to 2nd order, it is to be expected that *our variational equations will only depend on the normal vector field* $[v]$ *associated to the infinitesimal variation* v.

(I.b.5) PROPOSITION. *Suppose that* (I.b.1) *is an immersion and let* θ *be a differential form on* X. *Then as differential forms on* $N = N \times \{0\}$ *we have*

$$(L_{\partial/\partial t}(F^*\theta))_N = f^*(v \lrcorner d\theta + d(v \lrcorner \theta)) .$$

Remark. In case f is an embedding the right hand side of this equation is interpreted as follows: Let $V \in C^{\infty}(X,T(X))$ be any vector field that induces by restriction the infinitesimal variation $v \in C^{\infty}(N,f^*(T(X)))$. Then $V \lrcorner d\theta + d(V \lrcorner \theta)$ is a form on X and we set

$$f^*(v \lrcorner d\theta + d(v \lrcorner \theta)) = f^*(V \lrcorner d\theta + d(V \lrcorner \theta)) .$$

The point is that the right hand side does not depend on the extension V of v. Since the formula is local around a point of N, it will suffice to treat only the embedding case.

Proof.(5) Assuming that f is an embedding we let V be any vector field on X such that

$$V(F(x,t)) = F_*(\partial/\partial t) \in T_{F(x,t)}(X) .$$

The whole point of the proposition is the following computation: On $N \times [0,\varepsilon]$ the H. Cartan formula (0.a.1) implies that

$$\begin{aligned} L_{\partial/\partial t}(F^*\theta) &= d(\partial/\partial t \lrcorner F^*\theta) + \partial/\partial t \lrcorner dF^*\theta \\ &= dF^*(V \lrcorner \theta) + F^*(V \lrcorner d\theta) \\ &= F^*(d(V \lrcorner \theta) + V \lrcorner d\theta) . \end{aligned}$$

Both sides of this equation are differential forms on $N \times [0,\varepsilon]$, and the proposition follows by restricting both sides to $T(N) = T(N \times \{0\}) \subset T(N \times [0,\varepsilon])$ and noting (as above) that the right hand side depends only on v. Q.E.D.

(I.b.6) DISCUSSION (for later use--may be omitted now)

An interesting example of (I.b.5) occurs when $X = G$ is a Lie group with basis $\{\omega^i\}$ for the left invariant Maurer-Cartan forms and structure equations (cf. (0.a.3))

$$d\omega^i + \frac{1}{2} c^i_{jk} \omega^j \wedge \omega^k = 0 .$$

We recall the following well-known facts (cf. [35] for a proof):

(I.b.7) *Assuming that* N *is connected, a smooth mapping*

$$f: N \to G$$

is uniquely determined, up to a left translation, by the 1-forms

$$\psi^i = f^*\omega^i ; \qquad (I.b.8)$$

(I.b.9) *Given 1-forms* ψ^i *on* N *there exists mappings*

$$f: U \to G$$

defined in a neighborhood U *of any point* $x \in N$ *and satisfying* (I.b.8) *if, and only if, the pulled back Maurer-Cartan equations*

$$d\psi^i + \frac{1}{2} c^i_{jk} \psi^j \wedge \psi^k = 0 \qquad (I.b.10)$$

are satisfied.

Suppose now that $N = \{a \leqq s \leqq b\}$ is 1-dimensional. Then (I.b.10) is vacuous and giving $f: N \to G$, up to a left translation, is equivalent to giving the 1-forms

$$\psi^i = f^i(s)ds$$

where the $f^i(s)$ are *any* functions. Suppose we give a variation

$$f_t : N \to G \qquad 0 \leqq t \leqq \varepsilon$$

by requiring that

$$f_t^*\omega^i = f^i(s,t)ds$$

where $f^i(s,t)$ are functions satisfying $f^i(s,0) = f^i(s)$. Then we have

$$F: N \times [0,\varepsilon] \to G$$

defined by $F(s,t) = f_t(s)$. Writing

$$F^*\omega^i = f^i(s,t)ds + g^i(s,t)dt ,$$

(I.b.7) implies that the $g^i(s,t)$ are uniquely determined by $f^i(s,t)$ and $g^i(0,t)$.

To see this, by left translation we identify all the tangent spaces to G with the Lie algebra $\mathfrak{g}$ and let $e_i \in \mathfrak{g}$ be the vector fields dual to $\omega^i \in \mathfrak{g}^*$. Then the variation vector field of F is

$$v = g^i(s,0)e_i .$$

Now on the one hand (setting $f^i_t = \partial f^i/\partial t$, etc.)

$$L_{\partial/\partial t}(F^*\omega^i)\big|_N = f^i_t(s,0)ds ,$$

while on the other hand

$$f^*(v \lrcorner\, d\omega^i + d(v \lrcorner\, \omega^i)) = (-c^i_{jk} g^j(s,0) f^k(s,0) + g^i_s(s,0))ds .$$

Then (I.b.5) gives the O.D.E. system

$$g^i_s = f^i_t + c^i_{jk} g^j f^k \qquad (I.b.11)$$

In particular, this implies that $g^i(s,0)$ is uniquely determined by $f^i(s,0)$, $f^i_t(s,0)$, and $g^i(a,0)$ (and similarly for all $g^i(s,t)$).

Of course, (I.b.11) is more easily derived as a consequence of (I.b.10), but it serves to illustrate proposition (I.b.5).

We shall now derive the variational equations for integral manifolds of a Pfaffian system (I,ω) with one independent variable locally given by (I.a.1). For this we assume that

$$N = \{a \leqq s \leqq b\}$$

and let

$$f_t : N \to X \quad , \qquad 0 \leqq t \leqq \varepsilon \ , \qquad (I.b.12)$$

give a variation of an integral manifold (I.a.2) of (I,ω) (thus $f = f_0$). We assume that (I.b.12) is an integral manifold of (I,ω), so that by (I.a.3)

$$f^*_t \theta^\alpha = 0 \quad .^{(6)} \qquad (I.b.13)$$

Thinking of the f_t as giving a map (I.b.2) where

$$F(s,t) = f_t(s) \in X \ ,$$

(I.b.13) is equivalent to

$$F^* \theta^\alpha = g^\alpha(s,t)dt \quad .$$

Thus

$$(L_{\partial/\partial t}(F^*\theta^\alpha))_N = (g^\alpha_t(s,0)dt)_N = 0 \ .$$

By (I.b.5) this gives

$$f^*(v \lrcorner d\theta^\alpha + d(v \lrcorner \theta^\alpha)) = 0 \qquad (I.b.14)$$

where v is the infinitesimal variation associated to f_t.

Equations (I.b.14) may be thought of as giving a first approximation to $T_{(N,f)}(V(I,\omega))$. To put them in more palatable form we shall drop reference to f and consider integral manifolds of (I,ω) as submanifolds

$$N \subset X \ .$$

An infinitesimal variation of N in X is then given by tangent vectors $v(s) \in T_s(X)$ $(s \in N)$. Given such a v we extend it to a vector field on X, still denoted by v.(7) Then $v \lrcorner d\theta^\alpha + d(v \lrcorner \theta^\alpha)$ is a 1-form on X and (I.b.14) is equivalent to

$$(v \lrcorner d\theta^\alpha + d(v \lrcorner \theta^\alpha))_N = 0 .$$

For any 1-form σ on X we write

$$\sigma \equiv 0 \quad \text{mod } N$$

to mean that $\sigma_N = 0$. Then (I.b.14) is

$$\boxed{v \lrcorner d\theta^\alpha + d(v \lrcorner \theta^\alpha) \equiv 0 \quad \text{mod } N \qquad \alpha=1,..,s \;. \qquad (\text{I.b.15})}$$

Remarks: i) As noted in the proof of (I.b.5), these equations depend only on the infinitesimal variation $v \in C^\infty(N,T(X))$ and not on its extension to a vector field on all of X.

ii) Referring to the remark centered around (I.b.4) it is clear that (I.b.15) should only depend on the *normal* vector field $[v] \in C^\infty(N,T(X)/T(N))$ induced by v. In fact, suppose that $v(s) \in T_s(N)$ is tangent to N. Then again letting $v \in C^\infty(X,T(X))$ be any extension, $v \lrcorner \theta^\alpha$ is a function on X and

$$v \lrcorner \theta^\alpha = 0 \quad \text{on} \quad N$$

$$\Rightarrow d(v \lrcorner \theta^\alpha) \equiv 0 \quad \text{mod } N ,$$

while trivially

$$v \lrcorner d\theta^\alpha \equiv 0 \quad \text{mod } N$$

since $\langle v \lrcorner d\theta^\alpha, v\rangle = \langle d\theta^\alpha, v \wedge v\rangle = 0$.

iii) It is also clear that the equations (I.b.15) should be independent of the choice of basis $\theta^1,..,\theta^s$ for W^*. Thus let

$$\theta = \lambda_\alpha \theta^\alpha$$

where the λ^α are functions on X. Then

$$\begin{aligned} v \lrcorner d\theta + d(v \lrcorner \theta) &= v \lrcorner (d\lambda_\alpha \wedge \theta^\alpha) + \lambda_\alpha (v \lrcorner d\theta^\alpha) + d\lambda_\alpha (v \lrcorner \theta^\alpha) + \lambda_\alpha d(v \lrcorner \theta^\alpha) \\ &\equiv -d\lambda_\alpha (v \lrcorner \theta^\alpha) + d\lambda_\alpha (v \lrcorner \theta^\alpha) \quad \text{mod } N \\ &\equiv 0 \quad \text{mod } N \end{aligned}$$

where the middle step uses (I.b.15) and $\theta^\alpha \equiv 0 \mod N$.

We may therefore make the following

Definition. With $E = T(X)/T(N)$ denoting the normal bundle to N in X and $w_\alpha \in C^\infty(X,W)$ being a dual basis to the $\theta^\alpha \in C^\infty(X,W^*)$, we define the first order linear differential operator

$$L: C^\infty(N,E) \to C^\infty(N, W \otimes T^*(N)) \tag{I.b.16}$$

by

$$L([v]) = w_\alpha \otimes L^\alpha(v)$$

where

$$L^\alpha(v) = (v \lrcorner d\theta^\alpha + d(v \lrcorner \theta^\alpha))_N \quad .$$

Then we shall say that

$$L([v]) = 0$$

are the *variational equations of* N *as an integral manifold of* (I,ω).

We also sometimes refer to the "*tangent space*" $T_N(V(I,\omega))$ as being given by the solutions to the linear O.D.E. $L([v]) = 0$. This will be explained more precisely in Chapter I, Section c), and then in Chapter IV, Section a) we shall discuss "endpoint conditions" A,B and define the corresponding subspace $T_N(V(I,\omega;[A,B])) \subset T_N(V(I,\omega))$.

c) Differential Systems in Good Form; the Derived Flag and Cauchy Characteristics.

We retain the notations from the preceding sections and shall investigate the variational equations (I.b.15) for an important class of differential systems. We begin with the following

(I.c.1) Example. On $X = J^1(\mathbb{R},\mathbb{R}^m)$ with coordinates $(x;y^1,\ldots,y^m;\dot{y}^1,\ldots,\dot{y}^m)$ we consider the canonical Pfaffian system given by

$$\begin{cases} \theta^\alpha = dy^\alpha - \dot{y}^\alpha dx = 0 \\ \omega = dx \neq 0 \end{cases} \quad .$$

The structure equations are

$$d\theta^\alpha = -d\dot{y}^\alpha \wedge dx \quad . \tag{I.c.2}$$

It is convenient to use $\{\omega;\theta^1,..,\theta^m;d\dot{y}^1,..,d\dot{y}^m\}$ as a coframe on X, and we denote the dual tangent frame by $\{\partial/\partial\omega;\partial/\partial\theta^1,..,\partial/\partial\theta^m;\partial/\partial\dot{y}^1,..,\partial/\partial\dot{y}^m\}$. For a vector field

$$v = A\partial/\partial\omega + B^\alpha\partial/\partial\theta^\alpha + C^\alpha\partial/\partial\dot{y}^\alpha$$

on X, using (I.c.2) we find that

$$v \lrcorner d\theta^\alpha + d(v \lrcorner \theta^\alpha) = dB^\alpha + Ad\dot{y}^\alpha - C^\alpha dx \tag{I.c.3}$$

Let $x \to (x, y(x), \frac{dy(x)}{dx})$ be an integral manifold of the canonical system on $J^1(\mathbb{R},\mathbb{R}^m)$. Then using (I.c.3) the variational equations (I.b.15) are

$$\frac{dB^\alpha(x)}{dx} = C^\alpha(x) - \frac{d^2y^\alpha(x)}{dx^\alpha} A(x) \tag{I.c.4}$$

where now

$$v(x) = A(x)\partial/\partial\omega + B^\alpha(x)\partial/\partial\theta^\alpha + C^\alpha(x)\partial/\partial\dot{y}^\alpha \in C^\infty(N,T(X))$$

is a tangent vector to X defined along $N \subset X$. We may add to v a tangent vector field to N so as to make $A(x) = 0$, and then using (I.c.4) we see that

$$v(x) = B^\alpha(x)\partial/\partial\theta^\alpha + \left(\frac{dB^\alpha(x)}{dx}\right)\partial/\partial\dot{y}^\alpha . \tag{I.c.5}$$

In particular, a general $[v] \in T_N(V(I,\omega))$ may be said to "depend on m arbitrary functions (the $B^\alpha(x)$) of one variable."(8)

Of course, we may replace $\mathbb{R}^m$ by any manifold M in this example.

This example, and in fact all Pfaffian differential systems considered in this monograph, belong to a remarkable special class of Pfaffian systems that we now define in the case of one independent variable.

Definition. Let (I,ω) be an exterior differential system given locally by Pfaffian equations (cf. (I.a.1))

$$\begin{cases} \theta^\alpha = 0 & \alpha = 1,..,s \\ \omega \neq 0 & \end{cases}$$

on a manifold X. Denote by $\{\theta^\alpha\} \subset A^*(X)$ the *algebraic* ideal generated by the θ^α's. Then (I,ω) is said to be a Pfaffian system *in good form* in case there exist 1-forms π^α such that

$$d\theta^\alpha \equiv -\pi^\alpha \wedge \omega \bmod \{\theta^\alpha\}. \qquad (I.c.6)$$

Remarks. Some of the reasons for this terminology are explained in [11] and [12]. The concept was clearly isolated in Kähler's book [47]. One reason why Pfaffian systems in good form constitute a natural and theoretically unrestricted class is given in the discussion of prolongation at the end of this section.

To put the definition in intrinsic form we consider a Pfaffian system as given by a sub-bundle $W^* \subset T^*(X)$, and denote by

$$W^* \wedge T^*(X) \subset \Lambda^2 T^*(X)$$

the image of $W^* \otimes T^*(X)$ under exterior multiplication. Then the exterior derivative

$$d: C^\infty(X, W^*) \to C^\infty(X, \Lambda^2 T^*(X))$$

induces a mapping

$$\delta: W^* \to \Lambda^2 T^*(X)/W^* \wedge T^*(X) \qquad (I.c.7)$$

that is linear over the functions, and is therefore an algebraic mapping.(9)

Definitions. i) δ is called the *derived mapping* associated to the Pfaffian system W^*;

ii) Assuming that $\ker \delta = W_1^* \subset W^*$ is a constant rank sub-bundle of W^*, it defines a Pfaffian system called the *1st derived system* of W^*.(10)

Recalling that $L^* = \operatorname{span}\{\theta^1, .., \theta^s, \omega\}$, the structure equations (I.c.6) show that the condition that (I,ω) be in good form is

$$\delta(W^*) \subset L^* \wedge T^*(X)/W^* \wedge T^*(X). \qquad (I.c.8)$$

We now give some examples

(I.c.9) Example. From (I.c.2) it follows that the canonical system on $J^1(\mathbb{R}, M)$ is in good form.

In general, if a Pfaffian system (I,ω) on a manifold X is in good form, and if $Y \subset X$ is a submanifold such that $\omega_Y \neq 0$, then the restriction to Y of (I,ω) is again a Pfaffian system in good form.

Applying this to a submanifold $X \subset J^1(\mathbb{R},M)$ we see that the canonical system with constraints is a Pfaffian system in good form.

Remark. A natural question to ask is how much more general Pfaffian systems in good form are than canonical systems with constraints. Shifting notation slightly, let $\tilde{X} \subset J^1(\mathbb{R},M)$ be a submanifold and denote by $(\tilde{I},\tilde{\omega})$ the restriction of the canonical system on $J^1(\mathbb{R},M)$ to $\tilde{X}$. Let

$$\tilde{W}^* \subset \tilde{L}^* \subset T^*(\tilde{X})$$

be the corresponding sub-bundles. Since in standard local coordinates $(x;y^1,\dots,y^m;\dot{y}^1,\dots,\dot{y}^m)$ on $J^1(\mathbb{R},M)$ we have that $\tilde{\omega} = dx|\tilde{X}$, it is clear that *for canonical systems with constraints the Pfaffian system* $\tilde{L}^*$ *is completely integrable.* Conversely, it may be shown that (cf. [12]):

(I.c.10) *Let* (I,ω) *be a Pfaffian system in good form on a manifold* X *with corresponding sub-bundles* $W^* \subset L^* \subset T^*(X)$. *Then in order that there exist locally defined mappings* f *that induce* (I,ω) *from the canonical system on* $J^1(\mathbb{R},\mathbb{R}^m)$ *it is necessary and sufficient that* L^* *be completely integrable.*

Explanations. "Locally defined mappings" means that each point $p \in X$ has a neighborhood U on which f may be defined; i.e., we have

$$\begin{array}{l} X \\ \cup \\ U \xrightarrow{\ f\ } J^1(\mathbb{R},\mathbb{R}^m) \end{array} .$$

Using the standard coordinates $(x,y,\dot{y})$ on $J^1(\mathbb{R},\mathbb{R}^m)$ we set $\tilde{\theta}^\alpha = dy^\alpha - \dot{y}^\alpha dx$, $\tilde{\omega} = dx$, and

$$\begin{cases} \tilde{W}^* = \operatorname{span}\{\tilde{\theta}^1,\dots,\tilde{\theta}^m\} \\ \tilde{L}^* = \operatorname{span}\{\tilde{\theta}^1,\dots,\tilde{\theta}^m,\omega\} \end{cases} .$$

To say that f *induces* (I,ω) *from the canonical system* means that the differential f_* has maximal rank and that

$$\begin{cases} f^*(\tilde{W}^*) = W^* \\ f^*(\tilde{L}^*) = L^* \end{cases} .$$

Definition. A Pfaffian system in good form for which L^* is completely integrable is said to be *locally embeddable*.(11)

(I.c.11) Example. The canonical system on $J^2(\mathbb{R},\mathbb{R}^m)$ is given by

$$\begin{cases} \theta^\alpha = dy^\alpha - \dot{y}^\alpha dx = 0 \\ \dot{\theta}^\alpha = d\dot{y}^\alpha - \ddot{y}^\alpha dx = 0 \\ \omega = dx \neq 0 \end{cases}.$$

The structure equations are

$$\begin{cases} d\theta^\alpha = -\dot{\theta}^\alpha \wedge \omega \\ d\dot{\theta}^\alpha = -d\ddot{y}^\alpha \wedge \omega \end{cases} \qquad (I.c.12)$$

Thus the system is in good form. Since $\mathrm{span}\{\omega;\theta^\alpha;\dot{\theta}^\alpha\} = \mathrm{span}\{dx;dy^\alpha;d\dot{y}^\alpha\}$ the canonical system on $J^2(\mathbb{R},\mathbb{R}^m)$ is locally embeddable. In fact, the canonical inclusion (cf. Chapter 0, Section b))

$$J^2(\mathbb{R},\mathbb{R}^m) \subset J^1(\mathbb{R},J^1(\mathbb{R},\mathbb{R}^m))$$

gives a global embedding.

Again, it is clear that $\mathbb{R}^m$ may be replaced by any manifold M. Moreover, by considering the restriction of the canonical system to submanifolds $X \subset J^2(\mathbb{R},M)$ it is clear that any classical 2nd order system with constraints is an embeddable Pfaffian system in good form. Finally, it is also clear that these considerations extend to k^{th} order classical systems with constraints given by submanifolds $X \subset J^k(\mathbb{R},M)$.

Returning to the general discussion, we assume given a Pfaffian system (I,ω) in good form with derived system given by a sub-bundle $W_1^* \subset W^*$. The basic invariant of (I,ω) is given by the following

Definition. The *Cartan integer* $s_1 = s_1(I,\omega)$ is the rank of W^*/W_1^*.

Equivalently, s_1 is the number of independent 1-forms $\pi^\alpha \in C^\infty(X,T^*(X)/L^*)$ appearing in the structure equation (I.c.6).(12) In fact, let us say that the basis $\{\theta^\alpha\}$ for W^* is *adapted* to $W_1^* \subset W^*$ in case

$$\begin{cases} W_1^* = \mathrm{span}\{\theta^1,\dots,\theta^{s-s_1}\} \\ W^* = \mathrm{span}\{\theta^1,\dots,\theta^s\} \end{cases}.$$

If we use the additional ranges of indices

$$\begin{cases} 1 \leqq \rho,\ \sigma \leqq s - s_1 \\ s - s_1 + 1 \leqq \mu,\ \nu \leqq s \ , \end{cases}$$

then (I.c.6) becomes what we shall call the *refined structure equations*

$$\begin{cases} d\theta^\rho \equiv 0 \bmod \{\theta^\alpha\} \\ d\theta^\mu \equiv -\pi^\mu \wedge \omega \bmod \{\theta^\alpha\} \\ \pi^\mu \quad \text{linearly independent modulo} \quad L^* \end{cases} \tag{I.c.13}$$

for Pfaffian systems in good form.

(I.c.14) <u>Example</u>. Suppose that $X \subset J^1(\mathbb{R},\mathbb{R}^m)$ is defined by constraints

$$g^\rho(x,y,\dot{y}) = 0, \qquad 1 \leqq \rho \leqq s - s_1 \ ,$$

where

$$\operatorname{rank} \| \partial g^\rho(x,y,\dot{y}) / \partial \dot{y}^\alpha \| = s - s_1 \ .$$

Then since on X we have

$$0 = dg^\rho \equiv \frac{\partial g^\rho}{\partial \dot{y}^\alpha} \, d\dot{y}^\alpha \bmod \operatorname{span}\{\theta^\alpha,\omega\}$$

it follows that the restriction to X of the canonical system on $J^1(\mathbb{R},\mathbb{R}^m)$ has Cartan integer s_1. This is the setting for the Lagrange problem in Chapter IV, Section e).

(I.c.15) <u>Example</u>. From (I.c.12) it follows that for the canonical system on $J^2(\mathbb{R},\mathbb{R}^m)$ we have

$$\begin{cases} W_1^* = \operatorname{span}\{\theta^\alpha\} \\ s_1 = m \ . \end{cases}$$

If $X \subset J^2(\mathbb{R},\mathbb{R}^m)$ is defined by constraints

$$g^\rho(x,y,\dot{y},\ddot{y}) = 0 \qquad 1 \leqq \rho \leqq m - s_1$$

with

$$\operatorname{rank} \| \partial g^\rho / \partial \ddot{y}^\alpha \| = m - s_1 \ ,$$

then the restriction to X of the canonical system on $J^2(\mathbb{R},\mathbb{R}^m)$ has Cartan integer s_1.

It is straightforward to generalize these examples to submanifolds $X \subset J^k(\mathbb{R}, M)$ for any manifold M.

The goal of this section is to compute the variational equations (I.b.15) for a Pfaffian system in good form and thereby conclude a rigorous version of the following heuristic statement (cf. just below (I.c.5)):

(I.c.16) *A general* $[v] \in T_N(V(I,\omega))$ *depends on* s_1 *functions of one variable.*

Before doing this it is convenient to digress and discuss two completely integrable Pfaffian systems canonically associated to an arbitrary Pfaffian system I (without independence condition).

(I.c.17) Discussion. Let $W^* \subset T^*(X)$ be a sub-bundle defining a differential ideal $I \subset A^*(X)$ and consider the 1st derived map (I.c.7). Assuming that $W_1^* = \ker \delta$ is a sub-bundle we consider the 2nd derived map

$$\delta_1 : W_1^* \to \Lambda^2 T^*(X) / W_1^* \wedge T^*(X) \quad .$$

If we set $W_2^* = \ker \delta_1$, and continue assuming at each stage constant rank we obtain the so-called *derived flag* $W^* \supset W_1^* \supset W_2^* \supset W_3^* \supset \cdots$.

Definition. $W_\infty^* \equiv \cap_k W_k^*$ is called the *derived system* associated to the Pfaffian system W^*. Of course we must have

$$W_{k_0}^* = W_{k_0+1}^* = \cdots = W_\infty^*$$

for sufficiently large k_0. The derived system W_∞^* is completely integrable, and it is easy to see that W_∞^* *is the largest completely integrable sub-system of* W^*.

An interesting result concerning the derived system is due to Chow [24]. (This result is discussed in [7],[43], and [46].) Given the Pfaffian system $W^* \subset T^*(X)$, we say that two points $p,q \in X$ are *accessible* in case there is a sequence N_i of connected integral manifolds N_i, $i = 1,..,m$ of W^* and points $p_i \in N_i \cap N_{i+1}$ such that $p \in N_1$ and $q \in N_m$.

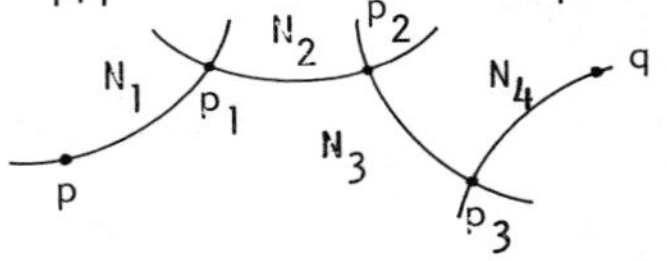

(I.c.18) THEOREM (Chow). *Two points* $p,q \in X$ *are accessible if and only if they lie on the same connected leaf of the foliation defined by the derived system* W_∞^*.

Clearly this is equivalent to showing that every point in a neighborhood of p lies on an integral curve of I through p.

(I.c.19) Example. For $J^2(\mathbb{R},\mathbb{R}^m)$ the derived flag is

$$\begin{array}{ccccc} \operatorname{span}\{\theta^\alpha;\dot\theta^\alpha\} & \supset & \operatorname{span}\{\theta^\alpha\} & \supset & (0) \\ \| & & \| & & \| \\ W^* & \supset & W_1^* & \supset & W_\infty^* \end{array} .$$

For $J^3(\mathbb{R},\mathbb{R}^m)$ the derived flag is (with the obvious notation)

$$\begin{array}{ccccccc} \operatorname{span}\{\theta^\alpha;\dot\theta^\alpha;\ddot\theta^\alpha\} & \supset & \operatorname{span}\{\theta^\alpha;\dot\theta^\alpha\} & \supset & \operatorname{span}\{\theta^\alpha\} & \supset & (0) \\ \| & & \| & & \| & & \| \\ W^* & \supset & W_1^* & \supset & W_2^* & \supset & W_\infty^* \end{array}$$

In general, roughly speaking the derived flag keeps track of "how many derivatives" are implicit in the Pfaffian system viewed as a system of differential equations.

We now turn to the other completely integrable system associated to W^*.

(I.c.20) Discussion. Given a sub-bundle $W^* \subset T^*(X)$ generating a differential ideal $I \subset A^*(X)$ we define the *associated system* $A(I) \subset C^\infty(X,T(X))$ to be the set of vector-fields v satisfying

$$v \lrcorner I \subset I \quad . \qquad (I.c.21)$$

It is easy to check that as a consequence of the differential closure of I

$$v,w \in A(I) \Rightarrow [v,w] \in A(I) \quad . \qquad (I.c.22)$$

We assume that the values $v(x) \in T_x(X)$, $v \in A(I)$, span a sub-bundle which we denote by $A(I) \subset T(X)$. Thus $A(I) = C^\infty(X,A(I))$. It is clear that

$$A(I) \subset (W^*)^\perp$$

and we set

$$C(I) = A(I)^{\perp} \subset T^*(X) \quad .$$

Note that by (I.c.22) the Pfaffian system $C(I)$ is completely integrable and that $W^* \subset C(I)$.

Definition. $C(I)$ is the *Cauchy characteristic system* associated to I.

Its importance lies in the following result (cf. [11], [17]):

(I.c.23) THEOREM. *Locally on* X *we may choose coordinates* $(u,v) = (u^1,\ldots,u^k;v^1,\ldots,v^\ell)$ *such that* $C(I)$ *is generated by* $du^1,\ldots,du^k$. *Moreover,* I *is generated by* 1*-forms*

$$\theta^\alpha = f_1^\alpha(u)du^1 + \cdots + f_k^\alpha(u)du^k \ .$$

In other words, the Cauchy characteristic system locally "picks out a minimal set of variables needed to express I". One dimensional integral manifolds of I are locally of the form

$$x \to (u(x), v(x))$$

where $x \to u(x)$ is an integral curve of the Pfaffian system generated by the above 1-forms $\theta^\alpha = \theta^\alpha(u,du)$ and where $v(x)$ is an arbitrary function of x.

(I.c.24) Example (cf. Chapter 0, Section d)). Let Ψ be a closed 2-form on X. Then $\Psi(x) \in \Lambda^2 T_x^*(X)$ is an alternating bilinear form in each tangent space $T_x(X)$. We define the *rank* $\rho(x)$ by

$$\begin{cases} \Psi(x)^{\rho(x)} \neq 0 \\ \Psi(x)^{\rho(x)+1} = 0 \end{cases}$$

(the notation $\Psi(x)^r$ means $\overbrace{\Psi(x) \wedge \ldots \wedge \Psi(x)}^{r}$).

(I.c.25) Definition (cf. (0.d.6). The *Cartan system* $C(\Psi)$ is the Pfaffian system generated by all 1-forms

$$v \lrcorner \Psi$$

where v is a vector field on X.

The simplest situation occurs when the rank is a constant ρ. Then the well-known *theorem of Darboux* (cf. (0.d.8)), [2], [9]) states that there is the local normal form

$$\Psi = du_1 \wedge dv^1 + \cdots + du_\rho \wedge dv^\rho$$

where the functions $u_1,\ldots,u_\rho,v^1,\ldots,v^\rho$ are part of a coordinate system on X. The Cartan system and Cauchy characteristic system coincide and are given by

$$C(\Psi) = \operatorname{span}\{du_1,\ldots,du_\rho,dv^1,\ldots,dv^\rho\}$$

Later on (cf. Chapter I, Section e)) we shall be especially concerned with the Cartan system $C(\Psi)$ in cases when the rank $\rho(x)$ is not constant.

Returning to the general discussion we want to compute the variational equations of a general Pfaffian system (I,ω) in good form and use these to justify (I.c.16). It will simplify our notation to assume that there are no Cauchy characteristics (the general case is done the same way). Referring to the refined structure equations (I.c.13), we claim as a consequence that:

(I.c.26) *the* 1-*forms* $\{\omega;\theta^\alpha;\pi^\mu\}$ *give a coframe on* X.

Indeed, these forms are linearly independent and therefore span a subbundle $K^*\subset T^*(X)$. It follows immediately from the definition and structure equations (I.c.13) that $A(I) = (K^*)^\perp$. Thus our assumption of no Cauchy characteristics is equivalent to $K^* = T^*(X)$.

We denote by $\{\partial/\partial\omega;\partial/\partial\theta^\alpha;\partial/\partial\pi^\mu\}$ the dual frame of vector fields.

To compute the number of solutions to the variational equations (I.b.15) we recall our notation $L^* = \operatorname{span}\{\omega,\theta^1,\ldots,\theta^s\}$ and we denote by $L^*\wedge L^*\subset A^2(X)$ the 2-forms spanned over $C^\infty(X)$ by the $\theta^\alpha\wedge\theta^\beta$ and $\theta^\alpha\wedge\omega$. We shall first make the computation under the assumption that (I.c.13) is

$$\begin{cases} d\theta^\rho \equiv 0 \mod L^*\wedge L^* \\ d\theta^\mu \equiv -\pi^\mu\wedge\omega \mod L^*\wedge L^*, \end{cases} \tag{I.c.27}$$

and then we will explain why it is sufficient to treat this case.

We denote by $W^*\wedge W^*\subset A^2(X)$ the 2-forms spanned over $C^\infty(X)$ by the $\theta^\alpha\wedge\theta^\beta$. Replacing π^μ by a term $\pi^\mu + F^\mu_\alpha\theta^\alpha$ if necessary, we may rewrite (I.c.27) as

$$\begin{cases} d\theta^\rho \equiv -E^\rho_\alpha \theta^\alpha \wedge \omega & \text{mod } W^* \wedge W^* \\ d\theta^\mu \equiv -\pi^\mu \wedge \omega & \text{mod } W^* \wedge W^* \quad . \end{cases} \tag{I.c.28}$$

A typical vector field on X is

$$v = A\, \partial/\partial\omega + B^\rho\, \partial/\partial\theta^\rho + B^\mu\, \partial/\partial\theta^\mu + C^\mu\, \partial/\partial\pi^\mu \, .$$

Adding to v a tangent vector to N we may assume that $A = 0$. Using (I.c.28) and the fact that $\theta^\alpha \equiv 0 \bmod N$ (so that $v \lrcorner (\theta^\alpha \wedge \theta^\beta) \equiv 0 \bmod N$), the equations (I.b.15)

$$d(v \lrcorner \theta^\alpha) + v \lrcorner d\theta^\alpha \equiv 0 \quad \text{mod } N$$

are

$$\begin{cases} dB^\rho - E^\rho_\alpha B^\alpha \omega \equiv 0 \quad \text{mod } N \\ dB^\mu - C^\mu \omega \equiv 0 \text{ mod } N \end{cases} .$$

If $N = \{a \leqq x \leqq b\}$ has coordinate x and $v \in C^\infty(N, T(X))$, then these equations are

$$\begin{cases} \dfrac{dB^\rho(x)}{dx} + E^\rho_\alpha(x) B^\alpha(x) D(x) = 0 \\ \dfrac{dB^\mu(x)}{dx} + C^\mu(x) D(x) = 0 \end{cases} \tag{I.c.29}$$

where $D(x) = \langle \omega, \partial/\partial x \rangle \neq 0$. The solution to this linear O.D.E. system on $a \leqq x \leqq b$ is uniquely given by prescribing arbitrarily the functions $C^\mu(x)$ and the initial values $B^\alpha(a)$. Thus we may say that the "size" of $T_N(V(I,\omega))$ is given as follows:

A general $[v] \in T_N(V(I,\omega))$ *is specified by* s_1 *arbitrary functions of one variable plus a certain number of constants.*(13)

Remark. The *Cartan-Kähler Theorem* for Pfaffian systems in good form states roughly that "the integral manifolds $N \in V(I,\omega)$ depend on s_1 functions of one variable (plus a certain number of constants)" (cf. [11],[17],[47], and [62]). Usually the result is stated with the additional assumptions of *involutiveness* and *real-analyticity*, but neither of these is required in the case of one independent variable. In fact, although we shall not need it, it seems almost certain that a $[v]$ given by solving the O.D.E. system (I.c.28) is always the infinitesimal variation associated to an honest curve $\{N_t\} \subset V(I,\omega)$. This explains our calling $T_N(V(I,\omega))$ the "tangent space" to $V(I,\omega)$ at N.

To explain why we may assume (I.c.27) is best done by a separate discussion, quite interesting in its own right, of the concept of the prolongation of an exterior differential system.

(I.c.30) Discussion. Let X be a manifold, n an integer, and denote by

$$\pi: G_n(X) \to X$$

the Grassmann bundle whose fibres

$$\pi^{-1}(p) = G_n(T_p(X))$$

are the Grassmann manifolds of n-planes in the tangent spaces $T_p(X)$. We will describe a canonical Pfaffian differential system (H,ω) *over* $G_n(X)$. To do this we must give a filtration

$$W^* \subset L^* \subset T^*(G_n(X)) \tag{I.c.31}$$

where the number of independent variables is n; i.e.,

$$\operatorname{rank}(L^*/W^*) = n .$$

We denote points of $G_n(X)$ by (p,E) where $p \in X$ and $E \subset T_p(X)$ is an n-plane, and then we set (where π^* is the *co*differential of π)

$$\left\{\begin{array}{lcl} W^*_{(p,E)} & = & \pi^*(E^\perp) \\ \cap & & \\ L^*_{(p,E)} & = & \pi^*(T^*_p(X)) \end{array}\right. . \tag{I.c.32}$$

Definition. The Pfaffian system (H,ω) defined by (I.c.31), (I.c.32) is called the *universal prolongation* associated to the manifold X and the integer n.

Let us see what (H,ω) looks like in local coordinates. If $(x^1,\ldots,x^n;y^1,\ldots,y^m) = (x^i;y^\alpha)$ $(1 \leqq i,j \leqq n;\ 1 \leqq \alpha,\beta \leqq m)$ are local coordinates on X, then an open set $U \subset G_n(X)$ is given by n-planes E such that

$$dx^1 \wedge \ldots \wedge dx^n|E \neq 0 .$$

Any plane $E \in U$ has a unique basis of the form

$$e_i = \partial/\partial x^i + \ell^\alpha_i\, \partial/\partial y^\alpha ,$$

and $(x^i;y^\alpha;\ell^\alpha_i)$ gives a local coordinate system on $U \subset G_n(X)$. From

(I.c.32) we see that $(H,\omega)|U$ is generated by the Pfaffian equations

$$\begin{cases} \theta_i^\alpha = dy^\alpha - \ell_i^\alpha \, dx^i = 0 \\ \omega = dx^1 \wedge .. \wedge dx^n \neq 0 \end{cases} \qquad . \qquad (I.c.33)$$

An integral manifold of (H,ω)

$$f: N \to G_n(X)$$

is given locally by

$$(x^i) \to (x^i ; y^\alpha(x)) ; \ell_i^\alpha(x)) \qquad ,$$

and from (I.c.33) it follows that

$$\ell_i^\alpha(x) = \frac{\partial y^\alpha(x)}{\partial x^i} \qquad .$$

In other words, given an immersion

$$f: N \to X \qquad\qquad (I.c.34)$$

there is a *canonical lift*

$$
\begin{array}{ccc}
 & & G_n(X) \\
 & f_* \nearrow & \downarrow \\
N & \xrightarrow{\;f\;} & X
\end{array}
$$

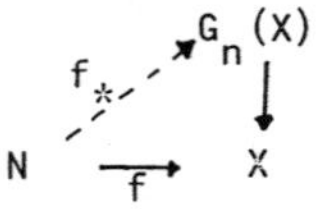

where by definition

$$f_*(p) = (f(p), f_* T_p(N)) \; .$$

Our remarks may be summarized by:

(I.c.35) *The integral manifolds of* (H,ω) *are the canonical lifts of maps* (I.c.34).

Now suppose that (I,ω) is an exterior differential system on X where $\deg \omega = n$. We do *not* assume that I is a Pfaffian system. The integral elements of (I,ω) form a subset $V(I,\omega) \subset G_n(X)$, which in practice will generally be an analytic variety. Assuming this to be the case, we denote by $X^{(1)} \subset V(I,\omega)$ the open dense set of smooth points.

__Definition.__ The 1st *prolongation* $(I^{(1)},\omega)$ of (I,ω) is given by the restriction to $X^{(1)} \subset G_n(X)$ of the canonical differential system (H,ω).

An immediate consequence of (I.c.35) and the definitions is:

(I.c.36) *The integral manifolds of* (I,ω) *and* $(I^{(1)},\omega)$ *are in one-to-one correspondence.*

Although easy to define the process of prolonging a differential system is somewhat subtle to understand, especially when $n > 1$. These matters are discussed in e.g., [12], [17], [60], and [62].

When $n = 1$ we see from (I.c.33) that locally (H,ω) is isomorphic to the canonical system on $J^1(\mathbb{R},\mathbb{R}^m)$. In particular, it follows that $(I^{(1)},\omega)$ is a Pfaffian system in good form. In fact we can say more. Suppose that (I,ω) is in good form with structure equations (I.c.13). The integral elements $V(I,\omega) \subset G_1(X)$ are lines in the tangent spaces to X defined by linear equations

$$\begin{cases} \theta^\alpha = 0 \\ \pi^\mu - \ell^\mu \omega = 0 \end{cases} . \tag{I.c.37}$$

Indeed, these are exactly the equations of a line in $T(X)$ on which $\theta^\alpha = 0$, $\omega \neq 0$. It follows that locally

$$X^{(1)} \cong X \times \mathbb{R}^{s_1}$$

where $\mathbb{R}^{s_1}$ has coordinates $\ell = (\ell^1,\ldots,\ell^{s_1}) = (\ell^\mu)$. The 1st prolongation $(I^{(1)},\omega)$ is the Pfaffian system on $X^{(1)}$

$$\begin{cases} \theta^\alpha = 0 \\ \Theta^\mu = \pi^\mu - \ell^\mu\omega = 0 \\ \omega \neq 0 \end{cases} . \tag{I.c.38}$$

Note that

$$\pi^\mu \equiv 0 \bmod L^{*(1)} \tag{I.c.39}$$

where

$$L^{(1)^*} = \operatorname{span}\{\omega;\theta^\alpha;\Theta^\mu\} .$$

Denote by $T^*(X) \wedge T^*(X)$ the subspace of $A^2(X^{(1)})$ spanned over $C^\infty(X^{(1)})$ by pullbacks of 2-forms on X. From (I.c.39) it follows that:

$$T^*(X) \wedge T^*(X) \subset L^{(1)^*} \wedge L^{(1)^*} .$$

On the other hand, clearly

$$d\theta^{\alpha} \equiv 0 \mod T^*(X) \wedge T^*(X)$$
$$d\Theta^{\mu} \equiv -d\ell^{\mu} \wedge \omega \mod T^*(X) \wedge T^*(X) \ .$$

Setting $\Pi^{\mu} = d\ell^{\mu}$ we infer that $(I^{(1)},\omega)$ has the structure equations

$$\begin{cases} d\theta^{\alpha} \equiv 0 \mod L^{(1)^*} \wedge L^{(1)^*} \\ d\Theta^{\mu} \equiv -\Pi^{\mu} \wedge \omega \mod L^{(1)^*} \wedge L^{(1)^*} \end{cases} . \qquad \text{(I.c.40)}$$

If we compare (I.c.38), (I.c.40) with (I.c.27), it follows that:

(I.c.41) *The Cartan integer for the 1st prolongation* $(I^{(1)},\omega)$ *of* (I,ω) *is the same as for* (I,ω). *Moreover, the structure equations* (I.c.27) *are valid for* $(I^{(1)},\omega)$.

On the basis of (I.c.36) and the previous discussion we may conclude that for any Pfaffian system in good form, a general $[v] \in T_N(V(I,\omega))$ depends on s_1-functions of one variable plus a certain number of constants. In fact, this can be made quite explicit by showing that the bijection

$$V(I,\omega) \to V(I^{(1)},\omega)$$

given by

$$\{f: N \to X\} \to \{f_*: N \to G_1(X)\}$$

induces an isomorphism

$$T_N(V(I,\omega)) \xrightarrow{\sim} T_N(V(I^{(1)},\omega)) \ ,$$

but we shall not do this here.

The process of prolongation of an exterior differential system (I,ω) on a manifold X with one independent variable (i.e., $\deg \omega = 1$) is closely related to the following construction, a special case of which will be of fundamental importance in Chapter I, Section e).

(I.c.42) <u>Construction</u>. Denote by $X_1 \subset X$ the image of the natural projection

$$\pi: V(I,\omega) \to X \ .$$

Any integral manifold of (I,ω) must certainly lie in X_1, but since the integral elements in $V(I,\omega)$ *may not be tangent to* X_1 (25)

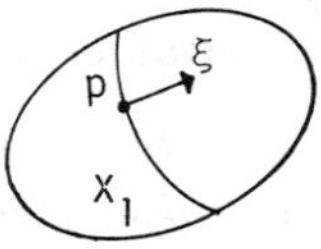

(here (p,ξ) is a typical element of $V(I,\omega)$), we cannot say there will be an integral manifold of (I,ω) through a general point of X_1. So we set $I_1 = I|X_1$, $\omega = \omega|X_1$, and repeat the construction for (I_1,ω) on X_1. In this way we obtain a "descending sequence" of exterior differential systems (I_k,ω) on X_k. Either for some k, $X_k = \not{c}$ and (I,ω) has no integral manifolds, or else the construction stabilizes at a non-empty X_k for which

$$V(I_k,\omega) \to X_k$$

is *surjective*. In this case, integral manifolds of (I,ω) through each point of X_k exist and coincide with those of (I_k,ω).

d) Derivation of the Euler-Lagrange Equations; Examples

We now return to the problem of determining the variational equations of the functional (I.a.4), only now we omit reference to f and simply write

$$\Phi(N) = \int_N \varphi \; . \tag{I.d.1}$$

For a 1-parameter family $N_t \subset X$ of integral manifolds of (I,ω) we want to compute the derivative at $t = 0$ of $\Phi(N_t)$. In other words, we want to evaluate

$$\frac{d}{dt}\left(\int_{N_t} \varphi\right)_{t=0} = \; ? \tag{I.d.2}$$

for a curve $\{N_t\} \subset V(I,\omega)$. Of course, without specifying "endpoint conditions" this doesn't make sense. In fact, due to the possibly complicated nature of (I,ω) it is not clear just what these endpoint conditions should be. Moreover, since the infinitesimal variations of $N \subset X$ are given by normal vector fields satisfying the O.D.E. system (I.b.15), the standard derivation (cf. [29]) of the Euler-Lagrange equations using arbitrary "test" variations will not carry over. Therefore, it would seem that adapting the usual formalism of the Euler-Lagrange equations requires digging into the structure theory of

(I,ω), and in particular into the derived flag (since this more or less tells "how many derivatives" are involved).

We shall proceed differently, arguing heuristically to find a set of equations on N that imply the vanishing of (I.d.2) under compactly supported variations. Of course, it follows from (I.c.27) that there will generally be no compactly supported $[v] \in T_N(V(I,\omega))$, but we simply *ignore* this and *formally* proceed to derive a set of equations that must hold if (I.d.2) is to vanish. This turns out to be a very beautiful system of equations in its own right, which we study in the remainder of this Chapter and in Chapters II, III. Then, in Chapter IV we will discuss endpoint conditions and justify calling them the Euler-Lagrange equations associated to the variational problem $(I,\omega;\varphi)$.

Let $N \subset X$ be an integral manifold of (I,ω) and $v \in C^\infty(N,T(X))$ an infinitesimal variation of N in X that satisfies the equations (I.b.15) (thus $[v] \in T_N(V(I,\omega))$. We assume that $N = \{a \leqq x \leqq b\}$ and, most importantly, that

$$v|\partial N = 0 ; \qquad \text{(I.d.3)}$$

i.e., $v(a) = v(b) = 0$. Let $N_t \subset X$ be a 1-parameter family of submanifolds with infinitesimal variation v. The basic computation (I.b.5) gives

$$\begin{aligned}\frac{d}{dt}\left(\int_{N_t} \varphi\right)_{t=0} &= \int_N v \lrcorner\, d\varphi + d(v \lrcorner\, \varphi) \\ &= \int_N v \lrcorner\, d\varphi + \int_{\partial N} v \lrcorner\, \varphi \\ &= \int_N v \lrcorner\, d\varphi\end{aligned}$$

by Stokes' theorem and our assumption (I.d.3). We consider $\Phi: V(I,\omega) \to \mathbb{R}$ formally as a function on an "infinite dimensional manifold" and *define* its *differential* $\delta\Phi(N)$ at $N \in V(I,\omega)$ to be the linear function

$$\delta\Phi(N): T_N(V(I,\omega)) \to \mathbb{R}$$

given by

$$\delta\Phi(N)[v] = \frac{d}{dt}\left(\int_{N_t} \varphi\right)_{t=0} .$$

Then for $v \in C^\infty(N,T(X))$ satisfying (I.b.15) and (I.d.3) we have

$$\delta\Phi(N)(v) = \int_N v \lrcorner d\varphi \quad . \tag{I.d.4}$$

Concerning this equation we make three important observations:

i) If we set

$$\varphi_1 = \varphi + \lambda_\alpha \theta^\alpha \tag{I.d.5}$$

where the λ_α are functions on X, then since $\theta^\alpha_N = 0$ the functional $\Phi(N)$ remains unchanged. Thus we must be able to show that

$$\int_N v \lrcorner d\varphi = \int_N v \lrcorner d\varphi_1 \quad . \tag{I.d.6}$$

<u>Proof of (I.d.6)</u>. Using the 2nd remark following (I.b.15) we have for $\theta = \lambda_\alpha \theta^\alpha$ that

$$v \lrcorner d\theta + d(v \lrcorner \theta) \equiv 0 \mod N \, .$$

Thus

$$\begin{aligned} \int_N v \lrcorner d(\varphi - \varphi_1) &= \int_N v \lrcorner d\theta \\ &= -\int_N d(v \lrcorner \theta) \\ &= -\int_{\partial N} v \lrcorner \theta \\ &= 0 \end{aligned}$$

by (I.d.3). This proves (I.d.6).

ii) The definition of $\delta\Phi(N)[v]$ makes sense; that is, the quantity

$$\frac{d}{dt}\left(\int_{N_t} \varphi\right)_{t=0}$$

has the properties: a) it depends only on $v \in C^\infty(N,T(X))$ and not on the extension of v to a vector field on all of X, and b) it vanishes in case v is tangent to N (and satisfies (I.d.3)), and therefore depends only on the section $[v]$ of the normal bundle determined by v.

This is proved in the same way as remarks i), ii) following (I.b.15).

iii) If we set

$$\varphi_2(v) = \varphi + d\eta(v) \tag{I.d.7}$$

where η depends on v, then because $d^2 = 0$ certainly

$$\int_N v \lrcorner d\varphi = \int_N v \lrcorner d\varphi_2(v) \quad .$$

Since

$$\int_N \varphi_2(v) = \int_N \varphi + \int_{\partial N} \eta(v)$$

we shall only want to consider substitutions (I.d.7) where $\eta(v)(p) = 0$ if $v(p) = 0$. This means that $\eta(v)$ *depends linearly on* v *over* $C^\infty(N)$; i.e.

$$\eta \in C^\infty(N, \mathrm{Hom}(T(X), \mathbb{R})) \quad .$$

For such an η we have

$$\eta_{\partial N} = 0 \qquad (I.d.8)$$

whenever (I.d.3) holds (recalling our notation for the restriction of differential forms to submanifolds, $\eta_{\partial N}$ means the restriction of the 0-form η to ∂N).

Using these remarks it follows that:
Whatever the equations

$$\delta\Phi(N)[v] = 0 \qquad (I.d.9)$$

turn out to be, they must be invariant under the substitutions (I.d.5) *and* (I.d.7).
Invariance under (I.d.7) means essentially that the equations (I.d.9) should be expressed in terms of $d\varphi$, and combining this with invariance under (I.d.5) gives the important conclusion:

The equations (I.d.9) *should be expressed in terms of* $d(\varphi + \lambda_\alpha \theta^\alpha)$ *where the* λ_α *are to be determined.*

With these two observations as guide we turn to our *heuristic* derivation of the Euler-Lagrange equations. Recalling the equations $L^\alpha(v) = 0$ that define $T_N(V(I,\omega))$ (cf. (I.b.15)), the condition that (I.d.9) hold may be phrased as:

$$\left\{\begin{matrix} L^\alpha(v) = 0 \text{ plus} \\ (I.d.3) \end{matrix}\right\} \Rightarrow \int_N v \lrcorner d\varphi = 0 \quad . \qquad (I.d.10)$$

By Stokes' Theorem the vanishing of the integral $\int_N v \lrcorner d\varphi$ for sufficiently many v's should mean that $v \lrcorner d\varphi \equiv d\eta \bmod N$ where $\eta_{\partial N} = 0$

(this is one place where "heuristic" comes in). Since also η is supposed to depend linearly on v we should have

$$v \lrcorner d\varphi \equiv d\eta(v) \bmod N$$

where $\eta(v)$ *depends linearly on* v *over* $C^\infty(N)$, and therefore satisfies (I.d.8) whenever v satisfies (I.d.3). With this understood and omitting reference to the endpoint conditions, (I.d.10) now reads:

$$L^\alpha(v) = 0 \quad \text{for all} \quad \alpha \Rightarrow v \lrcorner d\varphi \equiv d\eta(v) \bmod N \qquad \text{(I.d.11)}$$

Intuitively, this means that for <u>*any*</u> $v \in C^\infty(N,T(X))$ *the* 1-*form* $v \lrcorner d\varphi$ *must* (mod N) *be a linear combination of the* $L^\alpha(v)$ *plus* $d\eta(v)$ (this is the other place where "heuristic" comes in).[12] In particular, if the condition

$$v \in C^\infty(N,T(X)) \Rightarrow v \lrcorner d\varphi \equiv \lambda_\alpha L^\alpha(v) + d\eta(v) \bmod N \qquad \text{(I.d.12)}$$

is satisfied then our desired equation (I.d.9) will hold (where it is understood that (I.d.8) must be satisfied).

We now will determine what $\eta(v)$ must be in order that *no derivatives of* v *appear on the right hand side of* (I.d.12). Recalling that

$$L^\alpha(v) = v \lrcorner d\theta^\alpha + d(v \lrcorner \theta^\alpha) ,$$

if we set

$$\eta(v) = -\lambda_\alpha (v \lrcorner \theta^\alpha) \qquad \text{(I.d.13)}$$

then (I.d.8) is satisfied and

$$\lambda_\alpha L^\alpha(v) + d\eta(v) \equiv v \lrcorner d(\lambda_\alpha \theta^\alpha) \bmod N . \qquad \text{(I.d.13)}$$

For this choice of $\eta(v)$ and replacing λ_α by $-\lambda_\alpha$ we obtain for (I.d.12)

$$\boxed{v \lrcorner d(\varphi + \lambda_\alpha \theta^\alpha) \equiv 0 \bmod N \quad \text{for } all \quad v \in C^\infty(N,T(X))} \qquad \text{(I.d.14)}$$

Since no derivatives of v appear, these equations are *pointwise* along $N \subset X$. Moreover, they satisfy the conditions of being invariant under the substitutions (I.d.5) and (I.d.7); in fact, *they are the simplest such equations*. Consequently we give the following

<u>Definition</u>. The *Euler-Lagrange equations* associated to the variational problem $(I,\omega;\varphi)$ are the equations (I.d.14) imposed on the integral manifold N of (I,ω).

The somewhat mysterious role of the "functions λ_α to be determined" will be clarified in the next section (observe that in any case we only need determine the λ_α along N).

We shall now consider some examples.

(I.d.15) Example. We consider a classical variational problem (I.a.5). Since the Euler-Lagrange equations (I.d.14) are expressed in coordinate free terms, it will suffice to consider the case when $M = \mathbb{R}^m$ with coordinates $y = (y^1,..,y^m)$. Then using

$$\begin{cases} \varphi = L(x;y;\dot{y})\,dx \\ \theta^\alpha = dy^\alpha - \dot{y}^\alpha dx \end{cases}$$

we have

$$d(\varphi + \lambda_\alpha\theta^\alpha) = (L_{\dot{y}^\alpha} - \lambda_\alpha)\,d\dot{y}^\alpha \wedge dx + (d\lambda_\alpha - L_{y^\alpha}dx) \wedge \theta^\alpha$$

In computing the equations (I.d.14) we may use *any* $v \in C^\infty(N, T(X))$. Of course, it will be sufficient to use a set of vectors v that span $T_x(X)$ for each $x \in N$. More importantly we may use "one less" vector, as shown by the following

(I.d.16) LEMMA. *Let* T *be a vector space and* $\Psi \in \Lambda^2 T^*$ *an alternating bilinear form on* T. *Suppose that* $v_1,..,v_m \in T$ *are vectors that span a hyperplane* H *and* $w \in T$ *is a vector not in* H. *Then if*

$$\langle v_i \lrcorner \Psi, w\rangle = 0 \qquad i = 1,..,m ,$$

it follows that

$$\langle v \lrcorner \Psi, w\rangle \qquad \text{for } \underline{all} \ v \in T.$$

Proof. Any $v \in T$ is of the form

$$v = \alpha^i v_i + \beta w ,$$

and then

$$\begin{aligned} \langle v \lrcorner \Psi, w\rangle &= \alpha^i \langle v_i \lrcorner \Psi, w\rangle + \beta \langle w \lrcorner \Psi, w\rangle \\ &= 0 \end{aligned}$$

by assumption and since $\langle w \lrcorner \Psi, w\rangle = \langle \Psi, w \wedge w\rangle = 0$. Q.E.D.

This lemma will be used, *often without comment*, in computing examples throughout this text.

Another useful lemma, which is formalized in (II.b.4), is that *in computing Euler-Lagrange equations we may always work modulo linear combinations of terms* $\theta^\alpha \wedge \theta^\beta$. Utilizing these two lemmas greatly facilitates the computation of examples.

Returning to example (I.d.15), using the lemma we may compute the Euler-Lagrange equations by taking respectively

$$v = \partial/\partial\dot{y}^\alpha, \qquad v = \partial/\partial\theta^\alpha$$

(the w of the lemma is $\partial/\partial x$). Then the Euler-Lagrange equations (I.d.14) are

$$\begin{cases} (L_{\dot{y}^\alpha} - \lambda_\alpha)dx \equiv 0 \bmod N \\ d\lambda_\alpha - L_{y^\alpha}\,dx \equiv 0 \bmod N \end{cases} .$$

Since $dx \neq 0$ on N the 1st equations give

$$L_{\dot{y}^\alpha} = \lambda_\alpha ,$$

and then the 2nd equations give

$$dL_{\dot{y}^\alpha} - L_{y^\alpha}dx \equiv 0 \mod N \quad .$$

Thinking of N as a 1-jet $x \to (x, y(x), \frac{dy(x)}{dx})$ with independent variable x, these become the usual Euler-Lagrange equations

$$\frac{d}{dx}\,(L_{\dot{y}^\alpha}) = L_{y^\alpha} \tag{I.d.17}$$

found in any textbook (e.g. [29]). When written out this is a 2nd order O.D.E. in the coordinates $y^\alpha(x)$.

(I.d.18) Special Case of (I.d.15). We consider a mechanical system (I.a.7). In local coordinates

$$L(y,\dot{y}) = T(y,\dot{y}) - U(y)$$

where T is given by (I.a.8). The Euler-Lagrange equations (I.d.17) are

$$\frac{d}{dx}\left(g_{\alpha\beta}(y(x))\,\frac{dy^{\beta}(x)}{dx}\right) = \frac{1}{2}\left(\frac{\partial g_{\beta\gamma}(y(x))}{\partial y^{\alpha}}\,\frac{dy^{\beta}(x)}{dx}\,\frac{dy^{\gamma}(x)}{dx}\right) - \frac{\partial U(y(x))}{\partial y^{\alpha}} \tag{I.d.19}$$

Letting $\|g^{\alpha\beta}(y)\|$ be the inverse matrix to $\|g_{\alpha\beta}(y)\|$ and introducing the usual *Christoffel symbols*

$$\Gamma^{\alpha}_{\beta\gamma} = \frac{1}{2}\, g^{\alpha\delta}\left(\frac{\partial g_{\delta\beta}}{\partial y^{\gamma}} + \frac{\partial g_{\delta\gamma}}{\partial y^{\beta}} - \frac{\partial g_{\beta\gamma}}{\partial y^{\delta}}\right)$$

the equations (I.d.19) are equivalent to

$$\frac{d^2y^{\alpha}(x)}{dx^2} + \Gamma^{\alpha}_{\beta\gamma}(y(x))\,\frac{dy^{\beta}(x)}{dx}\,\frac{dy^{\gamma}(x)}{dx} = -U^{\alpha}(y(x)) \tag{I.d.20}$$

where $U^{\alpha} = g^{\alpha\beta}\partial U/\partial y^{\beta}$. In the more customary notation of Riemannian geometry (cf. [6], [53], [64]) we let γ be the curve $x \to y(x)$ in M and

$$v(x) = \frac{\partial y^{\alpha}(x)}{\partial x}\,\partial/\partial y^{\alpha}$$

$$= \text{tangent vector to } \gamma;$$

$$\frac{Dv(x)}{dx} = \left(\frac{d^2y^{\alpha}(x)}{dx^2} + \Gamma^{\alpha}_{\beta\gamma}(y(x))\,\frac{dy^{\beta}(x)}{dx}\,\frac{dy^{\gamma}(x)}{dx}\right)\partial/\partial y^{\alpha}$$

where D is the covariant differential of the Riemannian connection

$$= \textit{acceleration vector} \text{ to } \gamma; \text{ and}$$

$$\nabla U(y) = g^{\alpha\beta}(y)\,\frac{\partial U(y)}{\partial y^{\beta}}\,\frac{\partial}{\partial y^{\alpha}}$$

$$= \text{gradient of the potential function } U.$$

Then equations (I.d.20) are *Newton's law*

$$\frac{Dv(x)}{dx} + \nabla U(y(x)) = 0 \qquad (I.d.21)$$

for motion relative to the conservative force field $F = -\nabla U$.

In $\mathbb{R}^m$ with the flat metric these reduce to the familiar form

$$\frac{d^2y(x)}{dx^2} + \nabla U(y(x)) = 0 \quad .$$

If $M \subset \mathbb{R}^3$ is the cylinder

$$\begin{cases} z = u \\ x = v\cos\varphi, \quad y = v\sin\varphi \end{cases}$$

then the induced metric is

$$ds^2 = du^2 + v^2 d\varphi^2 \ ,$$

and when $U = 0$ we have

$$\begin{cases} L = \frac{1}{2}(\dot{u}^2 + v^2\dot{\varphi}^2) \\ \lambda_1 = L_{\dot{u}} = \dot{u} \\ \lambda_2 = L_{\dot{v}} = v^2\dot{\varphi} \end{cases} \quad .$$

As is well-known the solution to the equations (I.d.17) are the helices

$$\begin{cases} u = c_1 x \\ \varphi = c_2 x \end{cases} \quad .$$

We refer to [1] and [2] for numerous examples of classical mechanical systems. In Chapter III, Section a) we shall discuss in some detail the mechanical system corresponding to rigid body motion.

(I.d.22) Example. We consider a classical 2nd order variational problem (I.a.13). Again it will suffice to treat the case when $M = \mathbb{R}^m$. Then

$$\begin{cases} \varphi = L(x, y, \dot{y}, \ddot{y})dx \\ \theta^\alpha = dy^\alpha - \dot{y}^\alpha dx \\ \dot{\theta}^\alpha = d\dot{y}^\alpha - \ddot{y}^\alpha dx \end{cases}$$

and

$$d(\varphi + \lambda_\alpha\theta^\alpha + \dot{\lambda}_\alpha\dot{\theta}^\alpha) = (L_{\ddot{y}^\alpha} - \dot{\lambda}_\alpha)d\ddot{y}^\alpha \wedge dx$$

$$+ (d\dot{\lambda}_\alpha + \lambda_\alpha dx - L_{\dot{y}^\alpha}dx) \wedge \dot{\theta}^\alpha + (L_{y^\alpha}dx + d\lambda_\alpha) \wedge \theta^\alpha \quad .$$

Taking respectively (cf. lemma (I.d.16))

$$v = \partial/\partial\ddot{y}^\alpha , \qquad v = \partial/\partial\dot{\theta}^\alpha , \qquad v = \partial/\partial\theta^\alpha$$

the Euler-Lagrange equations (I.d.14) are

$$\begin{cases} (L_{\ddot{y}^\alpha} - \dot{\lambda}_\alpha)\,dx \equiv 0 \mod N \\ d\dot{\lambda}_\alpha + (\lambda_\alpha - L_{\dot{y}^\alpha})\,dx \equiv 0 \mod N \\ d\lambda_\alpha + L_{y^\alpha}\,dx \equiv 0 \mod N \end{cases} .$$

Since $dx \neq 0$ on N the first of these gives

$$L_{\ddot{y}^\alpha} = \dot{\lambda}_\alpha ,$$

and then the remaining two become

$$\begin{cases} dL_{\ddot{y}^\alpha} + (\lambda_\alpha - L_{\dot{y}^\alpha})\,dx \equiv 0 \mod N \\ d\lambda_\alpha + L_{y^\alpha}\,dx \equiv 0 \mod N \end{cases} \qquad (15)$$

Thinking of N as a 2-jet $x \to (x, y(x), \frac{dy(x)}{dx}, \frac{d^2y(x)}{dx^2})$ these combine to give

$$\frac{d^2}{dx^2}(L_{\ddot{y}^\alpha}) - \frac{d}{dx}(L_{\dot{y}^\alpha}) + L_{y^\alpha} = 0 \qquad \text{(I.d.23)}$$

(cf. [29]). When written out this is a 4th order O.D.E. in the coordinates $y^\alpha(x)$.

For a trivial special case we take $m = 1$ (i.e., $X = J^2(\mathbb{R},\mathbb{R})$) and

$$L = \frac{1}{2}(\ddot{y})^2 .$$

The equations (I.d.3) are

$$\frac{d^2}{dx^2}\left(\frac{d^2y(x)}{dx^2}\right) = 0$$

with general solution

$$y(x) = \alpha x^3 + \beta x^2 + \gamma x + \delta .$$

A more interesting higher order example will be discussed at the end of Chapter II, Section a) (cf. (II.a.50)).

(I.d.24) <u>Remark</u>. The natural boundary conditions in Example (I.d.15) are clearly

$$y(a) = A, \quad y(b) = B \;,$$

while those in Example (I.d.22) are

$$\begin{cases} y(a) = A\;, & y(b) = B \\ \dfrac{dy}{dx}(a) = \dot{A}\;, & \dfrac{dy}{dx}(b) = \dot{B} \;. \end{cases}$$

We observe that the corresponding conditions on the infinitesimal variation $v \in T_N(V(I,\omega))$ are respectively

$$(v \lrcorner \omega)(x) = (v \lrcorner \theta^\alpha)(x) = 0 \qquad \text{for} \quad x = a \quad \text{and} \quad x = b \tag{I.d.25}$$

$$(v \lrcorner \omega)(x) = (v \lrcorner \theta^\alpha)(x) = (v \lrcorner \dot{\theta}^\alpha)(x) = 0 \qquad \text{for} \quad x = a \quad \text{and} \quad x = b \tag{I.d.26}$$

However, this is misleading since it will *not* in general turn out that the correct boundary conditions are

$$(v \lrcorner \psi)(x) = 0 \qquad \text{for} \quad x \equiv a,b \quad \text{and for all} \quad \psi \in L^* \;,$$

as is the case in (I.d.25), (I.d.26).

(I.d.27) Example. Let (S, ds^2) be a two-dimensional Riemannian manifold. A curve $\gamma \subset S$ will be described parametrically by $s \to p(s) \in S$ where s is its arclength, and we denote by $\kappa(s)$ the *geodesic curvature* of γ (cf. [22], [28], and [56]). We shall investigate the Euler-Lagrange equations for the functional

$$\Phi(\gamma) = \frac{1}{2} \int_\gamma \kappa^2(s)\, ds \;. \tag{I.d.28}$$

When $S = \mathbb{E}^2$ with the flat metric the extremals of this functional have the following physical interpretation (cf. [25]): Imagine in the plane 2 pairs α, α' and β, β' of blocks of wood and a thin wire or hacksaw blade γ in the configuration (the wood blocks are glued in place):

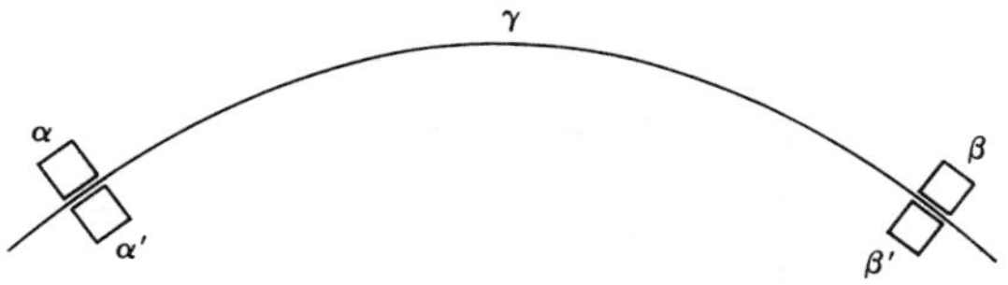

It is important that there be a small gap between α, α' and β, β' so that γ has variable length; only the *direction* of γ between α, α' and β, β' should be fixed. Then the position γ assumes will minimize the integral (I.d.28). If we actually *clamp* γ by closing the gap between α,α' and β,β'

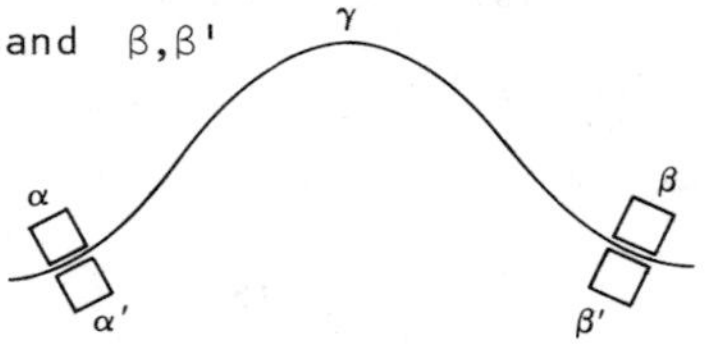

then the position γ assumes will minimize (I.d.28) subject to the constraint (cf. Part a) of the Appendix)

$$\int_\gamma ds = \ell = \text{constant} . \qquad (I.d.29)$$

The same physical interpretation remains true for a general surface S when γ is constrained to lie on S.

Since the geodesic curvature is a 2nd order invariant, this example is clearly a special case of Example (I.d.22). The Euler-Lagrange equations (I.d.23) are then a pair of 4th order equations in terms of any local coordinate system on S. It is worthwhile (but not necessarily recommended) to write these out to see the mess that ensues.

We shall therefore approach the problem more intrinsically by using frames. For this we give the following

Review (cf. [15], [22], [56], or [63]). Denote by $F(S)$ the bundle of orthonormal frames $(p;e_1,e_2)$ for S. Here $p \in S$ and $e_1,e_2 \in T_p(S)$ give an orthonormal basis. Clearly $\dim F(S) = 3$, and on $F(S)$ there is a coframing $\{\omega^1,\omega^2,\omega_1^2\}$ uniquely characterized by the conditions:

i) the 1-forms ω^1,ω^2 are horizontal for the fibering $F(S) \overset{\pi}{\to} S$ and satisfy

$$\langle\omega^i, v\rangle = (e_i, \pi_* v) \qquad v \in T(F(S)) ;$$

and

ii) the *structure equations*

$$\begin{cases} d\omega^1 = -\omega^2 \wedge \omega_1^2 \\ d\omega^2 = \omega^1 \wedge \omega_1^2 \\ d\omega_1^2 = -R\omega^1 \wedge \omega^2 \end{cases} \qquad (I.d.30)$$

are valid. Here R is the pullback to $F(S)$ of the *Gaussian curvature* on S.

Continuing with our discussion of the functional (I.d.28) we set

$$X = F(S) \times \mathbb{R}$$

where $\mathbb{R}$ has coordinate κ, and on X we consider the differential system (I,ω) given in the form (I.a.1) by

$$\begin{cases} \theta^1 = \omega^2 = 0 \\ \theta^2 = \omega_1^2 - \kappa\omega = 0 \\ \omega = \omega^1 \neq 0 \end{cases} . \tag{I.d.31}$$

Any curve $\gamma \subset S$ determines a lift $N \subset X$ given by

$$s \to (p(s), e_1(s), e_2(s), \kappa(s))$$

where $s \to p(s)$ gives γ, $e_1(s)$ is the unit tangent to γ, and $\kappa(s)$ is the geodesic curvature of γ (here, the normal $e_2(s)$ is only determined up to ± 1 and we make the choice given by

$$\frac{De_1}{ds} = \kappa(s)e_2(s)$$

where D is the covariant differential). It is clear that N is an integral manifold of (I,ω) on X, and that conversely any integral manifold of (I,ω) arises in this way from a curve $\gamma \subset S$. In fact, this just says that

$$\begin{cases} (\omega^1)_\gamma = ds \\ (\omega^2)_\gamma = 0 \\ (\omega_1^2)_\gamma = \kappa(s)ds , \end{cases}$$

which is well-known in the theory of surfaces (loc. cit.).

As a preliminary to analyzing the Euler-Lagrange equations associated to (I.d.28) we shall discuss the simpler arclength functional

$$\Theta(N) = \int_N \omega , \tag{I.d.32}$$

defined on integral manifolds $N \subset X$ of (I.d.31). It is clear that $\Theta(N) = \int_\gamma ds$, and the arclength functional is a variational problem $(I,\omega;\varphi)$ where $\varphi = \omega$. By the structure equations (I.d.30)

$$\begin{aligned} d(\varphi + \lambda_1\theta^1 + \lambda_2\theta^2) \equiv (d\lambda_1 + (\kappa - \lambda_2(R+\kappa^2))\omega) \wedge \theta^1 + (d\lambda_2 + \lambda_1\omega) \wedge \theta^2 \\ - \lambda_2 d\kappa \wedge \omega \quad \text{mod} \quad \theta^1 \wedge \theta^2 \end{aligned} . \tag{I.d.33}$$

By contracting respectively with (cf. lemma (I.d.16))

$$v = \partial/\partial\kappa, \qquad v = \partial/\partial\theta^2, \qquad v = \partial/\partial\theta^1$$

the Euler-Lagrange equations (I.d.14) are

$$\begin{cases} \lambda_2\omega \equiv 0 & \text{mod } N \\ d\lambda_2 + \lambda_1\omega \equiv 0 & \text{mod } N \\ d\lambda_1 + (\kappa-\lambda_2(R+\kappa^2))\omega \equiv 0 & \text{mod } N \end{cases}$$

Since $\omega \neq 0$ on N these give respectively

$$\lambda_2 = 0, \qquad \lambda_1 = 0, \qquad \kappa = 0 \quad .$$

This is the familiar characterization of *geodesics* as being curves whose geodesic curvature is zero.

We now consider a general variational problem $(I,\omega;\varphi)$ where, for some function $L(\kappa)$ of one variable,

$$\varphi = L(\kappa)\omega \quad .$$

The functional (I.d.28) corresponds to $L(\kappa) = \kappa^2/2$. Using (I.d.30) and (I.d.31) we find that

$$d(\varphi + \lambda_1\theta^1 + \lambda_2\theta^2) \equiv (L'(\kappa)-\lambda_2)d\kappa \wedge \omega + (d\lambda_1+(\kappa L(\kappa)-\lambda_2(R+\kappa^2))\omega)\wedge\theta^1$$
$$+ (d\lambda_2+\lambda_1\omega) \wedge \theta^2 \quad \text{mod} \quad \theta^1\wedge\theta^2$$

(this reduces to (I.d.33) when $L = 1$). Contracting this 2-form with $v = \partial/\partial\kappa$, $v = \partial/\partial\theta^1$, $v = \partial/\partial\theta^2$ as before, the Euler-Lagrange equations (I.d.14) are

$$\begin{cases} (L'(\kappa) - \lambda_2)\omega \equiv 0 & \text{mod } N \\ d\lambda_2 + \lambda_1\omega \equiv 0 & \text{mod } N \\ d\lambda_1 + (\kappa L(\kappa)-\lambda_2(R+\kappa^2))\omega \equiv 0 & \text{mod } N \end{cases} \quad . \tag{I.d.34}$$

The first equation gives

$$\lambda_2 = L'(\kappa) \quad ,$$

while the 2nd and 3rd yield the differential equations

$$\lambda_1 = -L''(\kappa)\frac{d\kappa}{ds}$$

(16)

$$\frac{d}{ds}\left(L''(\kappa)\frac{d\kappa}{ds}\right) - (\kappa L(\kappa) - L'(\kappa)(R+\kappa^2)) = 0. \qquad (\text{I.d.35})$$

In any local coordinate system (u,v) on S, (I.d.35) is a 4th order equation for the coordinates $u(x)$, $v(x)$ of γ. When $L = \kappa^2/2$ it is

$$\frac{d^2\kappa}{ds^2} + \left(\frac{\kappa^3}{2} + \kappa R\right) = 0 \quad . \qquad (\text{I.d.36})$$

When R is constant we may say more. Namely, from (I.d.36) it follows that

$$\frac{d}{ds}\left(\left(\frac{d\kappa}{ds}\right)^2 + \left(\frac{\kappa^4}{4} + \kappa^2 R\right)\right) = 0$$

along any solution curve γ to the Euler-Lagrange equations associated to (I.d.28). Along such a curve this gives

$$\kappa'^2 + \left(\frac{\kappa^4}{4} + \kappa^2 R\right) = c \qquad (\text{I.d.37})$$

where $\kappa' = d\kappa/ds$.

We recall that the complete, simply-connected surface S of constant curvature R is said to be a *space form* (of dimension two). It is known that S is either the sphere of radius $1/R$ in $\mathbb{E}^3$ $(R > 0)$, $\mathbb{E}^2$ $(R = 0)$, or a hyperbolic plane $(R < 0)$. In each case S admits a transitive simple Lie group of isometries, and it is a consequence of (I.b.7) (cf. Chapter III, Section b) for a general discussion) that a curve $\gamma \subset S$ is uniquely determined up to rigid motion by its geodesic curvature as a function of arclength. Because of (I.d.37) the Euler-Lagrange equations associated to the functional (I.d.26) are said to be quasi-integrable by quadratures (there is a general discussion of this notion in Chapter II, Section a)). We shall explain what this means in the present case.

We first do this analytically. Upon separation of variables (I.d.37) gives

$$\frac{d\kappa}{\sqrt{c - \kappa^2 R - \frac{\kappa^4}{4}}} = ds$$

Then we *define* the function $\kappa(s)$ by

$$\int_{\kappa_0}^{\kappa(s)} \frac{d\xi}{\sqrt{c - \xi^2 R - \frac{\xi^4}{4}}} = s + c_1$$

(this is similar to defining $\sin x$ by $\int_0^{\sin x} \frac{d\xi}{\sqrt{1-\xi^2}} = x$). Then it may be proved that $\kappa(s)$ is the restriction to an interval on $\mathbb{R} \subset \mathbb{C}$ of a doubly-periodic meromorphic function defined in the whole complex s-plane (these are called *elliptic functions*; cf. [27]). In this way it may be said that we have "solved" the Euler-Lagrange equations associated to (I.d.28). Actually, since so much is known about elliptic functions this is not too much of a misnomer.

Algebro-geometrically we consider the complex algebraic curve C defined in $\mathbb{C}^2$ with coordinates (x,y) by

$$y^2 + \frac{x^4}{4} + Rx^2 - c = 0 \quad .$$

(As usual we should consider the compactified curve in the complex projective plane obtained by adding the points at infinity. For present purposes we shall only worry about the affine curve C.) With the following exceptions, C is a smooth curve of genus one (elliptic curve):

<u>Case i)</u> $c = 0$. Then $(0,0)$ is an ordinary double point

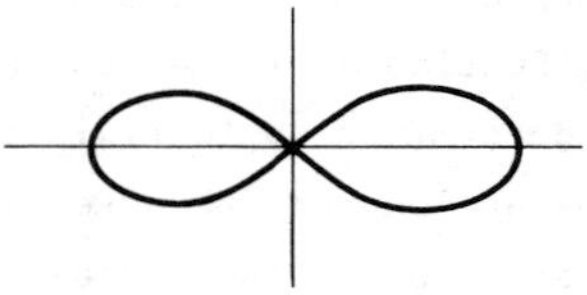

<u>Case ii)</u> $c = -R^2$. Then C has double points at $(0, \pm\sqrt{-2R})$. In fact, in this case C decomposes into a pair of imaginary parabolas having $\pm(0,\sqrt{-2R})$ as its only real points $(R < 0)$

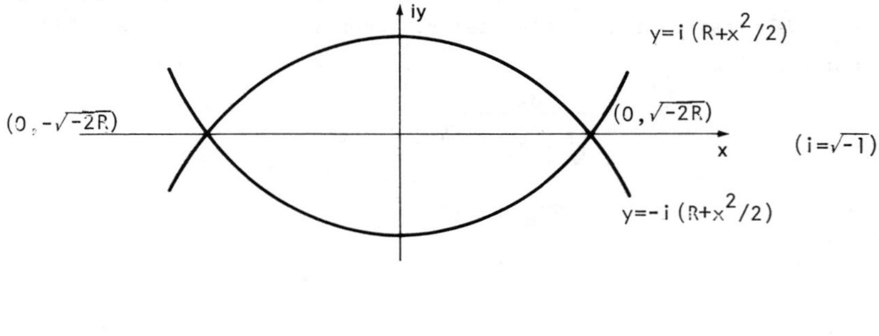

Aside from these cases, C is an elliptic curve whose points are given parametrically by

$$s \to (\kappa(s), \kappa'(s))$$

where $\kappa(s)$ is an elliptic function whose modulus depends on R and c (cf. [36] and [59]). In the exceptional cases each component of C is rational and is given parametrically by

$$s \to (\kappa(s), \kappa'(s))$$

where $\frac{d\kappa(s)}{ds}$ is a *rational* function of s. Thus, in these exceptional cases the Euler-Lagrange equations of (I.d.28) may be integrated by elementary functions.

It is also instructive to study (I.d.36) by the method of phase portraits. We do this here for the case $R = 0$ and refer to Section a) of the Appendix for the case $R \neq 0$.

First we note that (I.d.36) may be considered as Newton's equation of motion

$$\ddot{x} = -\frac{x^3}{2} \tag{I.d.38}$$

for a 1-dimensional mechanical system with time parameter s and potential energy

$$U(x) = \frac{x^4}{8} \quad .$$

Equation (I.d.38) may be thought of as a "higher order spring". In the (x,y) phase plane we consider the O.D.E. system

$$\begin{cases} \dot{x} = y \\ \dot{y} = -x^3/2 \end{cases}$$

given by the vector field

$$v = y\,\partial/\partial x - x^3/2\,\partial/\partial y \quad .$$

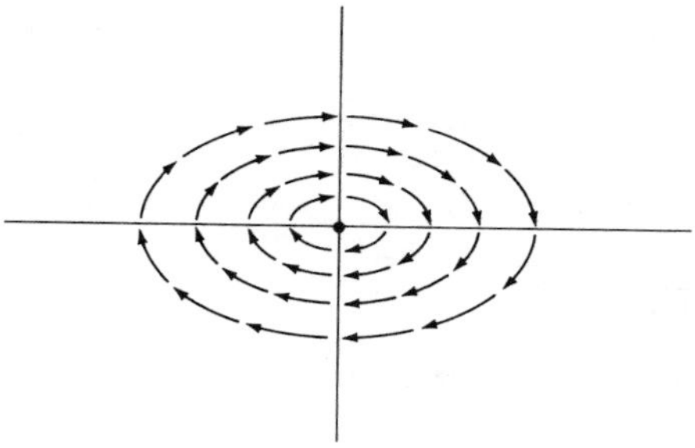

This vector field has a unique critical point (*equilibrium position*) at (0,0) corresponding to the extremal for (I.d.28) given by straight lines. The integral curves of v are the level sets of the total energy function

$$H = \frac{y^2}{2} + \frac{x^4}{8} = c > 0$$

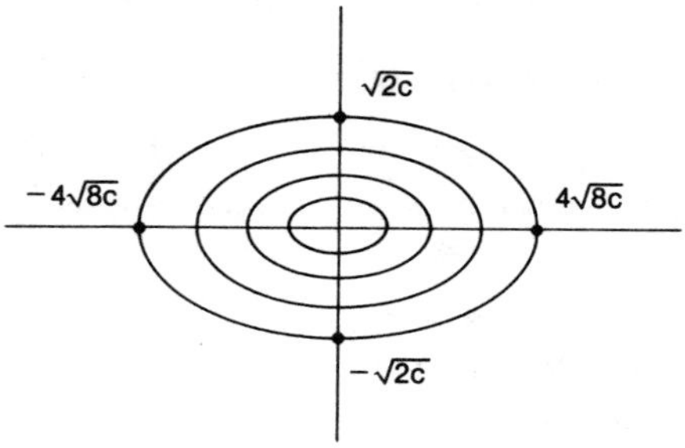

In fact these level sets are just the real points of the algebraic curve $y^2 + x^4/4 = 2c$ considered above.

Due to the absence of critical points of v on the x-axis other than (0,0) we infer that *circles are* <u>*not*</u> *extremals of* (I.d.28) (however, cf. (A.a.10)).

The elliptic function $\kappa(s)$ has the real period

$$\ell = \int_{-\sqrt[4]{8c}}^{+\sqrt[4]{8c}} \frac{dx}{\sqrt{2c - \frac{x^4}{4}}} .$$

However we remark that a periodic curvature function $\kappa(s)$ does not necessarily give a closed curve in $\mathbb{E}^2$ (cf. Gluck [30]). Suppose in fact that

$$\begin{cases} \dfrac{dx(s)}{ds} = e_1(s) & x(s) \in \mathbb{E}^2 \\ \dfrac{de_1(s)}{ds} = \kappa(s)e_2(s) & \\ \dfrac{de_2(s)}{ds} = -\kappa(s)e_1(s) & \end{cases}$$

are the Frenet equations of $\gamma \subset \mathbb{E}^2$, and that

$$\kappa(s + \ell) = \kappa(s) .$$

If $e_1(s) = (\cos\theta(s), \sin\theta(s))$ then

$$\theta'(s) = \kappa(s) .$$

Hence, a necessary condition that γ be a closed curve of period ℓ (i.e., that $x(s+\ell) = x(s)$) is

$$\int_0^\ell \kappa(s)\, ds = 2\pi w \qquad \text{(I.d.39)}$$

where $w \in \mathbb{Z}$ is to be the *rotation index* of γ.(17) Assuming (I.d.39) we have

$$\begin{cases} \theta(s+\ell) = \theta(s) + 2\pi w \\ e_1(s+\ell) = e_1(s) \end{cases} .$$

It then follows from the 1st Frenet equation that

$$x(s+\ell) - x(s) = C$$

is constant. The condition $C = 0$ is equivalent to

$$\int_0^\ell e_1(s)\, ds = 0 \quad ;$$

i.e. we must have

$$\int_0^\ell \cos\theta(s)\, ds = 0 = \int_0^\ell \sin\theta(s)\, ds . \qquad \text{(I.d.40)}$$

Since $\cos\theta(s)$, $\sin\theta(s)$ are periodic functions of $s \in \mathbb{R}/\ell\mathbb{Z}$, (I.d.40) gives two more conditions on $\kappa(s)$.

The solution $\kappa(s)$ to (I.d.36) contains two constants of integration. However, three conditions are required on $\kappa(s)$ to yield a closed curve γ. If we fix the length of γ, then it may be shown that the only closed extremals of (I.d.28) of this length are circles.

e) The Euler-Lagrange Differential System; Non-degenerate Variational Problems; Examples

Given a variational problem $(I,\omega;\varphi)$ on a manifold X we want to write the corresponding Euler-Lagrange equations (I.d.14) as a differential system (J,ω) on an associated manifold Y. Our construction will be functorial and will hopefully help to clarify the role of the "functions λ_α to be determined" in (I.d.14). (In classical cases they do *not* reduce to Lagrange multipliers; c.f. Chapter IV, Section e).)

We begin with the following general

(I.e.1) <u>Construction</u>.[18] On a manifold Z suppose we are given the data: i) a closed 2-form Ψ with associated Cartan system $C(\Psi)$ (cf. (I.c.25)) generated by the set of 1-forms

$$\{v \lrcorner \Psi : v \in C^{\infty}(Z, T(Z))\} ;$$

and ii) a 1-form ω.

<u>Remarks</u>: i) Thus $C(\Psi)$ is the differential ideal generated by the collection of 1-forms $\varphi = v \lrcorner \Psi$ where v is a smooth vector field on X.

ii) We do *not* assume that Ψ has constant rank (cf. footnote (25) and the remark below (I.e.5)); in fact, in our cases of interest this will generally not be the case. On the other hand, in examples everything will be real-analytic so that the loci where the rank of Ψ is constant will be analytic subvarieties.

We consider the Pfaffian differential system $(C(\Psi),\omega)$ on Z. Denoting by $\mathbb{P}T(Z) = G_1(Z)$ the *projectivized tangent bundle* consisting of tangent lines to Z, the set of integral elements (cf. Chapter 0, Section e))

$$V(C(\Psi),\omega) \subset \mathbb{P}T(Z)$$

is given by pairs (p,ξ) where $p \in Z$ and $\xi \subset T_p(Z)$ is a line spanned by a vector v satisfying

$$v \lrcorner (\Psi(p)) = 0, \quad v \lrcorner (\omega(p)) \neq 0.^{(19)} \qquad \text{(I.e.2)}$$

We denote by $Z_1 \subset Z$ the image of the natural projection

$$V(C(\Psi),\omega) \xrightarrow{\pi} Z \quad ,$$

and we assume that Z_1 is a manifold (this is not strictly necessary: in the real analytic case Z_1 will be an analytic subvariety and we may restrict our attention to the open dense set of smooth points of Z_1).

For any submanifold $N \subset Z$ we define its 1st *prolongation* to be the subset of $G_n(Z)$ $(n = \dim N)$ given by

$$p_1(N) = \{(p, T_p(N)) : p \in N\} \quad .$$

The condition that N (of dimension one) be an integral manifold of $(C(\Psi),\omega)$ is expressed by

$$p_1(N) \subset V(C(\Psi),\omega) \quad .$$

In particular this implies that $N \subset Z_1$.

We set

$$\begin{cases} C_1(\Psi) = C(\Psi)_{Z_1} = \{\varphi_{Z_1} : \varphi \in C(\Psi)\} \subset A^*(Z_1) \\ \omega = \omega_{Z_1} \in A^1(Z_1) \\ \Psi_1 = \Psi_{Z_1} \in A^2(Z_1) \end{cases} .$$

Thus $(C_1(\Psi),\omega)$ is the restriction to Z_1 of $(C(\Psi),\omega)$. We note that

$$C(\Psi_1) \subset C_1(\Psi)$$

but that equality may not hold (this somewhat subtle point has to do with the *position* of $T_p(Z_1)$ in $T_p(Z)$ relative to the 2-form $\Psi(p)$ at points $p \in Z_1$). By what we have said *the integral manifolds of* $(C(\Psi),\omega)$ *and* $(C_1(\Psi),\omega)$ *are the same.* Hence we consider

$$V(C_1(\Psi),\omega) \subset \mathbb{P}T(Z_1)$$

and denote by $Z_2 \subset Z_1$ the image of the projection

$$V(C_1(\Psi),\omega) \overset{\pi}{\to} Z_1 \quad .$$

Another somewhat subtle point is that

$$V(C_1(\Psi),\omega) \subset V(C(\Psi),\omega)$$

but again equality may not hold, since a vector v satisfying (I.e.2) may not be tangent to Z_1 (cf. the picture in construction (I.c.42)).

We assume that Z_2 is a manifold and repeat the construction to obtain the Pfaffian system $(C_2(\Psi),\omega)$ on Z_2 where

$$\begin{cases} C_2(\Psi) = C(\Psi)_{Z_2} = \{\varphi_{Z_2} : \varphi \in C_1(\Psi)\} \subset A^*(Z_2) \\ \omega = \omega_{Z_2} \in A^1(Z_2) \\ \Psi_2 = (\Psi_1)_{Z_2} = \Psi_{Z_2} \in A^2(Z_2) \end{cases} .$$

Again it is the case that

$$\begin{cases} C(\Psi_2) \subset C_2(\Psi) \\ V(C_2(\Psi)) \subset V(C_1(\Psi)) \end{cases}$$

but in either case equality may not hold. However, as before the integral manifolds of $(C_1(\Psi),\omega)$ and $(C_2(\Psi),\omega)$ are the same.

We inductively repeat this construction obtaining a sequence of Pfaffian systems $(C_k(\Psi),\omega)$ on manifolds Z_k where $Z_k \supset Z_{k+1}$. For some k_0 we must then have $Z_{k_0} = Z_{k_0+1}$ (we exclude the case where the number of components of Z_k becomes infinite), and this implies that $Z_{k_0+1} = Z_{k_0+2} = \cdots$. We then set

$$\begin{cases} Y = \bigcap_k Z_k = Z_{k_0} \subset Z \\ J = C_{k_0}(\Psi) \subset A^*(Y) \\ \omega = \omega_Y \in A^1(Y) \end{cases} \qquad \text{(I.e.3)}$$

We remark that $C(\Psi_Y) \subset J$, but in general equality will not hold.

By construction the Pfaffian differential system (J,ω) has the two properties (these uniquely characterize it):

(I.e.4) *The projection* $V(J,\omega) \to Y$ *is surjective; and*

(I.e.5) *The integral manifolds of* $(C(\Psi),\omega)$ *on* Z *coincide with those of* (J,ω) *on* Y.

Remark. In practice the construction of Y will proceed as follows: For $p \in Z$ we define $\Psi^{\perp}(p) \subset T_p(Z)$ by

$$\Psi^{\perp}(p) = \{v \in T_p(Z) : v \lrcorner \Psi(p) = 0 \text{ in } T_p^*(Z)\}$$

and $\omega^{\perp}(p) \subset T_p(Z)$ by

$$\omega^{\perp}(p) = \{v \in T_p(Z) : v \lrcorner \omega(p) = 0\} .$$

Then at a general point of Z we will have

$$\Psi^{\perp}(p) \subset \omega^{\perp}(p) . \qquad \text{(I.e.6)}$$

However, over a submanifold $Z_1 \subset Z$ (along which Ψ *must* jump in rank) $\Psi^{\perp}(p)$ and $\omega^{\perp}(p)$ will meet transversely; i.e., (I.e.6) will not hold. Thus for $p \in Z_1$ there exist vectors $v \in T_p(Z)$ satisfying (I.e.2), but at a general point of Z_1 no such v will lie in $T_p(Z_1) \subset T_p(Z)$. Then $Z_2 \subset Z_1$ is defined by the condition that there exists $v \in T_p(Z_1)$ satisfying (I.e.2), etc.

To motivate this construction we consider a variational problem $(I,\omega;\varphi)$ on a manifold X. Recall that $I \subset A^*(X)$ is the differential ideal generated by the sections of a sub-bundle $W^* \subset T^*(X)$, and that $\theta^1,\ldots,\theta^s$ denotes a set of 1-forms that locally give a framing for W^*. In $\mathbb{R}^s$ we use coordinates $\lambda = (\lambda_1,\ldots,\lambda_s)$ and set

$$\begin{cases} Z = X \times \mathbb{R}^s \\ \psi = \varphi + \lambda_\alpha \theta^\alpha \\ \omega = \omega \end{cases} \tag{I.e.7}$$

(I.e.8) <u>REMARK</u> (of more importance than most remarks). *Intrinsically* we have $Z = W^*$ where

$$\begin{array}{ccc} W^* & \subset & T^*(X) \\ \pi \downarrow & & \downarrow \pi \\ X & = & X \end{array}$$

On $T^*(X)$ there is the canonical 1-form θ (cf. Chapter 0, Section a)), and

$$\theta = \lambda_\alpha \pi^* \theta^\alpha$$

is simply the restriction to W^* of this canonical 1-form. Moreover,

$$\psi = \pi^* \varphi + \theta$$
$$\omega = \pi^* \omega$$

in (I.e.7). Having established these intrinsic identifications we shall prefer to work in coordinates and drop the π^*'s.[20]

The main observation is the following

(I.e.9) THEOREM. *The solutions to the Euler-Lagrange equations* (I.d.14) *are in a natural one-to-one correspondence with the integral manifolds* $N \subset Y$ *of* $(C(\Psi),\omega)$ *where* $\Psi = d\psi$.

<u>Proof</u>. Let $N \subset X$ be an integral manifold of (I,ω) that satisfies the Euler-Lagrange equations (I.d.14), and let s be a coordinate along N. Then we may determine functions $\lambda_\alpha(s)$ such that for all tangent vectors v *to the ambient manifold* X at points of N

$$v \lrcorner (d(\varphi + \lambda_\alpha \theta^\alpha))_N = 0 \tag{I.e.10}$$

(<u>Note</u>: Strictly speaking we must extend the functions $\lambda_\alpha(s)$ from N to X and then compute $d(\varphi + \lambda_\alpha \theta^\alpha)$. The point is that since $\theta^\alpha_N = 0$,

$v \lrcorner (d\varphi + \lambda_\alpha \theta^\alpha))_N$ is independent of the extension of the λ_α's.)

Associated to $N \subset X$ and the functions $\lambda_\alpha(s)$ is an obvious curve $\tilde{N} \subset X \times \mathbb{R}^s$, and we claim that $\tilde{N}$ is an integral manifold of $(C(\Psi),\omega)$ where

$$\Psi = d\psi = d\varphi + d\lambda_\alpha \wedge \theta^\alpha + \lambda_\alpha d\theta^\alpha \quad . \tag{I.e.11}$$

Using the isomorphism resulting from $Z = X \times \mathbb{R}^s$

$$T(Z) \cong T(X) \oplus T(\mathbb{R}^s)$$

and (I.e.10), it will suffice to show that

$$(\partial/\partial\lambda_\alpha \lrcorner \Psi)_{\tilde{N}} = 0 \quad . \tag{I.e.12}$$

But clearly

$$\partial/\partial\lambda_\alpha \lrcorner \Psi = \theta^\alpha \quad , \tag{I.e.13}$$

so that (I.e.12) follows from $\theta^\alpha_N = 0$. (To be precise, (I.e.13) should be $\partial/\partial\lambda_\alpha \lrcorner \Psi = \pi^*\theta^\alpha$, and then for $\tilde{v} \in T(Z)|_{\tilde{N}}$

$$\tilde{v} \lrcorner \pi^*\theta^\alpha = \pi_*\tilde{v} \lrcorner \theta^\alpha = 0 \ . \ )$$

Conversely, let $\tilde{N} \subset Z$ be an integral manifold of $(C(\Psi),\omega)$ with projection $\pi(\tilde{N}) = N \subset X$. Then from (I.e.13) it follows that N is an integral manifold of (I,ω). Moreover, the conditions

$$(v \lrcorner \Psi)_{\tilde{N}} = 0 \qquad \text{for} \qquad v \in T(X) \subset T(Z)$$

are just the Euler-Lagrange equations (I.d.14) for N. Q.E.D.

On the basis of (I.e.4), (I.e.5), which implies that through each point of Y there passes an integral manifold of (I,ω), and Theorem (I.e.9) we give the following

<u>Definitions</u>: i) We will call (J,ω) on Y the *Euler-Lagrange differential system* associated to the variational problem $(I,\omega;\varphi)$ on X.

ii) We shall call Y the *momentum space* associated to $(I,\omega;\varphi)$.

This latter terminology will be explained below; note that $Y \subset W^* \subset T^*(X)$ so that at first glance Y seems to be in the "right place" for a space of momenta. However, for a classical variational problem posed on a manifold M, $X = J^1(M) \cong \mathbb{R} \times T(M)$ so that

$$T^*(X) \cong T^*(\mathbb{R}) \times T^*(T(M)) \ ,$$

whereas the usual momentum space is $T^*(M)$ (or $\mathbb{R}\times T^*(M)$ for time dependent Lagrangians). Under suitable non-degeneracy assumptions on the Lagrangian Y will reduce to the usual momentum space, but the above construction of Y works under completely general circumstances.

In Section ii) of Chapter II, Section b) there is an example where the construction of Y through the sequence of intermediate manifolds $\{Z_k\}$ takes *five* steps. Roughly speaking these steps correspond to the derived flag of I and amount to successively determining more of the λ_α's. At first glance this might seem complicated, but in fact the successive determination of the Z_k's is *algorithmic* and never seems to be difficult to carry out in practice *once one has computed the structure equations of the original Pfaffian differential system* I.

We shall use our standard notations

$$\begin{cases} \psi_Y = \psi|Y \\ \Psi_Y = \Psi|Y \qquad \text{where} \quad \Psi = d\psi \end{cases} .$$

According to Remark (I.e.6) there is a natural inclusion

$$Y \subset T^*(X)$$

and

$$\Psi_Y = \pi^*\Phi + \Omega_Y$$

where $\Phi = d\varphi$, $\pi: Y \to X$ is the projection, and where Ω is the canonical symplectic form on $T^*(X)$. In general the Euler-Lagrange equations associated to any functional defined on integral manifolds of (I,ω) will involve the position of $W^* \subset T^*(X)$ relative to the 2-form Ψ.

Now it is possible to construct examples where the Euler-Lagrange equations have no solutions (this means that $Y = \emptyset$), and it is easy to give examples where they have too many solutions; cf. footnote (16) to Chapter I. Somewhat remarkably there seems to be a rather general class of variational problems in which the Euler-Lagrange system (J,ω) has a familiar standard local form. This is given by the following

Definition. The variational problem $(I,\omega;\varphi)$ is said to be *non-degenerate* in case

$$\begin{cases} \dim Y = 2m+1 \\ \psi_Y \wedge (\Psi_Y)^m \neq 0 \end{cases} . \tag{28}$$

Remark. It frequently but not always turns out that even for very general variational problems, non-degeneracy means that a maximum number s_1 (= The Cartan integer) of the local coordinates of X may be eliminated in favor of the undetermined functions λ_α. Moreover, the Euler-Lagrange Pfaffian system will almost always appear simpler when we use as many of the 1-forms $d\lambda_\alpha$ as possible as part of a coframe on Y (this is in fact clear from (I.e.11)). In this sense the philosophy is *opposite* the traditional method of Lagrange multipliers.

For non-degenerate variational problems the rank of Ψ_Y is everywhere the maximum possible value m. Moreover, whereas in general we only have an inclusion $C(\Psi_Y) \subset J$, *for non-degenerate problems we have*

$$C(\Psi_Y) = J \quad . \tag{I.e.14}$$

Proof. By definition, over each $y \in Y$ there are integral elements of (J,ω). Moreover, it is always the case that $V(J,\omega) \subset V(C(\Psi_Y),\omega)$. On the other hand, since Ψ_Y has maximal rank

$$\dim\{v \in T_y(Y) : v \lrcorner (\Psi(y)) = 0\} = 1 \ .$$

Hence we must have

$$V(C(\Psi_Y),\omega) = V(J,\omega)$$

since each side consists of a line in $T_y(Y)$ lying over each point $y \in Y$. Q.E.D.[21]

If $(I,\omega;\varphi)$ is non-degenerate, then by the Theorem of Pfaff-Darboux (cf. (0.d.9)) we may locally choose coordinates $(x;u_1,\ldots,u_m;v^1,\ldots,v^m)$ on Y such that

$$\psi_Y = -H(x,u,v)dx + u_i dv^i \tag{I.e.15}$$

where H is a non-vanishing function. In many examples we will even be able to choose global coordinates (perhaps many-valued as in the case of polar coordinates) so that (I.e.15) holds.

The exterior derivative of (I.e.15) gives

$$\Psi_Y = -H_{u_i} du_i \wedge dx - H_{v^i} dv^i \wedge dx + du_i \wedge dv^i \ .$$

By contracting with $\partial/\partial u_i$, $\partial/\partial v^i$ respectively (cf. lemma (I.d.16)) we find that locally (J,ω) is the Pfaffian system

$$\begin{cases} dv^i - H_{u_i}\,dx = 0 \\ du_i + H_{v^i}\,dx = 0 \\ \omega \equiv F dx \mod\{du_i, dv^i\}, \quad F \neq 0 \end{cases} \qquad (I.e.16)$$

Taking x as coordinate along the integral manifolds of (I.e.16) we see that the solution curves associated to the Euler-Lagrange equations of a non-degenerate variational problem on X have locally on Y the familiar *Hamiltonian form*

$$\begin{cases} \dfrac{dv^i}{dx} = \dfrac{\partial H}{\partial u_i} \\ \dfrac{du_i}{dx} = -\dfrac{\partial H}{\partial v^i} \end{cases} .$$

It is a remarkable fact that the Hamiltonian formalism appears, at least locally and in many cases globally, in such generality (cf. the various examples below).

In general it turns out to be somewhat subtle to predict the dimension of the momentum space associated to $(I,\omega;\varphi)$. In a little while we shall give a general theorem along these lines (cf. (I.e.34)); for the moment we shall give a criterion that $Z_k = Y$ in the construction (I.e.1).

(I.e.17) PROPOSITION. *Suppose that for some* k *we have*

$$\begin{cases} \text{(i)} & C_k(\Psi) = C(\Psi_{Z_k}) \\ \text{(ii)} & \dim Z_k = 2m+1 \\ \text{(iii)} & \psi_{Z_k} \wedge (\Psi_{Z_k})^m \neq 0 \\ \text{(iv)} & \omega \neq 0 \quad \text{on} \quad \Psi^{\perp}_{Z_k} \subset T(Z_k) \end{cases} .$$

(Both (iii) and (iv) should hold at each point of Z_k.)
Then $Y = Z_k$ *and the variational problem* $(I,\omega;\varphi)$ *is non-degenerate.*

<u>Proof</u>. In (iv) $\Psi^{\perp}_{Z_k}$ denotes the line sub-bundle of $T(Z_k)$ whose fibre over $p \in Z_k$ is $\{v \in T_p(Z_k) : v \lrcorner \Psi_{Z_k}(p) = 0\}$. It follows from (i)-(iv) that $V(C_k(\Psi),\omega) \to Z_k$ is surjective, so that $Y = Z_k$. Non-degeneracy follows from (ii) and (iii). Q.E.D.

(I.e.18) Example. Continuing with example (I.d.15) we consider the functional

$$\Phi = \int L\left(x, y(x), \frac{dy(x)}{dx}\right) dx \qquad y = (y^1, \ldots, y^m) \quad \text{(I.e.19)}$$

corresponding to a classical variational problem. We shall say that the functional (I.e.19) is *non-degenerate* in case

$$\det \| L_{\dot{y}^\alpha \dot{y}^\beta} \| \neq 0$$

(to be precise we should specify the open set on which this condition is assumed to hold, but we don't do this). If we write the Euler-Lagrange equations of (I.e.19) in the form (cf. (I.d.17))

$$L_{\dot{y}^\alpha \dot{y}^\beta} \frac{d^2 y^\beta}{dx^2} = g_\alpha\left(x, y(x), \frac{dy(x)}{dx}\right),$$

then non-degeneracy means that we can solve for d^2y^β/dx^2. Accordingly it is not surprising that

(I.e.20) *If* (I.e.19) *is a non-degenerate functional, then the variational problem* $(I, \omega; \varphi)$ *is non-degenerate. Moreover,* $\dim Y = 2m + 1$.

Proof. We shall use the proof as an opportunity to illustrate writing the Euler-Lagrange equations (I.d.17) as the Pfaffian system (J, ω). The manifold Z is $J^1(\mathbb{R}, \mathbb{R}^m) \times \mathbb{R}^m$ with coordinates $(x; y^\alpha; \dot{y}^\alpha; \lambda_\alpha)$, and on Z the forms ψ and Ψ are (cf. (I.e.7), (I.e.11))

$$\begin{cases} \psi = L(x, y, \dot{y})\, dx + \lambda_\alpha \theta^\alpha \\ \Psi = (L_{\dot{y}^\alpha} - \lambda_\alpha) d\dot{y}^\alpha \wedge dx + (d\lambda_\alpha - L_{y^\alpha}\, dx) \wedge \theta^\alpha \end{cases} .$$

On Z we take $\{dx; \theta^\alpha; d\dot{y}^\alpha; d\lambda_\alpha\}$ as a coframe and compute the Cartan system by contracting with $\partial/\partial\lambda_\alpha$, $\partial/\partial\dot{y}^\alpha$, $\partial/\partial\theta^\alpha$, respectively. This gives for $C(\Psi)$ the generating set of Pfaffian equations (cf. (I.d.16))

$$\begin{cases} \partial/\partial\lambda_\alpha \lrcorner\, \Psi = \theta^\alpha = 0 \\ \partial/\partial\dot{y}^\alpha \lrcorner\, \Psi = (L_{\dot{y}^\alpha} - \lambda_\alpha) dx = 0 \\ \partial/\partial\theta^\alpha \lrcorner\, \Psi = L_{y^\alpha} dx - d\lambda_\alpha = 0 \end{cases} \qquad \text{(I.e.21)}$$

We want integral manifolds of $C(Z)$ on which $dx \neq 0$, so following the algorithm in the construction (I.e.1), the 2nd equation in (I.e.21) implies that $Z_1 \subset Z$ is defined by $L_{\dot{y}^\alpha} = \lambda_\alpha$. In fact, at a point $p \in Z_1$ we have $T_p(Z) \cong \mathbb{R}^{3m+1}$ while the Cartan system is given by the vanishing of the $2m$ linear forms $\theta^\alpha(p)$, $(d\lambda_\alpha - L_{y^\alpha}dx)(p)$. Thus there certainly exist integral elements for $(C(\Psi),\omega)$ lying over p.

On Z_1 we have (since $\mathrm{span}\{dy^\alpha, dx\} = \mathrm{span}\{\theta^\alpha, dx\}$)

$$d\lambda_\alpha \equiv L_{\dot{y}^\alpha \dot{y}^\beta} d\dot{y}^\beta \mod \mathrm{span}\{\theta^\alpha, dx\}$$

(here *all forms are restricted to* Z_1). Since $\det\|L_{\dot{y}^\alpha\dot{y}^\beta}\| \neq 0$ we may take $\{dx;\theta^\alpha;d\lambda_\alpha\}$ as a coframe on Z_1. Then

$$\psi_{Z_1} = L dx + \lambda_\alpha \theta^\alpha$$

$$\Psi_{Z_1} = (d\lambda_\alpha - L_{y^\alpha} dx) \wedge \theta^\alpha \;;$$

from this and (I.e.21) it is clear that $C(\Psi_{Z_1}) = C_1(\Psi)$ and that

$$\psi_{Z_1} \wedge (\Psi_{Z_1})^m = m!\, L dx \wedge d\lambda_1 \wedge \theta^1 \wedge \cdots \wedge d\lambda_m \wedge \theta^m \neq 0 \;.$$

By Proposition (I.e.17) we conclude that $Y = Z_1$ and that the variational problem $(I,\omega;\varphi)$ is non-degenerate. Q.E.D. for (I.e.20)

(I.e.22) Remark. We have

$$\psi_Y = (L - \lambda_\alpha \dot{y}^\alpha) dx + \lambda_\alpha dy^\alpha \;.$$

Thus ψ_Y is globally of the form (I.e.15) with

$$H = \lambda_\alpha \dot{y}^\alpha - L$$

Here we are considering H as a function of the local coordinates $(x;y^\alpha;\lambda_\alpha)$ on Y.

It is also possible to take $(x;y^\alpha;\dot{y}^\alpha)$ as local coordinates on Y and write

$$H = \dot{y}^\alpha L_{\dot{y}^\alpha} - L \;,$$

which the reader may recognize as the Hamiltonian (Hamiltonians will be formally (and belatedly) introduced in Chapter II, Section a)).

In this discussion we may replace $\mathbb{R}^m$ by an arbitrary manifold M with local coordinates y^α. Then

$$X = J^1(\mathbb{R},M) \cong \mathbb{R} \times T(M)$$

and intrinsically (cf. Remark (I.e.8))

$$Z = W^* \subset T^*(X) \quad .$$

In fact it is easy to see from the formulas in local coordinates that

$$Z \subset \mathbb{R} \times \mathbb{R}\{dx\} \times \pi^* T^*(M) \subset T^*(X)$$

where we make the identification

$$T^*(X) \cong \mathbb{R} \times \mathbb{R}\{dx\} \times T^*(T(M))$$

and

$$\pi\colon T(M) \to M$$

is the projection. Since $Y \subset Z$ there is a map

$$L\colon Y \to \mathbb{R} \times T^*(M)$$

given in local coordinates by

$$L(x;y^\alpha;\dot{y}^\alpha) = (x;y^\alpha;L_{\dot{y}^\alpha}).$$

In fact, it is straightforward to check that in this case the λ_α are the fibre coordinates of $T^*(M)$ relative to the basis $\{dy^\alpha\}$, and the above map is frequently written as

$$\lambda_\alpha = L_{\dot{y}^\alpha} .$$

In this form we recognize it as the *Legendre transform* (cf. [29]), which by our description is intrinsically defined *as a map on* Y. Classically it is usually assumed that $L(x;y^\alpha;\dot{y}^\alpha)$ is a *convex function* of the $\dot{y}^\alpha$. In this case both the projection $Y \to \mathbb{R} \times T(M)$ and $L\colon Y \to \mathbb{R} \times T^*(M)$ are one-to-one and we may consider the Legendre transform (again intrinsically) as a map

$$L\colon \mathbb{R} \times T(M) \to \mathbb{R} \times T^*(M)$$

given by the same formula $\lambda_\alpha = L_{\dot{y}^\alpha}$.

Since in all cases, even when $\det\|L_{\dot{y}^\alpha\dot{y}^\beta}\| = 0$, there is a map

$$L\colon Y \to \mathbb{R} \times T^*(M)$$

we have chosen to call Y the momentum space associated to the variational problem under consideration.

In the case of a time independent mechanical system (cf. (I.a.7)) where the Lagrangian is

$$L = T(y,\dot{y}) - U(y) ,$$

the Riemannian metric associated to the kinetic energy gives an isomorphism

$$T(M) \overset{\sim}{\to} T^*(M)$$

which is the Legendre transform in this case (L is the identity on the $\mathbb{R}$-factor). We note that

$$\begin{aligned} H &= \dot{y}^\alpha g_{\alpha\beta}(y)\dot{y}^\beta - \frac{1}{2} g_{\alpha\beta}(y)\dot{y}^\alpha\dot{y}^\beta + U(y) \\ &= T(y,\dot{y}) + U(y) \end{aligned}$$

is the *total energy* of the mechanical system. Although we will have more to say about this later, the reader is advised to consult [1], [2].

(I.e.23) Example. Continuing with Example (I.d.22) we consider the functional

$$\Phi = \int L\left(x, y(x), \frac{dy(x)}{dx}, \frac{d^2y(x)}{dx^2}\right) dx \qquad y = (y^1,\dots,y^m) \qquad \text{(I.e.24)}$$

corresponding to a classical 2nd order variational problem. We will say that the functional (I.e.24) is *non-degenerate* in case

$$\det\|L_{\ddot{y}^\alpha\ddot{y}^\beta}\| \neq 0 .$$

Similarly to the previous example the Euler-Lagrange equations (I.d.23) have the form

$$L_{\ddot{y}^\alpha\ddot{y}^\beta} \frac{d^4y^\beta(x)}{dx^4} = g_\alpha\left(x, y(x), \frac{dy(x)}{dx}, \frac{d^2y(x)}{dx^2}, \frac{d^3y(x)}{dx^3}\right)$$

and non-degeneracy means that we can solve for $d^4y^\beta(x)/dx^4$. The analogue of (I.e.20) is

(I.e.25) *If* (I.e.24) *is a non-degenerate functional, then the variational problem* $(I,\omega;\varphi)$ *is non-degenerate. Moreover,* $\dim Y = 4m+1$.

Proof. Again we shall use the proof to illustrate writing the Euler-Lagrange equations as a Pfaffian system. On $Z = J^2(\mathbb{R},\mathbb{R}^m) \times \mathbb{R}^{2m}$ with coordinates $(x;y^\alpha;\dot{y}^\alpha;\ddot{y}^\alpha;\lambda_\alpha;\dot{\lambda}^\alpha)$ we have

$$\psi = Ldx + \lambda_\alpha \theta^\alpha + \dot{\lambda}_\alpha \dot{\theta}^\alpha$$

$$\Psi = (L_{\ddot{y}^\alpha} - \dot{\lambda}_\alpha) d\ddot{y}^\alpha \wedge dx + (d\dot{\lambda}_\alpha + (\lambda_\alpha - L_{\dot{y}^\alpha}) dx) \wedge \dot{\theta}^\alpha$$

$$+ (L_{y^\alpha} dx + d\lambda_\alpha) \wedge \theta^\alpha . \qquad (\text{I.e.26})$$

By contracting with $\partial/\partial\ddot{y}^\alpha$, $\partial/\partial\dot{\lambda}_\alpha$, $\partial/\partial\lambda_\alpha$, $\partial/\partial\dot{\theta}^\alpha$, $\partial/\partial\theta^\alpha$ we find that the Cartan system $C(\Psi)$ is generated by the Pfaffian equations (cf. (I.d.16))

$$\begin{cases} \partial/\partial\ddot{y}^\alpha \lrcorner \Psi = (L_{\ddot{y}^\alpha} - \dot{\lambda}_\alpha) dx = 0 \\ \partial/\partial\dot{\lambda}_\alpha \lrcorner \Psi = \dot{\theta}^\alpha = 0 \\ \partial/\partial\lambda_\alpha \lrcorner \Psi = \theta^\alpha = 0 \\ \partial/\partial\dot{\theta}^\alpha \lrcorner \Psi = -d\dot{\lambda}_\alpha - (\lambda_\alpha - L_{\dot{y}^\alpha}) dx = 0 \\ \partial/\partial\theta^\alpha \lrcorner \Psi = -d\lambda_\alpha - L_{y^\alpha} dx = 0 \end{cases} .$$

Then $Z_1 \subset Z$ is defined by $L_{\ddot{y}^\alpha} = \dot{\lambda}_\alpha$, and since *on* Z

$$L_{\ddot{y}^\alpha \ddot{y}^\beta} d\ddot{y}^\beta \equiv d\dot{\lambda}_\alpha \mod\{dx;\theta^\alpha;\dot{\theta}^\alpha\}$$

we may take $\{dx;\theta^\alpha;\dot{\theta}^\alpha;d\lambda_\alpha;d\dot{\lambda}_\alpha\}$ as a coframe on Z_1. It is then clear that there are integral elements of $(C(\Psi),\omega)$ *tangent to* Z_1. Using (I.e.26)

$$\psi_{Z_1} = Ldx + \lambda_\alpha \theta^\alpha + \dot{\lambda}_\alpha \dot{\theta}^\alpha$$

$$\Psi_{Z_1} = (d\dot{\lambda}_\alpha + (\lambda_\alpha - L_{\dot{y}^\alpha}) dx) \wedge \dot{\theta}^\alpha + (d\lambda_\alpha + L_{y^\alpha} dx) \wedge \theta^\alpha$$

$$\psi_{Z_1} \wedge (\Psi_{Z_1})^{2m} = (2m)!\, Ldx \wedge d\dot{\lambda}_1 \wedge \dot{\theta}^1 \wedge \ldots \wedge d\dot{\lambda}_m \wedge \dot{\theta}^m \wedge d\lambda_1 \wedge \theta^1 \wedge \ldots \wedge d\lambda_m \wedge \theta^m ,$$

and by Proposition (I.e.17) we obtain a proof of (I.e.25). Q.E.D.

Remark. Since

$$\psi_Y = (L - \dot{y}^\alpha \lambda_\alpha - \ddot{y}^\alpha \dot{\lambda}_\alpha) dx + \lambda_\alpha dy^\alpha + \dot{\lambda}_\alpha d\dot{y}^\alpha$$

we see that ψ_Y globally has the form (I.e.15) with 2nd order Hamiltonian

$$H = -L + \ddot{y}^\alpha \dot{\lambda}_\alpha + \dot{y}^\alpha \lambda_\alpha .$$

Note that, in contrast to the previous example, H is not defined on X but only up on Y. We will more to say about Hamiltonians in general in Chapter II.

We shall now give a general result that includes the two preceding examples as special cases. To explain this we assume that (I,ω) is a Pfaffian system in good form as defined in Chapter I, Section c) (cf. (I.c.6)). We then have the refined structure equations (I.c.13). It is obviously advantageous to allow changes of coframe on X that preserve these structure equations. Accordingly we give the following

Definition. *Admissable changes of coframe* of (I,ω) are given by invertible linear substitutions of the form

$$\begin{cases} \tilde{\theta}^{\rho} = A^{\rho}_{\sigma}\theta^{\sigma} \\ \tilde{\theta}^{\mu} = A^{\mu}_{\nu}\theta^{\nu} + A^{\mu}_{\rho}\theta^{\rho} \\ \tilde{\omega} = A\omega + B_{\alpha}\theta^{\alpha} \\ \tilde{\pi}^{\mu} = (A^{-1})^{\mu}_{\nu}\pi^{\nu} + B^{\mu}_{\alpha}\theta^{\alpha} + C^{\mu}\omega \end{cases} \quad . \tag{I.e.27}$$

Even in classical problems it is useful to allow such admissable changes of coframe. For example, suppose on a manifold M we have a mechanical system given by a kinetic energy function T on T(M) (for simplicity we assume the potential energy to vanish). In standard local coordinates $(x;y^{\alpha};\dot{y}^{\alpha})$ on $J^1(\mathbb{R},M)$

$$T = \frac{1}{2} g_{\alpha\beta}(y)\dot{y}^{\alpha}\dot{y}^{\beta}$$

and the canonical Pfaffian system is generated by the Pfaffian equations

$$\begin{cases} \theta^{\alpha} = dy^{\alpha} - \dot{y}^{\alpha}dx = 0 \\ \omega = dx \neq 0 \end{cases} \quad .$$

If we consider the Riemannian metric

$$ds^2 = g_{\alpha\beta}(y)\,dy^{\alpha}dy^{\beta}$$

associated to T, then in Riemannian geometry it is frequently advantageous for computations to use an orthonormal coframe

$$\omega^\alpha = A^\alpha_\beta(y)\,dy^\beta$$

on M whereby

$$ds^2 = \sum_\alpha (\omega^\alpha)^2 \quad .$$

If we identify $J^1(\mathbb{R},M)$ with $\mathbb{R}\times T(M)$ and introduce new coordinates $(x;y^\alpha;p^\alpha)$ where for $v=\dot{y}^\alpha\,\partial/\partial y^\alpha \in T(M)$

$$\langle\omega^\alpha,v\rangle = p^\alpha ,$$

then the canonical Pfaffian system is also generated by the Pfaffian equations

$$\begin{cases} \tilde{\theta}^\alpha = \omega^\alpha - p^\alpha dx = 0 \\ \tilde{\omega} = dx \neq 0 \end{cases} \quad .$$

Writing the two sets of structure equations respectively as

$$d\theta^\alpha = -\pi^\alpha\wedge\omega \qquad (\pi^\alpha = d\dot{y}^\alpha)$$

and

$$d\tilde{\theta}^\alpha \equiv -\tilde{\pi}^\alpha\wedge\tilde{\omega} \quad \text{modulo span}\{\omega;\tilde{\theta}^\alpha\},$$

it is straightforward to verify that $\{\tilde{\omega};\tilde{\theta}^\alpha;\tilde{\pi}^\alpha\}$ differs from $\{\omega;\theta^\alpha;\pi^\alpha\}$ by an admissable coframe change (I.e.27). In the new coframe

$$\tilde{\varphi} = \frac{1}{2}\sum (p^\alpha)^2\,dx \quad .$$

This particular change of coframe will be useful in Chapter III, Section a).

Returning to the general discussion we will make an assumption concerning the internal structure of the differential system (I,ω) and the 1-form φ in our variational problem $(I,\omega;\varphi)$ that will allow us to introduce a basic invariant given by a quadratic form on the vector bundle $(W^*/W_1^*)^*$. The assumption is stronger than what is really necessary but we are not sure what the exact correct hypothesis should be. In any case the assumption is satisfied in almost all of our examples, and by the discussion at the end of Chapter I, Section c) (cf. (I.c.30)) the hypotheses on (I,ω) are automatically satisfied on any prolongation (in this regard, cf. (I.c.40)).

Assumption. Concerning $(I,\omega;\varphi)$ we assume first that $L^* = \text{span}\{\omega;\theta^1,\dots,\theta^s\}$ is completely integrable, so that (I,ω) is locally embeddable (cf. (I.c.10)), and that (I,ω) is a Pfaffian system in good form. Moreover, we assume that there is an admissable coframe $\{\omega;\theta^\rho;\theta^\mu;\pi^\mu\}$ such that the relations

$$\begin{cases} \text{(i)} & d\theta^\rho \equiv 0 \mod L^*\wedge L^* \\ \text{(ii)} & d\theta^\mu \equiv -\pi^\mu\wedge\omega \mod L^*\wedge L^* \\ \text{(iii)} & d\omega \equiv 0 \mod L^*\wedge L^* \end{cases} \tag{I.e.28}$$

are valid, where $L^*\wedge L^*$ is the image of $L^*\otimes L^*\to\Lambda^2T^*(X)$. Concerning φ we assume that in this admissable coframe

$$d\varphi \equiv A_\mu\pi^\mu\wedge\omega \qquad \mod L^*\wedge L^* \ . \tag{I.e.29}$$

In practice we will always have

$$\varphi = f\omega \tag{I.e.30}$$

for some function f on X, and then (I.e.29) is a consequence of (iii) in (I.e.28).

Remark. What is actually needed is that for each $x\in X$ there should exist an admissable coframe *depending on* x such that (I.e.28), (I.e.29) are valid up *through* 1st order at x (this is somewhat analogous to choosing coordinates such that $g_{\alpha\beta}(y) = \delta_\alpha^\beta + O(y^2)$ in Riemannian geometry).

To a variational problem $(I,\omega;\varphi)$ satisfying (I.e.28), (I.e.29) we now intrinsically associate a quadratic form $\|A_{\mu\nu}(x)\|$ depending on $x\in X$. With the abused but hopefully clear notation

$$L^*\wedge L^* = \text{span over } C^\infty(X) \text{ of } \{\omega\wedge\theta^\alpha,\ \theta^\alpha\wedge\theta^\beta\}$$

the exterior derivative of (ii) in (I.e.28) gives

$$0 \equiv d(\pi^\mu\wedge\omega) \mod d(L^*\wedge L^*) \ .$$

Using (I.e.28) this implies that

$$d(\pi^\mu\wedge\omega) \equiv 0 \mod T^*(X)\wedge L^*\wedge L^* \ , \tag{I.e.31}$$

where $T^*(X)\wedge L^*\wedge L^*$ is the span over $A^1(X)$ of $L^*\wedge L^*$. The exterior derivative of (I.e.29) gives

$$0 \equiv dA_\mu \wedge \pi^\mu \wedge \omega + A_\mu d(\pi^\mu \wedge \omega) \mod d(L^* \wedge L^*) .$$

Using (I.e.28) and (I.e.31) this gives

$$dA_\mu \wedge \pi^\mu \wedge \omega \equiv 0 \mod T^*(X) \wedge L^* \wedge L^* . \qquad \text{(I.e.32)}$$

By a wonderful linear algebra result known as the *Cartan lemma* we then have

$$\begin{cases} dA_\mu \equiv A_{\mu\nu}\pi^\nu \mod L^* \\ A_{\mu\nu} = A_{\nu\mu} \end{cases} . \qquad \text{(I.e.33)}$$

<u>Proof</u>. Let $\{\omega;\theta^\alpha;\pi^\mu;\eta^j\}$ be a coframe on X and write

$$dA_\mu \equiv A_{\mu j}\eta^j + A_{\mu\nu}\pi^\nu \mod L^*$$

(actually, if we assume no Cauchy characteristics then we don't need to introduce the η^j since $\{\omega;\theta^\alpha;\pi^\mu\}$ already gives a coframe). Then by (I.e.32)

$$A_{\mu j}\eta^j \wedge \pi^\mu \wedge \omega + A_{\mu\nu}\pi^\mu \wedge \pi^\nu \wedge \omega \equiv 0 \mod T^*(X) \wedge L^* \wedge L^* .$$

By contracting with $\partial/\partial\eta^j$ we get

$$A_{\mu j}\pi^\mu \wedge \omega \equiv 0 \mod L^* \wedge L^* ,$$

which gives $A_{\mu j} = 0$. Contracting with $\partial/\partial\pi^\mu \wedge \partial/\partial\pi^\nu$ gives

$$(A_{\mu\nu} - A_{\nu\mu})\omega = 0$$

(since $(\partial/\partial\pi^\mu \wedge \partial/\partial\pi^\nu) \lrcorner T^*(X) \wedge L^* \wedge L^* = 0$). Taken together these two equations establish (I.e.33).

<u>Definitions</u>. i) We shall call $\|A_{\mu\nu}\|$ the *quadratic form* associated to the variational problem $(I,\omega;\varphi)$;[22] and ii) we shall say that $(I,\omega;\varphi)$ is *strongly non-degenerate* in case $\det\|A_{\mu\nu}\| \neq 0$.

For instance, in Example (I.e.18) strong non-degeneracy is equivalent to $\det\|L_{\dot{y}^\alpha\dot{y}^\beta}\| \neq 0$, while in example (I.e.23) it is equivalent to $\det\|L_{\ddot{y}^\alpha\ddot{y}^\beta}\| \neq 0$. In general strong non-degeneracy will mean that we need only consider the 1st derived system and not the whole derived flag.

(I.e.34) THEOREM. *If* $(I,\omega;\varphi)$ *is strongly non-degenerate, then it is non-degenerate and*

$$\dim Y = 2s + 1 .$$

Proof. Following the notation in construction (I.e.1) we have

$$\psi = \varphi + \lambda_\rho \theta^\rho + \lambda_\mu \theta^\mu$$

$$\Psi = d\varphi + d\lambda_\rho \wedge \theta^\rho + d\lambda_\mu \wedge \theta^\mu + \lambda_\rho d\theta^\rho + \lambda_\mu d\theta^\mu .$$

Using (I.e.28) and (I.e.29)

$$\Psi \equiv (A_\mu - \lambda_\mu)\pi^\mu \wedge \omega + (d\lambda_\alpha + \ell_\alpha \omega) \wedge \theta^\alpha \mod S \qquad \text{(I.e.35)}$$

where the ℓ_α are functions on Z and

$$S = W^* \wedge W^* \subset \Lambda^2 T^*(X) .$$

Since $\pi^\mu \wedge \omega$ is non-zero in $\Lambda^2 T^*(X)/S$ it follows that

$$Z_1 \subset Z$$

is defined by

$$A_\mu = \lambda_\mu . \qquad (23)$$

Then (I.e.33) gives *on* Z_1

$$d\lambda_\mu \equiv A_{\mu\nu}\pi^\nu \mod L^*$$

By our assumption of strong non-degeneracy we may take $\{\omega;\theta^\alpha;d\lambda_\alpha\}$ as a local coframe on Z_1. Then by (I.e.35)

$$\Psi_{Z_1} \equiv (d\lambda_\alpha + \ell_\alpha \omega) \wedge \theta^\alpha \mod S$$

$$\psi_{Z_1} = \varphi + \lambda_\alpha \theta^\alpha .$$

Now since $\dim Z_1 = 2s+1$, $\psi_{Z_1} \wedge (\Psi_{Z_1})^s$ is in any case a multiple of the non-zero volume form on Z_1

$$\Theta = \varphi \wedge d\lambda_1 \wedge \theta^1 \wedge \cdots \wedge d\lambda_s \wedge \theta^s . \qquad (27)$$

Since neither ψ_{Z_1} nor S contains any terms with a $d\lambda_\alpha$, it follows that we may neglect S in computing $\psi_{Z_1} \wedge (\Psi_{Z_1})^s$, and then this latter form is clearly equal to $s!\Theta$.

To complete the proof of the theorem, by Proposition (I.e.17) it will suffice to show that

$$C(\Psi_{Z_1}) = C_1(\Psi) .$$

We have

$$\Psi \equiv (d\lambda_\alpha + \ell_\alpha \omega) \wedge \theta^\alpha \quad \text{mod } S$$

$$\Psi_{Z_1} \equiv (d\lambda_\alpha + \ell_\alpha \omega) \wedge \theta^\alpha)_{Z_1} \quad \text{mod}(S_{Z_1}) \ .$$

Since $\{\omega;\theta^\alpha;d\lambda_\alpha\}$ gives a coframe on Z_1 it follows that the only way that the inclusion

$$C(\Psi_{Z_1}) \subset C_1(\Psi)$$

could fail to be an equality is due to the 1-forms

$$(v \lrcorner \eta)_{Z_1} \qquad \eta \in S,\ v \in C^\infty(N,T(X)) \ .$$

But since $W^* \wedge W^*$ it follows that

$$v \lrcorner \eta \in W^* \ .$$

On the other hand it is clear that $W^* \subset C(\Psi_{Z_1})$, and our result follows. Q.E.D.

(I.e.36) Examples. The theorem includes the classical examples (I.e.20) and (I.e.25) as special cases. More interestingly, referring to example (I.d.27) we consider the functional

$$\Phi(\gamma) = \int_\gamma L(\kappa)\, ds \tag{I.e.37}$$

where γ is a curve with arclength ds on a Riemannian surface S and $L(\kappa)$ is a smooth function of the geodesic curvature. Using the notations of that example we have

$$\begin{cases} d\theta^1 = -\theta^2 \wedge \omega \\ d\theta^2 = -\pi \wedge \omega \\ d\omega = -\kappa\theta^2 \wedge \omega - \theta^1 \wedge \theta^2 \end{cases}$$

where

$$\pi = d\kappa - (R^2 + \kappa^2)\theta^2 \ .$$

The derived flag (cf. Chapter I, Section c)) is

$$\begin{array}{ccccc} W_2^* & \subset & W_1^* & \subset & W^* \\ \| & & \| & & \| \\ (0) & & \text{span}\{\theta^1\} & & \text{span}\{\theta^1,\theta^2\} \ , \end{array}$$

and the Cartan integer $s_1 = 1$.

Moreover, for $\varphi = L(\kappa)\omega$ we have

$$d\varphi \equiv L'(\kappa)\pi \wedge \omega \bmod \mathrm{span}\{\theta^1 \wedge \theta^2, \theta^1 \wedge \omega, \theta^2 \wedge \omega\} .$$

Consequently both (I.e.28) and (I.e.29) are satisfied (there is one $\theta^\rho = \theta^1$ and one $\theta^\mu = \theta^2$), and since

$$dL'(\kappa) = L''(\kappa)\pi \bmod \mathrm{span}\{\omega;\theta^1,\theta^2\}$$

the condition of strong non-degeneracy is equivalent to

$$L''(\kappa) \neq 0 . \qquad \text{(I.e.38)}$$

In this regard we refer to footnote (16) for a discussion of the situation when $L'' \equiv 0$.

(I.e.39) Example. We will investigate what the quadratic form and strong non-degeneracy mean in the case of the Lagrange problem[24] (cf. Example (I.c.14); a specific non-degenerate Lagrange problem will be discussed in Example (II.a.40) below). We work in a neighborhood where $X \subset J^1(\mathbb{R},\mathbb{R}^m)$ is given by

$$g^\rho(x,y,\dot{y}) = 0 \qquad 1 \leqslant \rho \leqslant r \qquad \text{(I.e.40)}$$

and assume that in this neighborhood

$$\mathrm{rank} \left\| \frac{\partial g^\rho}{\partial \dot{y}^\sigma}(x,y,\dot{y}) \right\| = r \qquad 1 \leqslant \rho,\ \sigma \leqslant r . \qquad \text{(1.e.41)}$$

We take for (I,ω) the restriction to X of the canonical differential system on $J^1(\mathbb{R},\mathbb{R}^m)$

$$\begin{cases} \theta^\alpha = dy^\alpha - \dot{y}^\alpha dx = 0 \\ \omega = dx \neq 0 \end{cases} ,$$

and for φ we assume that

$$\varphi = L(x,y,\dot{y})dx .$$

To avoid confusion of notation, *in this example alone we take*

$$K^* = \mathrm{span}\{\theta^1,\ldots,\theta^n,\omega\}$$

$$= \mathrm{span}\{dy^1,\ldots,dy^n,dx\} ,$$

where it is understood that all forms are restricted to X. If we choose our coframe so that the refined structure equations (I.c.13) hold then we will have

$$d\dot{y}^\alpha \equiv \ell^\alpha_\mu \pi^\mu \bmod K^* .$$

Differentiation of (I.e.40) gives

$$g^{\rho}_{\dot{y}\beta}d\dot{y}^{\beta} \equiv 0 \mod K^*$$

and this implies that on X

$$g^{\rho}_{\dot{y}\beta}\ell^{\beta}_{\mu} = 0 \quad . \qquad \text{(I.e.42)}$$

Now recall that the quantities A_μ were defined by (cf. (I.e.29))

$$d\varphi \equiv A_\mu \pi^\mu \wedge \omega \mod K^* \wedge K^* \; .$$

Since

$$d\varphi \equiv L_{\dot{y}^\alpha} d\dot{y}^\alpha \wedge dx \mod K^* \wedge K^*$$

$$\equiv L_{\dot{y}^\alpha} \ell^\alpha_\mu \pi^\mu \wedge \omega \mod K^* \wedge K^*$$

it follows that

$$A_\mu = L_{\dot{y}^\alpha}\ell^\alpha_\mu \quad . \qquad \text{(I.e.43)}$$

In other words, if we set $p = (x,y)$ and

$$X_p = X \cap \pi^{-1}(p)$$

where $\pi\colon J^1(\mathbb{R},\mathbb{R}^m) \to \mathbb{R}\times\mathbb{R}$ is the projection, then $A_\mu \pi^\mu$ is the restriction of dL to the sub-bundle $T(X_p) \subset T(X)\big|_{X_p}$.

Now the $A_{\mu\nu}$ were defined by

$$dA_\mu \equiv A_{\mu\nu}\pi^\nu \mod K^* \; .$$

By (I.e.43)

$$dA_\mu \equiv L_{\dot{y}^\alpha \dot{y}^\beta}\ell^\alpha_\mu \ell^\beta_\nu \pi^\nu + L_{\dot{y}^\alpha}\ell^\alpha_{\mu\dot{y}^\beta}\ell^\beta_\nu \pi^\nu \; . \qquad \text{(I.e.44)}$$

To interpret this equation, we fix $p = (x,y)$ and consider X_p as a submanifold of $\mathbb{R}^m$. A change of coordinates $y^\alpha = y^\alpha(z)$ induces a *linear* automorphism of $\mathbb{R}^m$, so what is intrinsic is the *geometry of* X_p *in* $\mathbb{R}^m$. We may change the ℓ^μ_ν by an arbitrary substitution

$$\ell^\alpha_\mu \to \ell^\alpha_\mu B^\mu_\nu \, , \qquad \det\|B^\mu_\nu\| \neq 0 \; . \qquad \text{(I.e.45)}$$

This simply amounts to changing the coframe $\{\pi^\mu_{X_p}\}$. Using (I.e.41) it follows that

$$\det\|\ell^\mu_\nu\| \neq 0 \quad ,$$

so by a linear substitution (I.e.45) we may assume that $\ell^{\mu}_{\nu} = \delta^{\mu}_{\nu}$. Then (I.e.42) is

$$g^{\rho}_{\dot{y}^{\sigma}}\ell^{\sigma}_{\mu} + g^{\rho}_{\dot{y}^{\mu}} = 0 \quad . \tag{I.e.46}$$

Moreover we have

$$0 = d\ell^{\mu}_{\nu} = \left(\ell^{\mu}_{\nu\dot{y}^{\rho}}\ell^{\rho}_{\lambda} + \ell^{\mu}_{\nu\dot{y}^{\lambda}}\right)\pi^{\lambda} \quad . \tag{I.e.47}$$

We now evaluate at a point of X_p where

$$\begin{cases} g^{\rho}_{\dot{y}^{\sigma}} = \delta^{\rho}_{\sigma} \\ \ell^{\rho}_{\mu} = 0 \end{cases} \quad . \tag{I.e.48}$$

Then (I.e.47) is

$$\ell^{\mu}_{\nu\dot{y}^{\lambda}} = 0 \quad . \tag{I.e.49}$$

If we differentiate (I.e.46) *and then evaluate* at a point where (I.e.48) holds, we get

$$\ell^{\rho}_{\mu\dot{y}^{\nu}} + g^{\rho}_{\dot{y}^{\mu}\dot{y}^{\nu}} = 0 \quad . \tag{I.e.50}$$

Thus, at a point of X_p where the normalization (I.e.48) is valid, the coefficient of π^{ν} in the 2nd term in (I.e.44) is

$$\begin{aligned} L_{\dot{y}^{\alpha}}\ell^{\alpha}_{\mu\dot{y}^{\nu}} &= L_{\dot{y}^{\rho}}\ell^{\rho}_{\mu\dot{y}^{\nu}} + L_{\dot{y}^{\lambda}}\ell^{\lambda}_{\mu\dot{y}^{\nu}} \\ &= -L_{\dot{y}^{\rho}}g^{\rho}_{\dot{y}^{\mu}\dot{y}^{\nu}} \end{aligned}$$

by (I.e.49), (I.e.50). Consequently, at this point the quadratic form $\|A_{\mu\nu}\|$ is given by

$$A_{\mu\nu} = L_{\dot{y}^{\mu}\dot{y}^{\nu}} - L_{\dot{y}^{\rho}}g^{\rho}_{\dot{y}^{\mu}\dot{y}^{\nu}} \quad . \tag{I.e.51}$$

The first term is the *Hessian* $\|L_{\dot{y}^{\alpha}\dot{y}^{\beta}}\|$ restricted to the tangent space to X_p. The 2nd term is minus *the 2nd fundamental form* of X_p evaluated on dL (this makes sense in the affine linear case). The conclusion we may draw is this:

In the Lagrange problem the condition of strong non-degeneracy involves part of the Hessian $\|L_{\dot{y}^{\alpha}\dot{y}^{\beta}}\|$ *and 2nd fundamental form of the fibres of* $X \to \mathbb{R} \times \mathbb{R}^m$.

For example, the condition

$$\|A_{\mu\nu}\| > 0$$

means that, in a certain sense, *in the directions of* $T(X_p)$ L *should be more convex* (as a function of the $\dot{y}^\alpha$) *then* X_p *is.*

A special case occurs when the constraints are linear in the $\dot{y}^\alpha$. Then we have a distribution $S \subset T(\mathbb{R} \times \mathbb{R}^m)$ and the 2nd term in (I.e.51) drops out.

<u>Note</u>. There is extensive classical literature (and some confusion) concerning: i) the conditions for non-degeneracy in the Lagrange problem (in this regard [5] and [13] are helpful); ii) the property that a Lagrange problem be in Hamiltonian form (here [13] and [40] may be useful); and most seriously iii) the exact statement of when the classical Lagrange multiplier method gives a local minimum for a constrained functional (and, moreover, what the exact endpoint conditions for the competing curves should be). Here we have found that [4], [5], [13], and [40] are helpful (of course, there are certainly other useful sources). In Chapter IV, Section e) we shall briefly discuss these matters, and would venture a guess that the methods of Chapter IV, Sections a)-d) could be modified so as to clarify the points i)-iii) above.

FOOTNOTES FOR CHAPTER I

(1) This terminology has the following meaning, to be explained in detail later: Let $T_{(N,f)}(V(I,\omega))$ denote the tangent space to $V(I,\omega)$ at $f: N \to X$ and consider the differential of the functional (I.a.4) as a map

$$(\delta\Phi)(N,f): T_{(N,f)}(V(I,\omega)) \to \mathbb{R} .$$

The *variational* or *Euler-Lagrange equations* are the conditions

$$(\delta\Phi)(N,f) = 0$$

on $f: N \to X$. Integral manifolds (N,f) satisfying this equation are sometimes called *extremals* of Φ.

(2) According to Caratheodory (page (vii) of [11]) "the calculus of variations should be the servant of mechanics."

(3) Requiring that $dy(x)/dx \wedge d^2y(x)/dx^2 \neq 0$ is the analogue of requiring that $dy(x)/dx \neq 0$ in usual Euclidean geometry. Of course it may well turn out that the domain of the functional (I.a.16) should be enlarged; e.g., we might consider C^2-curves such that $dy(x)/dx \wedge d^2y(x)/dx^2 = 0$ at finitely many points. More importantly, we should perhaps allow certain C^1-curves as described in the remark below.

(4) The discussion in this section is valid for any number n of independent variables.

(5) In case N is 1-dimensional we may consider the variation as a map

$$F: [a,b] \times [0,\varepsilon] \to X$$

given by

$$(s,t) \to F(s,t) \in X .$$

Writing

$$F^*\theta = p(s,t)ds + q(s,t)dt$$

we have

$$(L_t(F^*\theta))_N = \frac{\partial p}{\partial t}(s,0)ds .$$

On the other hand

$$\frac{\partial}{\partial t} \lrcorner\ F^* d\theta = \frac{\partial}{\partial t} \lrcorner\ d(F^*\theta)$$

$$= \frac{\partial}{\partial t} \lrcorner \left(\frac{\partial p}{\partial t} - \frac{\partial q}{\partial s}\right) dt \wedge ds$$

$$= \left(\frac{\partial p}{\partial t} - \frac{\partial q}{\partial s}\right) ds \ ;$$

$$d\left(\frac{\partial}{\partial t} \lrcorner\ F^*\theta\right) = dq$$

$$= \frac{\partial q}{\partial s} ds + \frac{\partial q}{\partial t} dt.$$

Thus

$$\frac{\partial}{\partial t} \lrcorner\ (F^* d\theta) + d\left(\frac{\partial}{\partial t} \lrcorner\ F^*\theta\right) = \frac{\partial p}{\partial t} ds + \frac{\partial q}{\partial t} dt \ ,$$

and the restriction of this form to $N \times \{0\}$ is just $\frac{\partial p}{\partial t}(s,0)ds$. This computation is the basis of (I.b.5).

(6) Actually, all that is needed is that this equation holds up to order t^2.

(7) It now becomes convenient to drop the clumsy notation V for an extension of v to all of X.

(8) One should compare (I.c.5) with the following: A curve of integral manifolds of the canonical Pfaffian system on $J^1(\mathbb{R},\mathbb{R}^m)$ is given by

$$(x,t) \to \left(x; y^\alpha(x,t);\ dy^\alpha \frac{(x,t)}{dx}\right).$$

The tangent vector v to this curve at $t=0$ is

$$v(x) = \left(\frac{\partial y^\alpha(x,0)}{\partial t}\right) \partial/\partial y^\alpha + \left(\frac{\partial^2 y^\alpha(x,0)}{\partial x \partial t}\right) \partial/\partial \dot{y}^\alpha \ .$$

Then in the notations of example (I.c.1)

$$\begin{cases} B^\alpha(x) = \langle\theta^\alpha, v(x)\rangle = \dfrac{\partial y^\alpha(x,0)}{\partial t} \\ C^\alpha(x) = \langle d\dot{y}^\alpha, v(x)\rangle = \dfrac{\partial^2 y^\alpha(x,0)}{\partial x \partial t} \\ A(x) = \langle dx, v(x)\rangle = 0 \ , \end{cases}$$

and (I.c.4) is clear. We note that eliminating the A term corresponds to giving all the 1-jets a common parameter x.

(9) Passing to the quotient $\Lambda^2T^*(X)/W^* \wedge T^*(X)$ means working modulo the degree two part of the ideal $\{\theta^\alpha\}$. For a section $\theta = \lambda_\alpha\theta^\alpha \in C^\infty(X,W^*)$ we may write

$$\delta(\theta) \;=\; d\theta \bmod \{\theta^\alpha\} \;.$$

Then for f a function on X

$$\begin{aligned}\delta(f\theta) &= d(f\theta) \bmod \{\theta^\alpha\}\\ &= fd\theta \bmod \{\theta^\alpha\}\\ &= f\delta(\theta) \;.\end{aligned}$$

In the case of interest to us when $W^* \wedge T^*(X)$ is a sub-bundle of $\Lambda^2T^*(X)$, this linearity over $C^\infty(X)$ means that δ is a *bundle map*.

(10) The notion of derived mapping and 1st derived system, as well as for the whole derived flag to be discussed below, is valid for *any* Pfaffian system (i.e., it does not use the independence condition). If $W^{*\perp} = S \subset T(X)$ is the distribution corresponding to W^*, and if by $[S,S]$ we denote the subspace of vector fields generated by all brackets $[v,w]$ where $v,w \in C^\infty(X,S)$, then under the assumption that $W_1^* \subset W^*$ is a sub-bundle it is easy to see that

$$W_1^{*\perp} \;=\; [S,S] \;.$$

In particular, the condition that W^* be completely integrable is just $W_1^* = W^*$.

(11) With only a few exceptions the examples we consider will be locally embeddable. However, thus far this seems to be of little value in studying variational problems. Perhaps the reason is that admissable changes of coframe for (I,ω) (as explained in Chapter I, Section e)) do not in general correspond to diffeomorphisms of $J^1(\mathbb{R},M)$ induced by diffeomorphisms of M. The situation in theoretical mechanics and in the theory of Fourier integral operators is similar: Forget the fact that one is working on $T^*(M)$ and allow arbitrary contract transformations (perhaps satisfying a suitable transversality condition).

(12) We may also say that s_1 *measures the deviation of* W^* *from being completely integrable.* In this connection we remark that for the canonical system on $J^1(\mathbb{R},\mathbb{R}^s)$ the Cartan integer $s_1 = s$. Conversely, if (I,ω) is in good form $d\omega \equiv 0 \bmod\{I,\omega\}$, and $\dim X = 2s+1$, then if $s_1 = s$ it follows that (I,ω) is locally isomorphic to the canonical system on $J^1(\mathbb{R},\mathbb{R}^s)$ (this is a consequence of (I.c.10)). Thus the two extremes $s_1 = 0$ and $s_1 = s$ have local normal forms.

(13) Roughly speaking, we may say that the Cartan integer measures the "number" of integral manifolds of (I,ω) (cf. the preceding footnote).

(14) The intuitive principle is this: If on a vector space V we have linear functions $L^1,..,L^s$ and δ with the property that

$$L^1(v) = \cdot\cdot = L^s(v) = 0 \Rightarrow \delta(v) = 0 ,$$

then δ should be a linear combination of $L^1,..,L^s$. In the finite dimensional case this is obvious.

(15) It will be a frequent phenomenon that the number of undetermined functions" λ_α that may be eliminated is related to the Cartan integer s_1.

(16) The situation when $L''(\kappa) \equiv 0$ requires a special discussion. Then $L(\kappa) = \alpha + \beta\kappa$ and

$$\begin{aligned} \varphi &= (\alpha + \beta\kappa)\omega. \\ &\equiv \alpha\omega + \beta\omega_1^2 \text{ mod } \{\theta^\alpha\} . \end{aligned}$$

We have already examined the case when $\varphi = \omega$.

If $\varphi = \omega_1^2$ the equations (I.d.34) are

$$\begin{cases} (1 - \lambda_2)\omega \equiv 0 \text{ mod } N \\ d\lambda_2 + \lambda_1\omega \equiv 0 \text{ mod } N \\ d\lambda_1 + (\kappa^2 + \lambda_2(R - \kappa^2))\omega \equiv 0 \text{ mod } N . \end{cases}$$

These give $\lambda_2 = 1$, $\lambda_1 = 0$, and $R = 0$ respectively. *In general, then, no solutions to the Euler-Lagrange exist unless* $S = \mathbb{E}^2$.

In the case $S = \mathbb{E}^2$, *since* $d\omega_1^2 = 0$ *any curve is a solution of the Euler-Lagrange equations.* This has the following interpretation: Given any two line elements (p, e_1) and (p', e_1')

and *any* curve γ joining them, the total curvature integral is

$$\int_p^{p'} \kappa(s)\, ds = \ell$$

where ℓ is the length of the arc $\widehat{e_1 e_1'}$ on the unit circle.

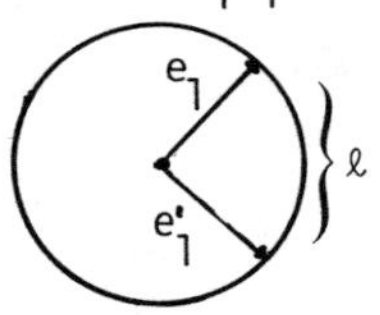

For essentially topological reasons, then, Φ is in some sense constant.

(17) In this regard, see the preceding footnote.

(18) This is a special case of the construction (I.c.42) given at the end of Chapter I, Section c).

(19) The notation means that we consider $\Psi(p) \in \Lambda^2 T_p^*(\mathbb{Z})$, $\omega(p) \in T_p^*(Z)$, and then $\lrcorner$ is the contraction operator

$$\lrcorner : T_p(Z) \otimes \Lambda^k T_p^*(Z) \to \Lambda^{k-1} T_p^*(Z) \quad .$$

(20) Even though we use coordinates we insist on all constructions being intrinsic and functorial. The failure to do this may account for some of the difficulties in systematically treating even the classical Lagrange problem.

(21) Given the 2-form Ψ_Y on the $(2m+1)$-dimensional manifold Y, then if $(\Psi_Y)^m \neq 0$ there is through each point of Y a unique *characteristic direction* (cf. Chapter 0, Section d), especially (0.d.5)). These characteristic directions give a line sub-bundle $\Psi_Y^\perp \subset T(Y)$, and the reason for equality in (I.e.14) is that we always have

$$\Psi_Y^\perp \supset J^\perp \quad ,$$

so that equality must hold since $\Psi_Y^\perp$ is a line bundle.

(22) As pointed out in the introduction, $\|A_{\mu\nu}\|$ is intrinsically given by a symmetric bilinear form

$$Q : (W^*/W_1^*)^* \otimes (W^*/W_1^*)^* \to \mathbb{R}$$

where $W_1^* = \mathrm{span}\{\theta^\rho\}$ is the 1st derived system of W^*.

(23) We may say that: For a strongly non-degenerate variational problem exactly s_1 (= the Cartan integer) of the functions λ_α are determined on X, while the remaining ones (corresponding to the 1st derived system of W^*) are only determined up on Y.

(24) It is *not* always the case that the structure relations (I.e.28) are satisfied for the Lagrange problem. Here we assume that (I.e.28) holds; the modifications necessary to treat the general case may easily be made using the discussion of the general Lagrange problem in Chapter IV, Section e) below.

(25) In this construction we may assume that (I,ω) is a Pfaffian system generated by Pfaffian equations

$$\theta^1 = \cdots = \theta^t = 0, \quad \omega \neq 0 \quad .$$

We do *not* assume that the $\theta^i(x) \in T_x^*(X)$ are linearly independent or that the rank of the linear equations

$$(*) \qquad \theta^i(x) = 0 , \qquad x \in X ,$$

is constant. Although this may at first sight seem to be an unpleasant situation, it will turn out to be quite manageable in examples, and in fact is the situation that is dictated by applications. What will happen is that at a general point x of X, for $v \in T_x(X)$

$$\langle\theta^i(x), v\rangle = 0 \quad \text{for all} \quad i \Rightarrow \langle\omega(x), v\rangle = 0 .$$

Thus, over such a general point there will be *no* integral elements of (I, ω). However, over a subvariety X_1 *along which the rank of the equations* (*) *must necessarily drop* there will, for each $x \in X_1$, be vectors $v \in T_x(X)$ such that

$$\langle\theta^i(x), v\rangle = 0 \quad \text{for all} \quad i \quad \text{and} \quad \langle\omega(x), v\rangle \neq 0 .$$

However, v may not be tangent to X_1, and this leads naturally to the construction in the text.

(26) We have not stated this result in its usual strong form, since in Chapter IV we will need the following variant: *On* $U \subset \mathbb{R}^{2\rho+1}$ *suppose we are given i) a 1-form* ψ *with* $\psi \wedge (d\psi)^\rho \neq 0$, *and ii) a completely integrable Pfaffian system* I *whose leaves have dimension* ρ *and where* $\psi \in I$. *Then the local normal form (0.d.9) is valid where* $I = \{dx, dv^1, \ldots, dv^\rho\}$.

(27) Here we must assume that $\{\varphi; \theta^\alpha; \pi^\mu\}$ gives a coframe on X, as is the case if (1.e.30) holds.

(28) On the somewhat sophisticated level, we let $V(J,\omega)$ denote the set of connected integral curves of (J,ω). Then we define a pre-symplectic structure Ω on $V(J,\omega)$ by setting

$$(*) \qquad \Omega(v,w) = \langle\Psi(p), v(p) \wedge w(p)\rangle$$

where $N \in V(J,\omega)$, v and $w \in T_N(V(J,\omega))$, and $p \in N$ is an arbitrary point. It may be shown that (*) is well-defined, and then $(I,\omega;\varphi)$ is non-degenerate in case Ω induces an honest symplectic structure on $V(J,\omega)$.

II. FIRST INTEGRALS OF THE EULER-LAGRANGE SYSTEM; NOETHER'S THEOREM AND EXAMPLES

a) First Integrals and Noether's Theorem; Some Classical Examples.

We consider a variational problem $(I,\omega;\varphi)$ given by the functional

$$\Phi(N) = \int_N \varphi \tag{II.a.1}$$

where $N \in V(I,\omega)$ is an integral manifold of the differential system (I,ω) on a manifold X. Associated to (II.a.1) is the Euler-Lagrange system (J,ω), which is a Pfaffian system defined on the momentum space Y (cf. Chapter I, Section e)) for the construction of (J,ω)).

Definition. A *1st integral* of the variational problem $(I,\omega;\varphi)$ is given by a function V defined on Y such that V is constant on the integral manifolds of (J,ω).

Remarks. The condition that V be constant on integral curves of (J,ω) is equivalent to

$$dV \equiv 0 \quad \text{mod } J \quad . \tag{II.a.2}$$

In practice, V will be the restriction to Y of a function (still denoted by V) defined on $Z = X \times \mathbb{R}^s$, and (II.a.2) may be verified by showing that

$$dV \equiv 0 \quad \text{mod } C(\Psi) \tag{II.a.3}$$

on Z.

Classically 1st integrals arose as *conserved quantities* for mechanical systems; accordingly the basic example is given by (I.d.15) where the Lagrangian

$$L = L(y,\dot{y})$$

is independent of x (to be more traditional, x should be replaced by a time parameter t). Then as follows from (I.d.17) and will be verified below (cf. example (II.a.12)), the Hamiltonian

$$H = \dot{y}^{\alpha} L_{\dot{y}^{\alpha}} - L$$

is a 1st integral. We will discuss this and several other examples later. For the moment we pause to give a further

Definition. Let $(I,\omega;\varphi)$ be a non-degenerate variational problem with momentum space Y of dimension $2m+1$. Given functions U,V defined on Y their *modified Poisson bracket* $[U,V]$ is the function on Y defined by

$$[U,V]\psi_Y \wedge (\Psi_Y)^m = m\,dU \wedge dV \wedge \psi_Y \wedge (\Psi_Y)^{m-1} . \tag{II.a.4}$$

Remark. We use the term "modified" Poisson bracket to distinguish it from the usual Poisson bracket $\{U,V\}$ to be discussed below. In the classical case the two are related.[20]

The importance of modified Poisson brackets lies in the following observation:

(II.a.5) *If* U,V *are each* 1st *integrals of* $(I,\omega;\varphi)$*, then so is their modified Poisson bracket* $[U,V]$.

Proof. Let v be the unique vector field on Y satisfying

$$\begin{cases} v \lrcorner \Psi_Y = 0 \\ v \lrcorner \psi_Y = 1 \end{cases} . \tag{II.a.6}$$

The integral manifolds of (J,ω) are locally just the integral curves of v, and to say that U is a 1st integral is equivalent to

$$L_v(U) = 0 . \tag{II.a.7}$$

By (II.a.6) and the H. Cartan formula (0.a.1)

$$\begin{cases} L_v(\Psi_Y) = 0 \\ L_v(\psi_Y) = 0 \end{cases} . \tag{II.a.8}$$

Since $L_v(dU) = dL_v(U)$, if U and V are each 1st integrals it follows from (II.a.4) and (II.a.7), (II.a.8) that

$$\begin{aligned} 0 &= L_v([U,V]\psi_Y \wedge (\Psi_Y)^m) \\ &= (L_v[U,V])\psi_Y \wedge (\Psi_Y)^m , \end{aligned}$$

which implies that $L_v[U,V] = 0$. Q.E.D.

A major source of 1st integrals is provided by symmetry groups of variational problems. To explain this we need the following

Definition. An *infinitesimal symmetry* of a variational problem $(I,\omega;\varphi)$ is given by a vector field v on X that satisfies

$$\begin{cases} L_v(I) \subset I \\ L_v(\varphi) \equiv 0 \bmod I \end{cases} \qquad \text{(II.a.9)}$$

These conditions are equivalent to

$$\begin{cases} \exp(tv)^* I = I \\ \exp(tv)^* \varphi \equiv \varphi \bmod I \ . \end{cases}$$

Since the independence condition is a transversality condition it does not appear in the definition of an infinitesimal symmetry.

We want to explain how v induces a vector field, for the moment denoted by $\tilde{v}$, on $Z = X \times \mathbb{R}^s$. The somewhat subtle point is that $\tilde{v}$ is *not* just the vector field induced by the product structure. To see what $\tilde{v}$ should be we recall that *intrinsically* (cf. remark (I.e.8))

$$Z = W^* \subset T^*(X)$$

and that

$$\psi = \pi^* \varphi + \theta$$

where θ is the canonical 1-form on $T^*(X)$ and $\pi : W^* \to X$ is the projection. Now $\exp(tv)^*$ operates as a 1-parameter group of *fibre-preserving transformations* on $T^*(X)$, and this action clearly leaves invariant the canonical 1-form on $T^*(X)$. Furthermore, by the 1st condition in (II.a.9), $\exp(tv)^*$ also leaves invariant the sub-bundle $W^* \subset T^*(X)$. Setting $Z = W^*$, by definition the 1-parameter group

$$\exp(tv)^* : Z \to Z$$

induces the vector field $\tilde{v}$ in question. The main result is (cf. [55])

(II.a.10) Noether's Theorem. *If* v *is an infinitesimal symmetry of* $(I,\omega;\varphi)$, *then the function*

$$V = \tilde{v} \lrcorner \psi$$

is a 1st *integral of the variational problem.*

Proof. Since $\exp(t\tilde{v})^*\theta = \theta$ and $\exp(tv)^*\varphi \equiv \varphi \bmod I$, it follows from the fact that $\exp(t\tilde{v})$ is a 1-parameter group of fibre preserving transformations covering the 1-parameter group $\exp(-tv)$ on X that

$$\exp(t\tilde{v})^*\psi \equiv \psi \bmod \pi^* I$$

where $\pi : Z \to X$ is the projection. Thus

$$L_{\tilde{v}}(\psi) \in \pi^* I \ .$$

By H. Cartan's formula (0.a.1)

$$dV = d(\tilde{v} \lrcorner \psi) = L_{\tilde{v}}(\psi) - \tilde{v} \lrcorner \Psi$$

where $\Psi = d\psi$. By the very definition of the Cartan system

$$\tilde{v} \lrcorner \Psi \in C(\Psi) \quad ,$$

while on the other hand since

$$\partial/\partial\lambda_\alpha \lrcorner \Psi = \pi^*\theta^\alpha$$

(cf. the proof of Theorem (I.e.9)) we have

$$\pi^* I \subset C(\Psi) \quad .$$

Combining these gives

$$dV \in C(\Psi) \quad ,$$

which is (II.a.3). Q.E.D.

Remark. In concrete terms, if

$$L_v \theta^\alpha = A^\alpha_\beta \theta^\beta \ ,$$

then in the product structure $Z = X \times \mathbb{R}^s$

$$L_{\tilde{v}} \lambda_\alpha = A^\beta_\alpha \lambda_\beta \quad .$$

Since $\psi = \varphi + \lambda_\alpha \theta^\alpha$ it follows that

$$\tilde{v} \lrcorner \psi = -v \lrcorner \psi \ ,$$

where the right hand side means the vector field v on $X \times \mathbb{R}^s$ induced by v on X in the product structure. Thus, although it is a slightly incorrect notation, for computational purposes we shall write the 1st integral in Noether's theorem simply as

$$V = v \lrcorner \psi \quad . \tag{II.a.11}$$

Due to the importance of this result we shall give a proof "down on X". The argument is less elegant but is perhaps more transparent than the one given above (of course, in essence the two are the same).

Proof. Setting $\theta = \lambda_\alpha \theta^\alpha$, by H. Cartan's formula

$$d(v \lrcorner (\varphi+\theta)) + (v \lrcorner d(\varphi+\theta)) = L_v(\varphi+\theta) \ .$$

By (II.a.9)

$$L_v(\varphi+\theta) \equiv 0 \bmod I \ .$$

On the other hand, the Euler-Lagrange equations (I.d.14) imply that

$$v \lrcorner d(\varphi+\theta) \equiv 0 \bmod N \ .$$

Combining these last two relations with H. Cartan's formula gives

$$\begin{aligned} dV &= -v \lrcorner d(\varphi+\theta) + L_v(\varphi+\theta) \\ &\equiv 0 \bmod N \end{aligned}$$

for any integral manifold N of (I,ω) that satisfies the Euler-Lagrange equations. Q.E.D.

(II.a.12) Example. We consider a classical variational problem (I.a.5) (cf. example (I.d.15)), given on the space $J^1(\mathbb{R},\mathbb{R}^m)$ with coordinates $(x;y^\alpha;\dot{y}^\alpha)$ by the data

$$\begin{cases} \theta^\alpha = dy^\alpha - \dot{y}^\alpha dx = 0 \\ \omega = dx \neq 0 \\ \varphi = L(x,y,\dot{y})dx \ . \end{cases}$$

Following the construction in Chapter I, Section e), on $Z = J^1(\mathbb{R},\mathbb{R}^m) \times \mathbb{R}^m$ we consider the 1-form

$$\psi = Ldx + \lambda_\alpha \theta^\alpha \ .$$

To put ψ in the normal form (0.d.9) of Pfaff-Darboux we write it as

$$\psi = -Hdx + \lambda_\alpha dy^\alpha$$

where

$$H = -L + \lambda_\alpha \dot{y}^\alpha \ .$$

This brings us to the famous (but long overdue)

Definition. The *Hamiltonian* associated to the classical variational problem is the restriction to the momentum space $Y \subset Z$ of the function

$$H = -L + \lambda_\alpha \dot{y}^\alpha \ .$$

Remark. In case the variational problem is non-degenerate, $Y = Z_1$ is defined by the relations

$$\lambda_\alpha = L_{\dot{y}^\alpha} \quad .$$

In any case Y is always *contained in* this locus, so that the Hamiltonian is sometimes written as

$$H = -L + \dot{y}^\alpha L_{\dot{y}^\alpha} \quad .$$

A small point here is that Hamiltonian is always defined even if the variational problem is degenerate (in the sense that $\det\|L_{\dot{y}^\alpha\dot{y}^\beta}\| \equiv 0$). A related small point is that in our formulation the properties, such as being globally one-to-one, of the Legendre transform $\lambda_\alpha = L_{\dot{y}^\alpha}$ never arise. It seems that this assumption merely provides an unnecessary convenience in the classical case.

Now suppose that

$$v = A\,\partial/\partial x + B^\alpha\,\partial/\partial y^\alpha + C^\alpha\,\partial/\partial \dot{y}^\alpha$$

is an infinitesimal symmetry of the variational problem $(I,\omega;\varphi)$ above. Then Noether's theorem gives the 1st integral (cf. (II.a.11))

$$\begin{aligned} V &= v \lrcorner (L\,dx + \lambda_\alpha\theta^\alpha) \\ &= A(L - \dot{y}^\alpha\lambda_\alpha) + B^\alpha\lambda_\alpha \\ &= -AH + B^\alpha L_{\dot{y}^\alpha} \end{aligned}$$

where H is the Hamiltonian.

For example, suppose that the Lagrangian $L(y,\dot{y})$ is independent of x. We claim that

$$v = \partial/\partial x$$

gives an infinitesimal symmetry of $(I,\omega;\varphi)$ (this is usually expressed by saying that the variational problem admits a *time shift*). In fact it is immediate that

$$\begin{cases} \mathcal{L}_v(\theta^\alpha) = 0 \\ \mathcal{L}_v(\varphi) = 0 \quad . \end{cases}$$

(<u>Note</u>: If, as usual, we use $\{\omega = dx; \theta^\alpha; d\dot{y}^\alpha\}$ as a coframe for $J^1(\mathbb{R},\mathbb{R}^m)$ then the time shift is given by

$$v = \partial/\partial\omega - \dot{y}^\alpha\,\partial/\partial\theta^\alpha \quad .)$$

According to Noether's theorem:
The Hamiltonian H *is a* 1st *integral* (note that H is also independent of x).

For mechanical systems (I.a.7) the Hamiltonian is easily seen to be

$$H = T(y,\dot{y}) + U(y)$$

and Noether's theorem reduces to the *law of conservation of total energy* (cf. remark i) in (I.e.22)).

We refer to the abundant excellent references including [1], [2], and [29] for numerous examples of 1st integrals for classical mechanical systems. We have little to add other than to make the point that in some problems (special integrals of Keplerian motion, Newtonian motion with two central forces each given by an inverse square law, Kowaleska systems, geodesics on a triaxial ellipsoid, and others--cf. [2], [66] and (added in proof) the article by J. Moser in *Dynamical Systems*, Progress in Math., no. 8, Birkhäuser (1980)) there are 1st integrals not accounted for by symmetry, at least in an obvious way.

Before concluding this particular discussion, for later use we want to point out one further easy source of 1st integrals.

Definition. In a classical variational problem suppose that the variable y^α does not appear in the Lagrangian (i.e., $L_{y^\alpha} = 0$). Then we say that y^α is a *cyclic coordinate*.

In this case the vector field

$$v = \partial/\partial y^\alpha$$

gives an infinitesimal symmetry of the corresponding classical variational problem with 1st integral

$$L_{\dot{y}^\alpha} = \lambda_\alpha .$$

Actually, for this we don't need Noether's theorem; it is clear from the classical form

$$\frac{d}{dx}(L_{\dot{y}^\alpha}) = L_{y^\alpha}$$

of the Euler-Lagrange equations.

(II.a.13) Example. In $\mathbb{R}^3$ with cylindrical coordinates (r,φ,z) the arclength on a surface of revolution given by

$$z = f(r)$$

is

$$ds^2 = (1 + f'(r)^2)dr^2 + r^2 d\varphi^2$$

$$= F(r)dr^2 + r^2 d\varphi^2$$

where the 2nd equation defines $F(r)$ (here for simplicity we assume that the surface is not part of a cylinder). The corresponding mechanical system in which the potential energy $U = 0$ has Lagrangian

$$L(r,\varphi,\dot{r},\dot{\varphi}) = \frac{1}{2}(F(r)\dot{r}^2 + r^2\dot{\varphi}^2)$$

in which φ is a cyclic coordinate. Thus

$$\lambda_2 = L_{\dot{\varphi}} = r^2\dot{\varphi} \tag{II.a.14}$$

is the 1st integral that corresponds to conservation of angular momentum about the vertical axis (cf. [2], [29]).

(II.a.15) <u>Example</u>. We consider a classical 2nd order variational problem (cf. example (I.a.13)) where $X = J^2(\mathbb{R},\mathbb{R})$ has coordinates $(x,y,\dot{y},\ddot{y})$ and the Lagrangian L is one of the following(1)

$$L = \frac{1}{2}(\ddot{y}^2 - \dot{y}^2) \tag{II.a.16}$$

$$L = \frac{1}{2}(\ddot{y}^2 - y^2) . \tag{II.a.17}$$

Both give non-degenerate variational problems on a 5-dimensional momentum space (cf. (I.e.25)).

<u>Remarks</u>. i) In general, if $L = L(x;y^\alpha;\dot{y}^\alpha;\ddot{y}^\alpha)$ is a 2nd order Lagrangian in which x does not appear, then taking $v = \partial/\partial x$ gives the 1st integral

$$H = \ddot{y}^\alpha\dot{\lambda}_\alpha + \dot{y}^\alpha\lambda_\alpha - L . \tag{II.a.18}$$

Note that here H is defined on the momentum space Y and not on X (or even on a manifold of the same dimension as X).

ii) Still in general, if one of the variables y^α, say y^1, does not appear in $L(x;y^\alpha;\dot{y}^\alpha;\ddot{y}^\alpha)$ (as in the 1st order case we say that y^1 is a cyclic coordinate), then $\partial/\partial y^1$ gives an infinitesimal symmetry of the corresponding variational problem with 1st integral

$$V = \lambda_1 . \tag{II.a.19}$$

If L does not contain either x or y^1, then the same is true of H in (II.a.18). From this it is easy to conclude that the modified Poisson bracket

$$[H,V] = 0 \qquad \text{(II.a.20)}$$

where V is given by (II.a.19).

Proof. On the momentum space Y

$$\psi_Y = -Hdx + \lambda_\alpha dy^\alpha + \dot{\lambda}_\alpha d\dot{y}_\alpha$$
$$\Psi_Y = -dH \wedge dx + d\lambda_\alpha \wedge dy^\alpha + d\dot{\lambda}_\alpha \wedge d\dot{y}_\alpha \ .$$

By exterior algebra

$$dH \wedge d\lambda_1 \wedge \psi_Y \wedge (\Psi_Y)^{m-1} = dH \wedge \psi_Y \wedge \Theta$$

where

$$\Theta = -(2m-1)!\, d\lambda_1 \wedge d\lambda_2 \wedge dy^2 \wedge \cdots \wedge d\lambda_m \wedge dy^m \wedge d\dot{\lambda}_1 \wedge d\dot{y}^1 \wedge \cdots \wedge d\dot{\lambda}_m \wedge d\dot{y}^m$$
$$= (H_x d_x + H_{y_1} dy^1) \wedge \psi_Y \wedge \Theta$$
$$= 0$$

which gives (II.a.20).

Turning to (II.a.16), the Euler-Lagrange differential system is (cf. the computation just below (I.e.26))

$$\begin{cases} d\dot{\lambda} + (\lambda + \dot{y})dx = 0 \\ d\lambda = 0 \qquad\qquad (\text{since } L_y = 0) \\ dy - \dot{y}dx = 0 \\ d\dot{y} - \dot{\lambda}dx = 0 \end{cases}$$

on the space of variables $(x,y,\dot{y},\lambda,\dot{\lambda})$. The above remarks give the two 1st integrals (using that $\ddot{y} = \dot{\lambda}$ on Y)

$$\begin{cases} H = \frac{1}{2}(\dot{\lambda}^2 + \dot{y}^2) + \dot{y}\lambda \\ \lambda \end{cases} \qquad \text{(II.a.21)}$$

Of course, in this, as in any "constant coefficient" case, the Euler-Lagrange equations may be explicitly integrated and a general solution is

$$y = c_1 + c_2 x + c_3 \sin x + c_4 \cos x \ .$$

Thus in addition to (II.a.21) we have quantities such as

$$\dddot{y}^2 + \dot{\lambda}^2 \qquad (\dot{\lambda} = \ddot{y}) \qquad \text{(II.a.22)}$$

that are not functions on Y but are constants of motion, and are thus 1st integrals in some extended sense.

As for (II.a.17) the Euler-Lagrange differential system is (cf. just below (I.e.26))

$$\begin{cases} d\dot{\lambda} + \lambda dx = 0 \\ d\lambda - y dx = 0 \\ dy - \dot{y} dx = 0 \\ d\dot{y} - \dot{\lambda} dx = 0 \end{cases}$$

on the space of variables $(x,y,\dot{y},\lambda,\dot{\lambda})$. This is equivalent to the 4th order O.D.E.

$$\frac{d^4\lambda}{dx^4} + \lambda = 0 \quad .$$

(II.a.23) Example. Suppose that we have a classical 1st order system given by a Lagrangian L that is homogeneous of order μ in the tangent variables and that does not depend on x. In local coordinates $(x;y^\alpha;\dot{y}^\alpha)$ on $J^1(\mathbb{R},M)$

$$\begin{cases} L = L(y^\alpha;\dot{y}^\alpha) \\ L(y^\alpha;\eta\dot{y}^\alpha) = \eta^\mu L(y^\alpha;\dot{y}^\alpha) \end{cases} \quad .$$

Then by example (II.a.12) and *Euler's homogeneity relation*(2)

$$\begin{aligned} H &= \dot{y}^\alpha L_{\dot{y}^\alpha} - L \\ &= (\mu - 1)L \end{aligned}$$

is a 1st integral. Consequently, *if* $\mu \neq 1$ *then* L *is constant on solution curves to the Euler-Lagrange equations for* L.

In particular, suppose that the quadratic Lagrangian (for computational convenience we omit the usual $\frac{1}{2}$ factor)

$$L = g_{\alpha\beta}(y)\dot{y}^\alpha\dot{y}^\beta \qquad \text{(II.a.24)}$$

gives a Riemannian metric ds^2 on a manifold M. Set

$$\tilde{L} = \sqrt{L}$$

so that

$$\tilde{\Phi}(\gamma) = \int_{\gamma} ds$$

gives the length functional. Along a solution curve to the Euler-Lagrange equations for L we have (cf. (I.d.17))

$$\begin{cases} \frac{d}{dx}(L_{\dot{y}^\alpha}) = L_{y^\alpha} \\ \frac{d}{dx}(L) = 0 \end{cases} .$$

Along such curves it follows that

$$\begin{aligned} \frac{d}{dx}(\tilde{L}_{\dot{y}^\alpha}) &= \frac{1}{2}\frac{d}{dx}(L^{-1/2} L_{\dot{y}^\alpha}) \\ &= \frac{1}{2} L^{-1/2} \frac{d}{dx}(L_{\dot{y}^\alpha}) \\ &= \frac{1}{2} L^{-1/2}(L_{y^\alpha}) \\ &= \tilde{L}_{y^\alpha} \end{aligned}$$

which are just the Euler-Lagrange equations for $\tilde{L}$, and therefore we may conclude that:

The solution curves to the Euler-Lagrange equations of (II.a.24) give the geodesics for the corresponding Riemannian metric. (3)

For a special case, referring to example (II.a.13) we easily obtain a proof of

Clairaut's Theorem. *If* α *is the angle that a geodesic makes with the meridian curve on a surface of revolution, then*

$$r \sin\alpha = \text{constant}.$$

Proof. We have $r\dot{\varphi} = \|v\| \sin\alpha$ where v is the tangent vector to the geodesic γ. If we parametrize γ by arclength then (II.a.24) gives

$$r^2\dot{\varphi} = r\sin\alpha = \text{constant}$$

along γ. Q.E.D.

Note. See pages 85, 86 in [2] for a nice discussion of this example.

We now want to give one reasonably precise formulation of the classical notion that an O.D.E. (or dynamical system) should be "algebraically integrable by quadratures."[4] To begin with we shall say that $(I,\omega;\varphi)$ gives an *algebraic variational problem* in case X is a smooth (real) algebraic variety and I,ω,φ are all given by smooth rational forms.[5] This is the case in all examples we have encountered. As is clear from the construction of the Euler-Lagrange differential system (J,ω) on the momentum space Y (cf. Chapter I, Section e)), if $(I,\omega;\varphi)$ is an algebraic variational problem then Y is an algebraic variety and (J,ω) is an algebraic differential system.[6] By an *algebraic 1st integral* we shall mean a 1st integral V that is an algebraic function on Y (i.e., V should lie in a finite algebraic extension of the field of rational functions on Y, such as the function $\sqrt{1+t^2}$ on $\mathbb{R}$).

Definition. We shall say that the Euler-Lagrange equations associated to the algebraic variational problem $(I,\omega;\varphi)$ are *algebraically integrable by quadratures* in case there are algebraic 1st integrals V_j such that for general constants C_j the equations

$$V_j = C_j$$

determine an integral *curve* of (J,ω).

(II.a.25) Example. Of course the classical example is given by a conservative mechanical system with one degree of freedom; i.e., on a Zariski open subset $X \subset J^1(\mathbb{R},\mathbb{R})$ by the Euler-Lagrange equations associated to the Lagrangian

$$L = \frac{\dot{y}^2}{2} - U(y) \qquad \text{(II.a.26)}$$

where $U(y)$ is a rational function of y.
(Note. The reason that X may be only a Zariski open subset is to allow interesting potential functions such as $U(y) = k/y$.) The Hamiltonian

$$H = \frac{\lambda^2}{2} + U(y) \qquad (\lambda = \dot{y}) \qquad \text{(II.a.27)}$$

is an algebraic 1st integral on the momentum space. Each solution $y(x)$ of the Euler-Lagrange equations

$$\frac{d^2y(x)}{dx^2} + U'(y(x)) = 0 \qquad \text{(II.a.28)}$$

associated to (II.a.26) traces out a curve $\left(y(x), \frac{dy(x)}{dx}\right)$ in the phase space $\mathbb{R}^2$ with coordinates $(y,\dot{y})$, and these phase curves are just the level sets

$$H(y,\dot{y}) = c$$

where H is given by (II.a.27).(7)

<u>Remarks</u>: i) Any 2nd order O.D.E. of the form

$$\frac{d^2y(x)}{dx^2} + u(y(x)) = 0 \qquad \text{(II.a.29)}$$

arises as the Euler-Lagrange equation associated to a Lagrangian (II.a.26) where

$$U(y) = \int_{y_0}^{y} u(t)\, dt \quad .$$

Thus, if $U(y)$ is a rational function then (II.a.29) is algebraically integrable by quadratures.

ii) We may also take for L a Lagrangian

$$L = \frac{g(y)\dot{y}^2}{2} - U(y)$$

where $g(y)$ is a rational function that is positive at a general point $y \in \mathbb{R}$. The corresponding Hamiltonian is

$$H = \frac{g(y)\lambda^2}{2} + U(y) \quad ,$$

and the Euler-Lagrange equations are again algebraically integrable by quadratures.

The algebraic curves that turn up in this way are all of the form

$$\dot{y}^2 = R(y) \qquad \text{(II.a.30)}$$

where $R(y)$ is a rational function of y; i.e., they are always *hyper-elliptic algebraic curves* (cf. [36], [59]).

For reasons arising from our examples we want to give another concept of being integrable by quadratures. Let G be a Lie group with basis $\omega^i \in \mathfrak{g}^*$ for the Maurer-Cartan forms. Then by (I.b.7) and (I.b.9) any curve

$$f: N \to G \quad , \qquad N = \{t : a \leqq t \leqq b\} \ ,$$

is given uniquely up to left translations by specifying functions $q^i(t)$ such that (cf. (I.b.8))

$$f^*\omega^i = q^i(t)\, dt \quad .$$

Moreover, the functions $q^i(t)$ may be prescribed arbitrarily. In invariant terms, $q(t) = \{q^i(t)\}$ is a curve in $\mathfrak{g}$, and so we may say that $\underline{f}$ *is uniquely determined by* $q(t)$ *once we specify the* dim G *constants* $f(0) \in G$.

Now let $(I,\omega;\varphi)$ be a variational problem that may be posed on G by a left-invariant 1-form. The solution curves to the corresponding Euler-Lagrange equations are then curves in G uniquely given up to left translation by the corresponding curve $q(t)$ in $\mathfrak{g}^*$.

<u>Definition</u>: We shall say that $(I,\omega;\varphi)$ is *quasi-integrable by quadratures* in case there are functions $H^\rho(q)$ on $\mathfrak{g}$ such that for general constants c^ρ the equations

$$H^\rho(q) = c^\rho$$

determine a curve in $\mathfrak{g}$ with the property that these curves are exactly the curves corresponding to solutions of the Euler-Lagrange equations of $(I,\omega;\varphi)$.(19)

(II.a.31) <u>Example</u>. Let M be the simply connected n-dimensional complete Riemannian manifold of constant sectional curvature R (i.e., M is $\mathbb{E}^n$ in the case $R = 0$, the sphere $S^n(1/R)$ of radius $1/R$ in the case $R > 0$, or the hyperbolic space $H^n(1/R)$ in the case $R < 0$). To each curve $\gamma \subset M$ parameterized by arclength and with curvature function $\kappa(s)$ we associate the functional

$$\Phi = \frac{1}{2} \int_\gamma \kappa^2\, ds \quad . \tag{II.a.32}$$

As will be seen below this functional may, in a natural way, be posed as an invariant variational problem on the Lie group Aut(M) of isometries of M. During the course of this text we shall prove the

(II.a.33) THEOREM: *The Euler-Lagrange equations associated to the functional (II.a.32) are quasi-integrable by quadratures.*

In fact much more than this will come out, including the complete "phase portrait" in $\mathfrak{g}^*$ of the curves $\{\lambda(t)\}$ corresponding to to extremals of (II.a.32). Discussions with Robert Bryant have shown that the motion in Aut(M) is given by a *linear* flow on either a cylinder $\mathbb{R}\times\mathbb{R}/\mathbb{Z}$ or on a 2-torus $\mathbb{R}^2/\mathbb{Z}^2$; these curves are well understood, and have an especially nice picture in the torus case (cf. [2]). We shall only give the complete proof of (II.a.33) when $R \geqq 0$, as the case $R<0$ is formally the same as $R>0$.

To relate our terminology with that in standard references we want to discuss another notion of integrability for the Euler-Lagrange differential system associated to a variational problem $(I,\omega;\varphi)$. For this we make the following

(II.a.34) Assumptions. The variational problem $(I,\omega;\varphi)$ is non-degenerate and the Pfaff-Darboux normal form (I.e.15) holds *globally* on Y.

More precisely, there should be a diffeomorphism

$$Y \cong \mathbb{R}\times P \qquad \dim P = 2m$$

where $\mathbb{R}$ has coordinate x and a 1-form η on P such that

$$d\eta = \Omega$$

is a 2-form of maximal rank on P, and such that

$$\psi_Y = -H dx + \eta$$

for some function H on $\mathbb{R}\times P$. In particular, the pair (P,Ω) defines a *symplectic structure* on P.(8) Any point of P has a neighborhood with coordinates (u_i, v^i) such that (cf. [11], [64] for a proof)

$$\begin{cases} \eta = u_i dv^i \\ \Omega = du_i \wedge dv^i \end{cases} \qquad \text{(II.a.35)}$$

Examples include a classical variational problem of any order that is non-degenerate in the sense of examples (I.e.18) and (I.e.23).

Definition. The Euler-Lagrange equations associated to $(I,\omega;\varphi)$ are said to be (globally) in *Hamiltonian form* in case (II.a.34) holds where $H = H(u,v)$ is independent of x.

For functions U,V on a symplectic manifold (P,Ω) the (usual) *Poisson bracket* $\{U,V\}$ is the function defined by

$$\{U,V\}\Omega^m = m\,dU \wedge dV \wedge \Omega^{m-1} . \qquad \text{(II.a.36)}$$

In local coordinates for which (II.a.35) is valid,

$$\{U,V\} = \frac{\partial U}{\partial u_i}\frac{\partial V}{\partial v^i} - \frac{\partial U}{\partial v^i}\frac{\partial V}{\partial u_i} . \qquad \text{(II.a.37)}$$

The functions U,V are said to be *in involution* in case $\{U,V\} = 0$.

Note. We may also consider U,V as functions on $Y = \mathbb{R} \times P$ and then their modified Poisson bracket is defined by (II.a.4). It is not immediately clear what the relation between $[U,V]$ and $\{U,V\}$ is (other than the property that if both U,V are 1st integrals, then so is either bracket. However, see footnote[20]).

Definition. Let $(I,\omega;\varphi)$ be a variational problem whose Euler-Lagrange differential system is in Hamiltonian form. Then these equations are said to be *completely integrable* in case there are functions $H = V_1, V_2, \ldots, V_m$ on P such that $dV_1 \wedge \ldots \wedge dV_m \neq 0$ and such that all V_i, V_j are in involution; i.e., all

$$\{V_i, V_j\} = 0 .$$

Completely integrable Hamiltonian systems are discussed in many places; e.g., in [2]. Recently many interesting finite- and infinite-dimensional examples of completely integrable Hamiltonian systems have turned up (cf. the appendices to [2]).

It is well-known that the presence of a cyclic coordinate in a classical 1st order variational problem may be used to reduce from $J^1(\mathbb{R},\mathbb{R}^m)$ to $J^1(\mathbb{R},\mathbb{R}^{m-1})$. Moreover, when $m = 2$ and the Lagrangian $L = L(y,\dot{y})$ does not depend on x, the presence of one cyclic coordinate implies that the Euler-Lagrange equations are completely integrable (cf. page 272 of [2]).

(II.a.38) Example. We consider again a surface of revolution in $\mathbb{R}^3$ with the Lagrangian given by (1/2 times) the induced metric. In the notation of example (II.a.13)

$$L = \frac{1}{2}\left(F(r)\dot{r}^2 + r^2\dot{\varphi}^2\right) . \qquad \text{(II.a.39)}$$

Two 1st integrals are given by

$$\begin{cases} V_1 = H & (= L \text{ in this case}) \\ V_2 = r^2\dot{\varphi} , \end{cases}$$

and from (II.a.37) it follows that these functions are in involution.

Proof. The momentum space $Y = \mathbb{R} \times P$ has coordinates $(x, r, \lambda_1, \varphi, \lambda_2)$ where

$$\begin{cases} \lambda_1 = L_{\dot{r}} = F(r)\dot{r} \\ \lambda_2 = L_{\dot{\varphi}} = r^2\dot{\varphi} \end{cases}$$

It follows that P has coordinates $(r, \lambda_1, \varphi, \lambda_2)$ with symplectic form

$$\Omega = dr \wedge d\lambda_1 + d\varphi \wedge d\lambda_2 ,$$

and that in terms of these coordinates on P

$$\begin{cases} \dot{r} = \lambda_1/F(r), \quad \dot{\varphi} = \lambda_2/r^2 \\ V_1 = \frac{1}{2} (\lambda_1^2/F(r) + \lambda_2^2/r^2) \\ V_2 = \lambda_2 \end{cases}$$

The involutivity of V_1, V_2 then follows from (II.a.37) and $\partial V_1/\partial\varphi = 0$.

(II.a.40) Example. A classical example of what is essentially a completely integrable system is provided by a wheel of unit radius rolling without slipping on a plane:

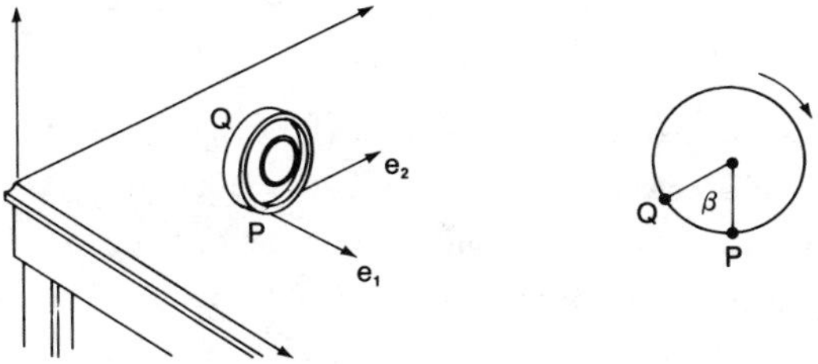

This is also the classic example of a mechanical system having a non-holonomic constraint.

To set the problem up we take the configuration manifold of the point Q to be

$$M = \mathcal{F}(\mathbb{E}^2) \times S^1 .$$

This has the following meaning: To each position of Q we associate the frame (P, e_1, e_2) where P is the point of contact of the wheel with the plane (to avoid later confusion in notation, in this example we denote the position vector by P rather than x), e_1 is the unit vector in the direction in which the wheel is rolling, $e_1 \wedge e_2 = 1$, and finally β is the angle in the picture. The fact that the wheel rolls without slipping is expressed by saying that trajectories $Q(t)$ are integral curves of the Pfaffian differential system

$$\begin{cases} \varphi^1 = \omega^1 - d\beta = 0 \\ \varphi^2 = \omega^2 = 0 \end{cases} \tag{II.a.41}$$

Denote this system by K. From the structure equations (0.c.3) of a moving frame we have

$$\begin{cases} d\varphi^1 = -\varphi^2 \wedge \omega_1^2 \\ d\varphi^2 = \omega^1 \wedge \omega_1^2 \end{cases}$$

Thus the 1st derived system K_1 is generated by φ^1 and the 2nd derived system $K_2 = (0)$. According to Chow's theorem discussed in (I.c.18) we may find an integral curve of (II.a.41) taking Q to Q' in the following figure

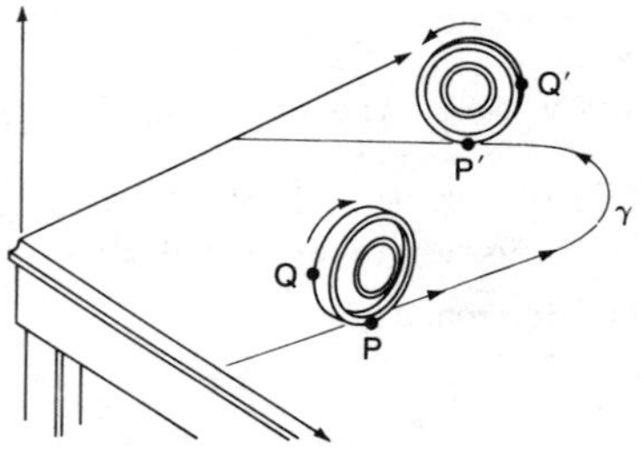

This may be done by choosing the path γ to have length $2\pi k + \widehat{QQ'}$ $(k \in \mathbb{Z})$ and to have the dotted lines as tangent lines at its endpoints.

To investigate the equations of motion of Q we use coordinates (x,y,α,β) on the configuration manifold M as depicted in the following figure:

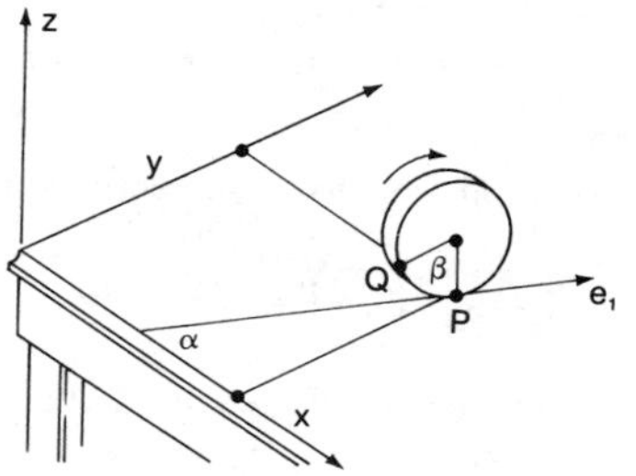

Thus

$$\begin{cases} P = (x,y,0) \\ Q = (x - \sin\beta\ \cos\alpha,\ y - \sin\beta\ \sin\alpha,\ 1 - \cos\beta)\ . \end{cases}$$

That the motion $Q(t)$ should be subject to the constraint (II.a.41) is expressed by

$$\begin{cases} \dfrac{dx}{dt} = \cos\alpha\ \dfrac{d\beta}{dt} \\[2ex] \dfrac{dy}{dt} = \sin\alpha\ \dfrac{d\beta}{dt} \end{cases} \tag{II.a.42}$$

Following our usual notations we consider

$$J^1(\mathbb{R},M) \cong \mathbb{R} \times T(M)$$

with coordinates

$$(t;x,y,\alpha,\beta;\dot{x},\dot{y},\dot{\alpha},\dot{\beta})\ ,$$

and then the constraining manifold $X \subset J^1(\mathbb{R},M)$ is given by

$$\begin{cases} \dot{x} - \dot{\beta}\cos\alpha = 0 \\ \dot{y} - \dot{\beta}\sin\alpha = 0\ . \end{cases} \tag{II.a.43}$$

The motion of Q is described by *some* of the solution curves to the Euler-Lagrange equations associated to the variational problem $(I,\omega;\varphi)$ where (I,ω) is the canonical Pfaffian differential system

$$\begin{cases} \theta^1 = dx - \dot{x}dt = dx - \dot{\beta}\cos\alpha\ dt = 0 \\ \theta^2 = dy - \dot{y}dt = dy - \dot{\beta}\sin\alpha\ dt = 0 \\ \theta^3 = d\alpha - \dot{\alpha}dt = 0 \\ \theta^4 = d\beta - \dot{\beta}dt = 0 \\ \omega = dt \neq 0 \end{cases}$$

on X and

$$\varphi = L dt$$

where

$$L = \frac{1}{2}(\dot{Q},\dot{Q})$$

is the kinetic energy of Q.

Remark. The reason that the motion of Q is not described by a *general* solution curve of the Euler-Lagrange equations associated to $(I,\omega;\varphi)$ is a somewhat subtle matter dealing with endpoint conditions and will be discussed in detail in Chapter IV, Section a) (cf. example (IV.a.26).

Returning to the general discussion, using (II.a.43) the restriction to X of $\dot{Q}$ is

$$\dot{Q} = (\dot{x} - \dot{\beta}\cos\beta\cos\alpha + \dot{\alpha}\sin\beta\sin\alpha, \dot{y} - \dot{\beta}\cos\beta\sin\alpha - \dot{\alpha}\sin\beta\cos\alpha, \dot{\beta}\sin\beta)$$

$$= (\dot{\beta}\cos\alpha(1 - \cos\beta) + \dot{\alpha}\sin\beta\sin\alpha, \dot{\beta}\sin\alpha(1 - \cos\beta) - \dot{\alpha}\sin\beta\cos\alpha, \dot{\beta}\sin\beta).$$

It follows that the restriction of L to X is

$$L = (1 - \cos\beta)\dot{\beta}^2 + \frac{\sin^2\beta}{2}\dot{\alpha}^2 . \tag{II.a.44}$$

This suggests that on the space $\tilde{X}$ with coordinates $(t,\alpha,\beta,\dot{\alpha},\dot{\beta})$ we consider the classical variational problem $(\tilde{I},\omega;\tilde{\varphi})$ where $(\tilde{I},\omega)$ is the Pfaffian differential system

$$\begin{cases} \tilde{\theta}^1 = d\alpha - \dot{\alpha}dt = 0 \\ \tilde{\theta}^2 = d\beta - \dot{\beta}dt = 0 \\ \omega = dt \neq 0 \end{cases}$$

and

$$\tilde{\varphi} = \left((1 - \cos\beta)\dot{\beta}^2 + \frac{\sin^2\beta}{2}\dot{\alpha}^2\right) dt .$$

We shall prove that:

(II.a.45). *The curves in* X *obtained from solution curves to* $(\tilde{I},\omega;\tilde{\varphi})$ *by integrating the equations (II.a.42) are solution curves to* $(I,\omega;\varphi)$. *These curves in* X *describe the motion of the wheel.*

Proof. The Euler-Lagrange system for $(\tilde{I},\omega;\tilde{\varphi})$ is generated by the Pfaffian equations (cf. (I.d.16))

$$\begin{cases} dL_{\dot{\alpha}} - L_{\alpha}dt = 0 \\ dL_{\dot{\beta}} - L_{\beta}dt = 0 . \end{cases} \tag{II.a.46}$$

This gives a pair of 2nd order equations for $\alpha(t)$, $\beta(t)$ that may be uniquely solved for small t with arbitrary initial values

$$\alpha(0),\ \beta(0),\ \alpha'(0),\ \beta'(0) . \tag{II.a.47}$$

Using (II.a.42) we may then uniquely determine $x(t)$, $y(t)$ with given initial values

$$x(0),\ y(0) . \tag{II.a.48}$$

Note that we cannot arbitrarily assign $x'(0)$ and $y'(0)$. Physically the initial values (II.a.47) and (II.a.48) amount to giving the wheel an initial position and push.

We now compute the Euler-Lagrange system for $(I,\omega;\varphi)$. Setting

$$\psi = \varphi + \lambda_1\theta^1 + \lambda_2\theta^2 + \lambda_3\theta^3 + \lambda_4\theta^4$$

we have for $\Psi = d\psi$ that

$$\Psi = (L_{\dot\alpha} - \lambda_3)d\dot\alpha \wedge dt + (L_{\dot\beta} - \lambda_4 - \lambda_1 \cos\alpha - \lambda_2 \sin\alpha)d\dot\beta \wedge dt$$

$$+ (d\lambda_3 - (L_\alpha + \dot\beta(\lambda_2 \cos\alpha - \lambda_1 \sin\alpha))dt) \wedge \theta^3 + (d\lambda_4 - L_\beta dt) \wedge \theta^4 + \sum_{i=1,2} d\lambda_i \wedge \theta^i$$

The Euler-Lagrange system is generated by the Pfaffian equations

$$\begin{cases} \text{(i)} & \partial/\partial\dot\alpha \lrcorner \Psi = (L_{\dot\alpha} - \lambda_3)dt = 0 \\ \text{(ii)} & \partial/\partial\dot\beta \lrcorner \Psi = (L_{\dot\beta} - \lambda_4 - \lambda_1 \cos\alpha - \lambda_2 \sin\alpha)dt = 0 \\ & \partial/\partial\lambda_i \lrcorner \Psi = \theta^i = 0 \\ \text{(iii)} & \partial/\partial\theta^3 \lrcorner \Psi = -d\lambda_3 + (L_\alpha + \dot\beta(\lambda_2 \cos\alpha - \lambda_1 \sin\alpha))dt = 0 \\ \text{(iv)} & \partial/\partial\theta^4 \lrcorner \Psi = -d\lambda_4 + L_\beta dt = 0 \\ \text{(v)} & \partial/\partial\theta^1 \lrcorner \Psi = -d\lambda_1 = 0 \\ \text{(vi)} & \partial/\partial\theta^2 \lrcorner \Psi = -d\lambda_2 = 0 \quad . \end{cases}$$

Equations (i), (ii), (v), (vi) give

$$\begin{cases} L_{\dot\alpha} = \lambda_3 \\ L_{\dot\beta} = \lambda_4 + \lambda_1 \cos\alpha + \lambda_2 \sin\alpha \\ \lambda_1 = \text{constant} \\ \lambda_2 = \text{constant} \quad . \end{cases}$$

Plugging these into (iii), (iv) gives

$$\begin{cases} dL_{\dot\alpha} - (L_\alpha + \dot\beta(\lambda_2 \cos\alpha - \lambda_1 \sin\alpha))dt = 0 \\ dL_{\dot\beta} + (\lambda_1 \dot\alpha \sin - \lambda_2\dot\alpha \cos\alpha - L_\beta)dt = 0 \quad . \end{cases} \qquad \text{(II.a.49)}$$

Given a solution $\tilde N$ to (II.a.46) we consider the curve N in X obtained from $\tilde N$ by integrating the equations (II.a.42). If along this curve we set

$$\lambda_3 = L_{\dot{\alpha}} , \quad \lambda_4 = L_{\dot{\beta}} , \quad \lambda_1 = \lambda_2 = 0 ,$$

then by comparing (II.a.46) with (II.a.49) we infer that N is a solution to the Euler-Lagrange system associated to $(I,\omega;\varphi)$. This proves the first part of (II.a.45).

The second part follows from the least action principle together with our remarks above concerning initial values. Q.E.D.

Remark. The somewhat mysterious relationship between the variational problems $(I,\omega;\varphi)$ on X and $(\tilde{I},\omega;\tilde{\varphi})$ on $\tilde{X}$ will be clarified in Chapter IV, Section a) (cf. example (IV.a.26)).

It remains to discuss the variational problem $(\tilde{I},\omega;\tilde{\varphi})$. Due to the presence of the cyclic coordinate α the Euler-Lagrange equations associated to $(\tilde{I},\omega;\tilde{\varphi})$ give a completely integrable Hamiltonian system (cf. corollary 1 on page 272 of [2]). We shall indicate briefly how the integration may be carried out.

Using (II.a.44) we have from the cyclic coordinate α and conservation of total energy the two 1st integrals

$$\begin{cases} \text{(i)} \quad L_{\dot{\alpha}} = (\sin^2 \beta)\dot{\alpha} \\ \text{(ii)} \quad L = (1 - \cos \beta)\dot{\beta}^2 + \dfrac{c_1^2}{2 \sin^2 \beta} = c_2 \end{cases} .$$

Setting $u = 1 - \cos \beta$ we have

$$\begin{cases} \cos \beta = 1 - u \\ \sin \beta = \sqrt{u(2-u)} \\ \dot{\beta} = \dfrac{\dot{u}}{\sqrt{u(2-u)}} \end{cases} .$$

Then (ii) is

$$\frac{\dot{u}^2}{2-u} + \frac{c_1^2}{2u(2-u)} = c_2 .$$

Relabelling constants and separating variables this is

$$\frac{du}{\sqrt{\dfrac{au^2+bu+c}{u}}} = dt .$$

This O.D.E. may be integrated by elliptic functions (cf. the note concerning surfaces of revolution having constant Gauss curvature in example (II.a.50)). From (i) we have

$$d\alpha = \frac{c_1\, dt}{2u(2-u)}$$

$$= \frac{c_1\, du}{2\dot{u}u(2-u)}\ ; \qquad \text{i.e.}$$

$$d\alpha = \frac{c_1}{2(2-u)} \frac{du}{\sqrt{u(au^2+bu+c)}} \ .$$

This equation may also be integrated using elliptic functions.

We note the special solution

$$\begin{cases} \alpha = \text{constant} \\ \beta = \arccos(ct^2-1) \end{cases}$$

where the wheel rolls in a straight line.

(II.a.50) Example. The basic invariant of an oriented submanifold $M^n \subset \mathbb{E}^{n+r}$ is its 2nd *fundamental form* II, which assigns to each point $p \in M$ a quadratic form $II(p)$ with values in the normal bundle to M at p (cf. [6], [63]). A natural functional to consider is

$$\Phi(M) = \frac{1}{k} \int_M \|II(p)\|^k \, dA(p) \tag{II.a.51}$$

where dA is the volume form of the induced Riemannian metric on M and $\|II(p)\|$ is the length of the 2nd fundamental form at p. When $n=1$ and $k=2$ this is just the functional (II.a.32). When $n=k=2$ and $r=1$, and $M \subset \mathbb{E}^3$ is a compact oriented surface, by the Gauss-Bonnet theorem the functional is

$$\frac{1}{2} \int_M (H^2 - 2K)\, dA = \frac{1}{2} \int_M H^2 dA - \frac{1}{2\pi} \chi(M) \tag{II.a.52}$$

where H is the *mean curvature* (= trace (II)) and $\chi(M)$ is the Euler-Poincaré characteristic of M. The quantity

$$\int_M H^2 \, dA$$

was introduced by Wilmore [67] and recently studied by Li-Yau [51].

In general

$$\frac{1}{2} \int_M \|II\|^2 \, dA \tag{II.a.53}$$

might be considered as a sort of "extrinsic Yang-Mills functional," while for dimension reasons the functional

$$\frac{1}{n} \int_M \|II\|^n \, dA$$

may be more natural. For example, it is conformally invariant (a fact exploited by Li-Yau in [51]). When $n=2$ the two functionals coincide.

Here we shall investigate the Euler-Lagrange equations of (II.a.53) for a surface of revolution $S \subset \mathbb{R}^3$. We first put them in Hamiltonian form

$$\begin{cases} \dfrac{dy}{dx} = H_\lambda \\ \dfrac{d\dot{y}}{dx} = H_{\dot{\lambda}} \\ \dfrac{d\lambda}{dx} = -H_y \\ \dfrac{d\dot{\lambda}}{dx} = -H_{\dot{y}} \end{cases}$$

where $H = H(y,\dot{y},\lambda,\dot{\lambda})$ is a function in the space of coordinates $(x,y,\dot{y},\lambda,\dot{\lambda})$. Since $H_x = 0$ we immediately obtain one 1st integral, which geometrically reflects the rotational symmetry of S about an axis. Next, the conformal vector field

$$v = x^i \, \partial/\partial x^i$$

in $\mathbb{R}^3$ will be shown via Noether's theorem to yield the 1st integral

$$V = -xH + \lambda y \, .$$

Notice that $V_x = -H \neq 0$. A computation during which an unfortunate miraculous cancellation occurs shows that the modified Poisson-bracket

$$[H,V] = cH \qquad (c = \text{constant}) \, . \tag{II.a.54}$$

Thus the algebra of 1st integrals generated by H, V, $[H,V]$, $[H,[H,V]]$, $[V,[H,V]]$, etc. is the same as the algebra generated by H, V.

Essentially because of this we are unable to anser the following:

<u>Question</u>. *Are there any further algebraic 1st integrals of the Euler-Lagrange equations associated to (II.a. 53)? Equivalently, are these equations algebraically integrable by quadratures?* (cf.(21))

Before computing the Euler-Lagrange equations we need to derive the basic structure equations for a surface of revolution $S \subset \mathbb{R}^3$, given in cylindrical coordinates (z,r,φ) by

$$\begin{cases} z = f(r) \\ x = r\cos\varphi \\ y = r\sin\varphi \end{cases}$$

where (x,y,z) are the usual rectangular coordinates in $\mathbb{R}^3$:

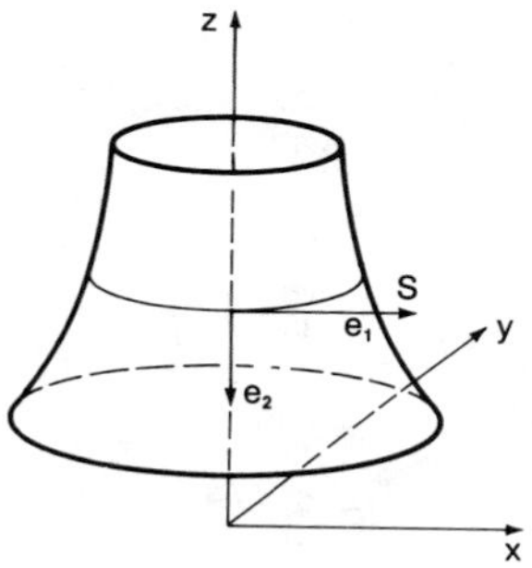

We set

$$F(r) = \sqrt{1+f'(r)^2}$$

and then

$$\begin{cases} \partial/\partial\varphi = (-r\sin\varphi, r\cos\varphi, 0); & \|\partial/\partial\varphi\| = r \\ \partial/\partial r = (\cos\varphi, \sin\varphi, f'(r)); & \|\partial/\partial r\| = F(r) \end{cases} .$$

Hence we take as frame field

$$\begin{cases} e_1 = \frac{1}{r}\,\partial/\partial\varphi \\ e_2 = \frac{1}{F(r)}\,\partial/\partial r \end{cases} .$$

The unit normal e_3 is given by

$$e_3 = e_1 \times e_2 \quad ;$$

i.e.,

$$e_3 = \frac{1}{F}\,(f'\cos\varphi, f'\sin\varphi, -1) .$$

This gives a field of *Darboux frames*

$$(r,\varphi) \to (e_1(r,\varphi),\ e_2(r,\varphi),\ e_3(r,\varphi)) ,$$

and the pullbacks via this map of the basic 1-forms ω^i, ω_i^j on $F(\mathbb{E}^3)$ (cf. Chapter 0, Section c)) are

$$\begin{cases} \omega^1 = r d\varphi \\ \omega^2 = F dr \\ \omega^3 = 0 \\ \omega_1^2 = -\dfrac{d\varphi}{F} & (=(de_1,e_2)) \\ \omega_1^3 = -\dfrac{f'}{F}\, d\varphi & (=(de_1,e_3)) \\ \omega_2^3 = -\dfrac{f''}{F^2}\, dr & (=(de_2,e_3)) . \end{cases} \tag{II.a.55}$$

The 2nd fundamental form is then

$$\mathrm{II} = \left(\frac{-f'}{rF}\right)(\omega^1)^2 + \left(\frac{-f''}{F^3}\right)(\omega^2)^2 \tag{II.a.56}$$

(in general, II is the quadratic differential form (cf. [15], [22])

$$\omega^1\omega_1^3 + \omega^2\omega_2^3 \qquad).$$

Remark. We will check these formulas. By the Gauss equation the Gaussian curvature

$$K = \det \mathrm{II} = \frac{f'f''}{rF^4}$$

(to check the sign take the case $z=\sqrt{1-r^2}$ of the unit sphere). On the other hand, by the structure equations of surface theory (loc. cit.)

$$d\omega_1^2 = -K\omega^1 \wedge \omega^2$$

$$d\omega_1^2 = \frac{F'}{F^2}\, dr \wedge d\varphi \qquad \text{by (II.a.55)}$$

$$= -\frac{F'}{rF^3}\, \omega^1 \wedge \omega^2$$

which gives

$$K = \frac{F'}{rF^3} = \frac{f'f''}{rF^4}$$

since $F' = f'f''/F$.

Note. These formulas lead to the well-known interesting fact that: *The surfaces of revolution of constant Gaussian curvature are described by elliptic functions.*

Proof. For simplicity we take the case $K = 1$. Then integration of

$$\frac{F'}{F^3} = r$$

gives

$$\frac{1}{F^2} = c - r^2 .$$

This is the same as

$$f'(r) = \sqrt{\frac{r^2 - c_1}{c - r^2}} , \qquad c_1 = c - 1 ,$$

which gives

$$f(r) = \int_{r_0}^{r} \frac{dx}{\sqrt{\frac{c-x^2}{x^2-c_1}}} + c_2 .$$

To evaluate this integral we consider the algebraic curve C given by

$$y^2 = \frac{c-x^2}{x^2-c_1} .$$

In homogeneous coordinates this is

$$y^2x^2 - c_1z^2y^2 + z^2x^2 - cz^4 = 0 .$$

This is a plane quartic with ordinary double points at $[0,1,0]$. $[0,0,1]$ as its only singularities. By the genus formula ([36], [59]) its desingularization is a curve of genus one. On this curve $\omega = dx/y$ is an abelian integral (in fact, it is *the* holomorphic differential). Hence $f = \int \omega$ is the inverse of an elliptic function (loc. cit.).

As a 2nd check, the mean curvature is

$$H = \text{Trace } \mathrm{II}$$

$$= -\frac{1}{F}\left(\frac{f'}{r} + \frac{f''}{F^2}\right)$$

$$= -\frac{1}{rF^3}\left((1+f'^2)f' + rf''\right) .$$

This gives for the minimal surface equation

$$H = 0 = \left((1+f'^2)f' + rf''\right) .$$

On the other hand it is well-known (cf. (IV.b.41) below) that the minimal surface equation is the Euler-Lagrange equation of the classical variational problem on $J^1(\mathbb{R},\mathbb{R})$ with coordinates (r,f,f') and Lagrangian $L = 2\pi r\sqrt{1+f'^2}$. The Euler-Lagrange equation

$$\frac{d}{dr}(L_{f'}) = L_f = 0$$

is then

$$0 = \frac{d}{dr}\left(\frac{rf'}{\sqrt{1+f'^2}}\right) = \frac{1}{\sqrt{1+f'^2}^{\,3}}\left(f'(1+f'^2) + rf''\right)$$

which checks with the above condition $H = 0$.

We return to the Euler-Lagrange equations for the functional (II.a.53). By (II.a.56)

$$\frac{1}{2}\|\mathrm{II}\|^2 = \frac{1}{2}\left(\frac{f'^2}{r^2F^2} + \frac{f''^2}{F^6}\right)$$

$$= \frac{1}{2F^2}\left(\left(\frac{f'}{r}\right)^2 + \left(\frac{f''}{F^2}\right)^2\right) .$$

Taking into account that

$$dA = \omega^1 \wedge \omega^2$$

$$= rF\,d\varphi \wedge dr$$

and ignoring the 2π-factor, the functional (II.a.53) is

$$\Phi = \int L(r,f(r),f'(r),f''(r))\,dr$$

where

$$L = \frac{r}{2F}\left(\left(\frac{f'}{r}\right)^2 + \left(\frac{f''}{F^2}\right)^2\right), \qquad F = \sqrt{1+f'^2} .$$

We note that f is a cyclic coordinate.[9]

Now for reasons of Hamiltonian symmetry it is preferable to have a time-independent Lagrangian. This suggests that we take $f = x$ as independent variable and $r = y$ as dependent variable. Thus we must set

$$\begin{cases} r = y \\ f = x \\ f' = \dfrac{1}{\dot{y}} \\ f'' = \dfrac{\ddot{y}}{\dot{y}^3} \\ dr = \dot{y}dx \end{cases} .$$

In these coordinates the Lagrangian turns out to be

$$L(x,y,\dot{y},\ddot{y}) = \frac{1}{2G}\left(\frac{1}{y} + \frac{y\ddot{y}^2}{G^4}\right), \quad G = \sqrt{1+\dot{y}^2} . \tag{II.a.57}$$

Note that L does not depend on x. In terms of the canonical coordinates $(x,y,\dot{y},\lambda,\dot{\lambda})$ with

$$\dot{\lambda} = L_{\ddot{y}} = \frac{y\ddot{y}}{G^5} ,$$

which may be inverted to give

$$\ddot{y} = \frac{\dot{\lambda}G^5}{y} ,$$

the Hamiltonian is

$$H = -L + \lambda\dot{y} + \dot{\lambda}\ddot{y} .$$

This is

$$H = \frac{1}{2y}\left(\dot{\lambda}^2 G^5 - \frac{1}{G}\right) + \lambda\dot{y} . \tag{II.a.58}$$

Discussion. Geometrically taking $y = r$ as dependent variable means that "we look at the ruled surface the other way."

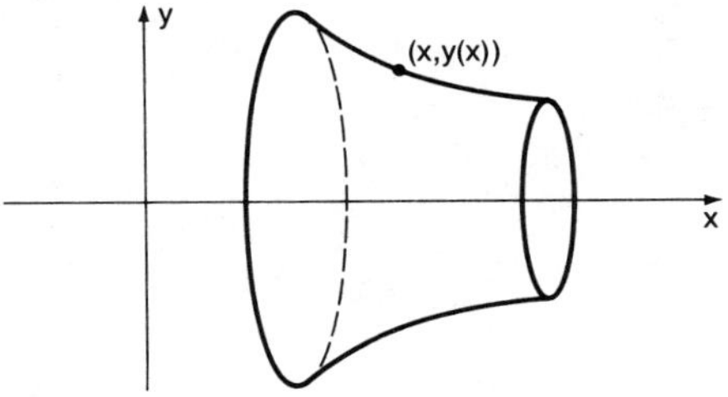

(Here the x axis is the same as the z-axis previously.) Just for fun we shall check (II.a.57). The surface is now given by

$$(x,y) \to (x, y(x)\cos\varphi, y(x)\sin\varphi) = X(x,\varphi)$$

then

$$\begin{cases} \partial/\partial x \to (1, \dot{y}\cos\varphi, \dot{y}\sin\varphi) \\ \partial/\partial\varphi \to (0, -y\sin\varphi, y\cos\varphi). \end{cases}$$

We take as orthonormal frame field

$$\begin{cases} e_1 = \frac{1}{G}(1, \dot{y}\cos\varphi, \dot{y}\sin\varphi) \\ e_2 = \frac{1}{y}(0, -y\sin\varphi, y\cos\varphi) \end{cases}$$

where

$$G = \sqrt{1+\dot{y}^2} .$$

Then a unit normal is

$$e_3 = e_1 \times e_2 = \frac{1}{G}(\dot{y}, -\cos\varphi, -\sin\varphi) ,$$

and

$$\begin{cases} \omega_1^3 = (de_1, e_3) = \frac{-\ddot{y}}{G^2}\, dx \\ \omega_2^3 = (de_2, e_3) = \frac{d\varphi}{G} \end{cases} .$$

From

$$dX(x,\varphi) = \omega^1 e_1 + \omega^2 e_2$$

it follows that

$$\begin{cases} \omega^1 = G\,dx \\ \omega^2 = y\,d\varphi \\ dA = y\sqrt{1+\dot{y}^2}\, dx \wedge d\varphi \end{cases}$$

(the latter is clearly the correct expression for dA). Finally

$$\begin{aligned} II &= \omega^1\omega_1^3 + \omega^2\omega_2^3 \\ &= \frac{-\ddot{y}}{G}\, dx^2 + \frac{y}{G}\, d\varphi^2 \\ &= \frac{-\ddot{y}}{G^3}(\omega^1)^2 + \frac{1}{Gy}(\omega^2)^2 , \end{aligned}$$

$$\frac{1}{2}\,\|II\|^2 = \frac{1}{2G^2}\left(\frac{\ddot{y}^2}{G^4} + \frac{1}{y^2}\right)$$

$$\frac{1}{2}\,\|II\|^2 dA = \frac{1}{2G}\left(\frac{\ddot{y}^2 y}{G^4} + \frac{1}{y}\right) dx \wedge d\varphi\ .$$

Thus the functional (II.a.53) is 2π times

$$\int \frac{1}{2G}\left(\frac{y\ddot{y}^2}{G^4} + \frac{1}{y}\right) dx\ ,$$

which agrees with (II.a.57).

Returning to the general discussion, an obvious computation using (II.a.58) and $G_{\dot{y}} = \dot{y}/G$ gives for the Euler-Lagrange equations in Hamiltonian form

$$\begin{cases} \dfrac{dy}{dx} = H_\lambda = \dot{y} \\[2mm] \dfrac{d\dot{y}}{dx} = H_{\dot{\lambda}} = \dfrac{\dot{\lambda}G^5}{y} \\[2mm] \dfrac{d\lambda}{dx} = -H_y = \dfrac{1}{2y^2}\left(\dot{\lambda}^2 G^5 - \dfrac{1}{G}\right) \\[2mm] \dfrac{d\dot{\lambda}}{dx} = -H_{\dot{y}} = -\lambda - \dfrac{\dot{y}}{2y}\left(5\dot{\lambda}^2 G^3 + \dfrac{1}{G^3}\right)\ . \end{cases} \tag{II.a.59}$$

The dilation vector field on $\mathbb{R}^3$ gives the infinitesimal symmetry

$$v = x\frac{\partial}{\partial x} + y\frac{\partial}{\partial y} - \ddot{y}\frac{\partial}{\partial \ddot{y}} \tag{II.a.60}$$

of the variational problem $(I,\omega;\varphi)$ given in $(x,y,\dot{y},\ddot{y})$-space by the data

$$\begin{cases} \theta = dy - \dot{y}dx \\ \dot{\theta} = d\dot{y} - \ddot{y}dx \\ \omega = dx \neq 0 \\ \varphi = Ldx \end{cases}$$

where L is given by (II.a.57). Let us check that this is the case. The 1-parameter group generated by v is

$$\begin{cases} x \to \mu x \\ y \to \mu y \\ \dot{y} \to \dot{y} \\ \ddot{y} \to \mu^{-1}\ddot{y} \ . \end{cases}$$

It follows that

$$\begin{cases} L_v\theta = \theta \\ L_v\dot{\theta} = 0 \\ L_v\varphi = 0 \end{cases}$$

(this may be also verified by the H. Cartan formula (0.a.1)). The induced vector field $\tilde{v}$ on the space Z of variables $(x,y,\dot{y},\ddot{y},\lambda,\dot{\lambda})$ is

$$\tilde{v} \;=\; x\frac{\partial}{\partial x} + y\frac{\partial}{\partial y} - \ddot{y}\frac{\partial}{\partial \ddot{y}} - \lambda\frac{\partial}{\partial \lambda} \ .$$

The 1st integral given by Noether's theorem is (cf. (II.a.11))

$$V \;=\; v \,\lrcorner\, (\varphi + \lambda\theta + \dot{\lambda}\dot{\theta}) \ .$$

Since $\varphi + \lambda\theta + \dot{\lambda}\dot{\theta} = -Hdx + \lambda dy + \dot{\lambda}d\dot{y}$ this gives

$$V = -xH + y\lambda \ . \tag{II.a.61}$$

On any integral curve of (II.a.59) both V and H are constant; this implies that

$$\frac{d^2}{dx^2}(y\lambda) = 0$$

on integral curves, a relation that is not obvious from (II.a.59).

To compute the modified Poisson bracket $[H,V]$ we must evaluate the exterior product

$$dH \wedge dV \wedge \psi_Y \wedge \Psi_Y \ . \tag{II.a.62}$$

Setting

$$\Omega = d\lambda \wedge dy + d\dot{\lambda} \wedge d\dot{y}$$

we have

$$\begin{cases} \psi_Y = -Hdx + \lambda dy + \dot{\lambda}d\dot{y} \\ \Psi_Y = -dH \wedge dx + \Omega \\ dV \equiv -Hdx + yd\lambda + \lambda dy \quad \mathrm{mod}\ \ dH \ . \end{cases}$$

Since $dH \wedge dH = 0$, (II.a.62) is

$$dH \wedge (-Hdx + yd\lambda + \lambda dy) \wedge (-Hdx + \lambda dy + \dot{\lambda}d\dot{y}) \wedge \Omega \ .$$

We write this as

$$dH \wedge (-Hdx + \eta) \wedge (-Hdx + \xi) \wedge \Omega$$

where $\eta = yd\lambda + \lambda dy$, $\xi = \lambda dy + \dot{\lambda}d\dot{y}$. Since $H_x = 0$ the only terms containing dx come from $-Hdx$. Thus (II.a.62) is

$$\begin{aligned} Hdx \wedge dH \wedge (\xi - \eta) \wedge \Omega &= Hdx \wedge dH \wedge (\dot{\lambda}d\dot{y} - yd\lambda) \wedge \Omega \\ &= Hdx \wedge dH \wedge (\dot{\lambda}d\lambda \wedge dy \wedge d\dot{y} - yd\lambda \wedge d\dot{\lambda} \wedge d\dot{y}) \\ &= H(\dot{\lambda}H_{\dot{\lambda}} + yH_y)dx \wedge d\lambda \wedge dy \wedge d\dot{\lambda} \wedge d\dot{y} \quad . \end{aligned}$$

A similar computation shows that

$$\psi_Y \wedge \psi_Y^2 = 2(\lambda H_\lambda + \dot{\lambda}H_{\dot{\lambda}} - H)dx \wedge d\lambda \wedge dy \wedge d\dot{\lambda} \wedge d\dot{y} \ .$$

Referring to (II.a.58), (II.a.59) an unfortunate miracle occurs; namely

$$\dot{\lambda}H_{\dot{\lambda}} + yH_y = \frac{\dot{\lambda}^2 G^5}{2y} + \frac{1}{2Gy} \quad ,$$

$$\begin{aligned} \lambda H_\lambda + \dot{\lambda}H_{\dot{\lambda}} - H &= \lambda\dot{y} + \frac{\dot{\lambda}^2 G^5}{y} - \left(\lambda\dot{y} + \frac{\dot{\lambda}^2 G^5}{2y} - \frac{1}{2yG}\right) \\ &= \frac{\dot{\lambda}^2 G^5}{2y} + \frac{1}{2Gy} \quad . \end{aligned}$$

By the definition of the modified Poisson bracket (Chapter II, Section a)) it follows that $[H,V]$ is a constant multiple of H, which proves (II.a.54).

b) Investigation of the Euler-Lagrange System for Some Differential-Geometric Variational Problems

i) We shall begin our discussion of the functional

$$\Phi(\gamma) = \frac{1}{2}\int_\gamma \kappa^2 \, ds \tag{II.b.1}$$

defined on curves $\gamma \subset \mathbb{E}^n$ in the simple case $n = 2$. We take $X = F(\mathbb{E}^2) \times \mathbb{R}$ where $\mathbb{R}$ has coordinate κ, and on X we consider the variational problem $(I,\omega;\varphi)$ given by the data (where for the moment $L(\kappa)$ is an arbitrary smooth function of the single variable κ with derivatives L', L'', etc.)

$$\begin{cases} \omega = \omega^1 \\ \theta^1 = \omega^2 \\ \theta^2 = \omega_1^2 - \kappa\omega \\ \varphi = L(\kappa)\omega \end{cases} \tag{II.b.2}$$

The integral manifolds of (I,ω) are given by curves

$$s \to (x(s), e_1(s), e_2(s), \kappa(s)) \in F(\mathbb{E}^2) \times \mathbb{R}$$

where $s \to x(s)$ describes a curve $\gamma \subset \mathbb{E}^2$ along which

$$\begin{cases} \omega = ds = \text{arclength on } \gamma \\ e_1(s) = x'(s) = \text{unit tangent to } \gamma \\ \kappa(s) = e_1'(s) = \text{curvature of } \gamma \end{cases} .$$

We will prove that:

(II.b.3) *The data (II.b.2) gives a non-degenerate variational problem if* $L''(\kappa) \neq 0$.(10) *In this case the momentum space has dimension five. If* $L(\kappa) = \kappa^2/2$ *then the Euler-Lagrange equations associated to* $(I,\omega;\varphi)$ *are quasi-integrable by quadratures.*

Proof. We follow the algorithm for the construction of (J,ω) in Chapter I, Section e). On $Z = X \times \mathbb{R}^2$ where $\mathbb{R}^2$ has coordinates (λ_1, λ_2) we consider the 1-form

$$\psi = L(\kappa)\omega + \lambda_1\theta^1 + \lambda_2\theta^2$$

with exterior derivative

$$\Psi = d(L(\kappa)\omega) + \lambda_1 d\theta^1 + \lambda_2 d\theta^2 + d\lambda_1 \wedge \theta^1 + d\lambda_2 \wedge \theta^2 .$$

We must compute i) the Cartan system $C(\Psi)$, ii) the momentum space Y with Euler-Lagrange system (J,ω) on Y, and iii) assuming that $\dim Y = 2m+1$, the top degree differential form

$$\psi_Y \wedge (\Psi_Y)^m$$

on Y. In computing the Cartan system there is a useful remark that is worth isolating:

(II.b.4) LEMMA. *Given a pair of 2-forms* Ψ, $\tilde{\Psi}$ *on* $Z = X \times \mathbb{R}^s$ *we write*

$$\Psi \equiv \tilde{\Psi}$$

in case

$$\Psi - \tilde{\Psi} \equiv 0 \bmod \operatorname{span}\{\theta^\alpha \wedge \theta^\beta\}. \qquad \text{(II.b.5)}$$

If this holds then

$$C(\Psi) = C(\tilde{\Psi}).$$

Proof. We have

$$\psi = \varphi + \lambda_\alpha \theta^\alpha$$

$$\Psi = d\psi = d\varphi + d\lambda_\alpha \wedge \theta^\alpha + \lambda_\alpha d\theta^\alpha,$$

and so

$$\frac{\partial}{\partial \lambda_\alpha} \lrcorner \Psi = \theta^\alpha \in C(\Psi).$$

It is then clear that

$$C(\Psi) = C(\Psi + h_{\alpha\beta}\theta^\alpha \wedge \theta^\beta)$$

for any functions $h_{\alpha\beta}$ on Z. Q.E.D.

From the structure equations (i)-(v) in (0.c.3) of a moving frame we have

$$\begin{cases} d\theta^1 = -\theta^2 \wedge \omega & \overset{=}{\equiv} -\theta^2 \wedge \omega \\ d\theta^2 = -\pi \wedge \omega + \kappa\theta^1 \wedge \theta^2 & \overset{=}{\equiv} -\pi \wedge \omega \qquad (\pi = d\kappa - \kappa^2\theta^1) \\ d\varphi = L'\pi \wedge \omega + (L'\kappa^2 - L\kappa)\theta^1 \wedge \omega - L\theta^1 \wedge \theta^2 & \\ & \overset{=}{\equiv} L'\pi \wedge \omega + (L'\kappa^2 - L\kappa)\theta^1 \wedge \omega . \end{cases}$$

In terms of the coframe $\{\omega; \theta^1, \theta^2, \pi, d\lambda_1, d\lambda_2\}$ on Z we therefore have

$$\Psi \overset{=}{\equiv} (L' - \lambda_2)\pi \wedge \omega + (d\lambda_1 - (L'\kappa^2 - L\kappa)\omega) \wedge \theta^1 + (d\lambda_2 + \lambda_1\omega) \wedge \theta^2 .$$

Letting $\equiv$ denote congruence modulo $\operatorname{span}\{\theta^\alpha\}$ and noting (II.b.4) that

$$\Psi \overset{=}{\equiv} \tilde{\Psi} \Rightarrow v \lrcorner \Psi \equiv v \lrcorner \tilde{\Psi} ,$$

we see that the Cartan system is generated by the Pfaffian equations

$$\begin{cases} \partial/\partial\lambda_\alpha \lrcorner \Psi = \theta^\alpha = 0 \\ \partial/\partial\pi \lrcorner \Psi \equiv (L' - \lambda_2)\omega = 0 \\ \partial/\partial\theta^2 \lrcorner \Psi \equiv -d\lambda_2 - \lambda_1\omega = 0 \\ \partial/\partial\theta^1 \lrcorner \Psi \equiv -d\lambda_1 + (L'\kappa^2 - L\kappa)\omega = 0 \end{cases} \qquad \text{(II.b.6)}$$

Thus $Z_1 \subset Z$ is given by the equation

$$L'(\kappa) = \lambda_2 .$$

Proof. At a point $p \in Z_1$ the Cartan system is given by the *four* linear equations on $T_p(Z)$

$$\begin{cases} \theta^1 = \theta^2 = 0 \\ d\lambda_2 + \lambda_1\omega = 0 \\ d\lambda_1 - (L'\kappa - L\kappa)\omega = 0 \,. \end{cases}$$

It is clear that they always have a solution $v \in T_p(Z)$ on which $\langle\omega, v\rangle \neq 0$. Q.E.D.

Remark. What is *not* clear, and in fact is not always true, is that these equations have a solution $v \in T_p(Z_1)$ on which $\langle\omega, v\rangle \neq 0$ (cf. footnote (16) to Chapter I).

On $Z_1 \subset Z$ we have (here all differential forms are viewed as multilinear functions restricted tc $T(Z_1) \subset T(Z)$)

$$\begin{aligned} d\lambda_2 &= L''(\kappa)d\kappa \\ &\equiv L''(\kappa)\pi \text{ mod span}\{\theta^1, \theta^2\} \,. \end{aligned}$$

If $L'' \neq 0$ then we may solve for π and take $\{\omega, \theta^1, \theta^2, d\lambda_1, d\lambda_2\}$ as a coframe on Z_1. When this is done we see that for each point $p \in Z_1$ the above four linear equations generating $C_1(\Psi)$ have a solution $v \in T_p(Z_1)$ on which $\langle\omega, v\rangle \neq 0$. Thus at each point of Z_1 there is an integral element of $C_1(\Psi)$ *tangent to* Z_1, and therefore $Z_1 = Y$. Note that $\dim Y = 5$.

To check the non-degeneracy condition

$$\psi_Y \wedge (\Psi_Y)^2 \neq 0$$

we cannot in general work modulo terms $\theta^\alpha \wedge \theta^\beta$ (a good example of this is given just below in (ii) of this section). In the present example we have

$$\begin{aligned} \psi_Y &= L\omega + \lambda_1\theta^1 + \lambda_2\theta^2 \\ \Psi_Y &= (d\lambda_1 - (L'\kappa^2 - L\kappa)\omega) \wedge \theta^1 + (d\lambda_2 + \lambda_1\omega) \wedge \theta^2 + (\lambda_2\kappa - L)\theta^1 \wedge \theta^2 \,. \end{aligned}$$

It is clear that we may also choose

$$\{\psi_Y, \theta^1, \theta^2, \eta_1 = d\lambda_1 - (L'\kappa^2 - L\kappa)\omega, \eta_2 = d\lambda_2 + \lambda_1\omega\}$$

as a coframe on Y, and then

$$\psi_Y \wedge (\Psi_Y)^2 = 2\psi_Y \wedge \eta_1 \wedge \theta^1 \wedge \eta_2 \wedge \theta^2 \neq 0$$

Consequently, proposition (I.e.17) applies and the variational problem is non-degenerate with the 5-dimensional momentum space Y.

To determine an integral curve of the Euler-Lagrange differential system on Y we must specify *four* constants. Now $\dim E(3) = .3$ and so Noether's theorem gives *three* 1st integrals V_1, V_2, V_3, which we won't write out explicitly. In the case $L(\kappa) = \kappa^2/2$ we shall determine another 1st integral V_4, and this will be sufficient to establish (II.b.3). We remark without proof that the level sets $V_\rho = c_\rho$ determine *surfaces* in Y, which are cylinders $\mathbb{R} \times \mathbb{R}/\mathbb{Z}$ with the property the solution curves to the Euler-Lagrange equations are projections of straight lines.

The last two equations in (II.b.6) are then (using that $\lambda_2 = \kappa$ on Y)

$$\begin{cases} d\lambda_1 - \dfrac{\lambda_2^3}{2}\,\omega = 0 \\ d\lambda_2 + \lambda_1\omega = 0 \end{cases} .$$

Writing these equations as

$$\frac{2d\lambda_1}{\lambda_2^3} = \omega = -\frac{d\lambda_2}{\lambda_1}$$

we may separate variables and obtain

$$d\left(\lambda_1^2 + \frac{\lambda_2^4}{4}\right) = 2\lambda_1 d\lambda_1 + \lambda_2^3 d\lambda_2 = 0$$

along solution curves to (J,ω). Thus

$$V_4 = \lambda_1^2 + \frac{\lambda_2^4}{4}$$

is our promised 1st integral. Q.E.D. for (II.b.3)

The phase portrait of this system has already been discussed in example (I.d.27). In particular we recall that the Euler-Lagrange equations may be integrated by elliptic functions.

ii) For our next example we denote by $d\sigma$ the element of *affine arclength* (cf. example (I.a.15)) defined on curves $\gamma \subset \mathbb{A}^2$ and consider the functional

$$\Phi(\gamma) = \int_\gamma d\sigma \, . \tag{II.b.7}$$

We will show that:

(II.b.8) *The solutions to the Euler-Lagrange equations associated to (II.b.7) are plane conics.*(12)

To carry out the computation it is convenient to use affine Frenet frames (cf. [34], [44]). We give $\gamma \subset \mathbb{A}^2$ by a map

$$\sigma \to x(\sigma)$$

where σ is an affine arclength parameter; i.e.,

$$|x'(\sigma) \wedge x''(\sigma)| = 1 \, .$$

We set

$$\begin{cases} \omega = d\sigma \\ dx = e_1\omega \\ de_1 = e_2\omega \end{cases} \, . \tag{II.b.9}$$

By the structure equations (0.c.3) for affine frames, on the curve $\sigma \to (x(\sigma), e_1(\sigma), e_2(\sigma))$ in $F(\mathbb{A}^2)$ we have

$$\begin{cases} \omega^1 = \omega, & \omega^2 = 0 \\ \omega_1^1 = 0, & \omega_1^2 = \omega \end{cases} \tag{II.b.10}$$

Since $\omega_1^1 + \omega_2^2 = 0$ this gives also that $\omega_2^2 = 0$, and we define the *affine curvature* $\kappa(\sigma)$ by

$$\omega_2^1 = \kappa(\sigma)\omega \, . \tag{II.b.11}$$

Equivalently since $\omega_2^2 = 0$

$$de_2 = \kappa(\sigma)e_1\omega \, . \tag{II.b.12}$$

We call $(x(\sigma), e_1(\sigma), e_2(\sigma))$ the *affine Frenet frame*, and observe that as a consequence of (I.b.7) the curve $\gamma \subset \mathbb{A}^2$ is uniquely determined up to affine motion by its affine curvature function $\kappa(\sigma)$.

Remark. The affine normal e_2 has the following geometric interpretation, to be encountered again in Chapter IV, Section a). First note that under the action of the affine group $A(2)$ we may i) measure area (and therefore determine the *center of gravity* of a plane region), and ii) say when 2 lines are parallel (i.e., they do not meet). Given

$p \in \gamma$ with tangent line T_p we consider the lines $T_p(\varepsilon)$ parallel to T_p and at distance ε (use any identification $\mathbb{A}^2 \cong \mathbb{R}^2$ to measure this distance--we are going to let $\varepsilon \to 0$, and the end result will not depend on the particular identification).

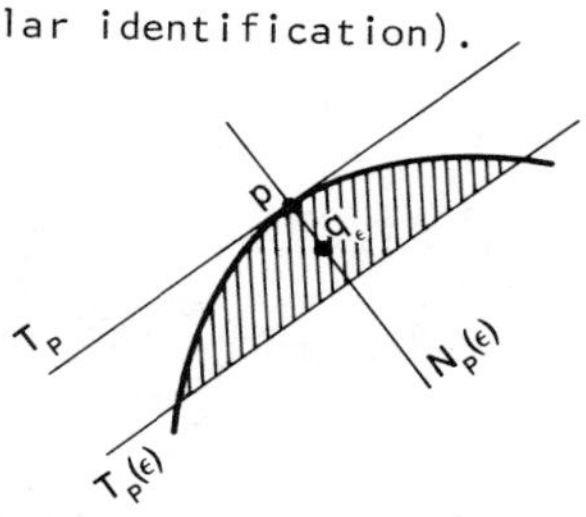

Let q_ε be the center of gravity of the region between γ and $T_p(\varepsilon)$ and draw the line $\overline{pq_\varepsilon} = N_p(\varepsilon)$. The limit $N_p = \lim_{\varepsilon \to 0} N_p(\varepsilon)$ is the *affine normal line*, and the *affine normal* e_2 is determined by the condition

$$\begin{cases} e_2 \in N_p \\ e_1 \wedge e_2 = 1 \end{cases}$$

where $e_1 = dx(\sigma)/d\sigma$.

In particular we consider curves with $\kappa(\sigma) \equiv 0$. By the structure equations (II.b.9), (II.b.11) this is equivalent to

$$x'''(\sigma) = 0 \ .$$

Consequently, $x(\sigma)$ is a quadratic polynomial in σ and describes a conic in $\mathbb{A}^2$. Our assertion (II.b.8) then follows from:

(II.b.13). *The solutions to the Euler-Lagrange equations associated to (II.b.7) have affine curvature zero.*

On $X = F(\mathbb{A}^2) \times \mathbb{R}$ where $\mathbb{R}$ has coordinate κ we consider the Pfaffian differential system (I,ω) generated by the equations

$$\begin{cases} \theta^1 = \omega^2 = 0 \\ \theta^2 = \omega_1^2 - \omega = 0 \\ \theta^3 = \omega_1^1 = 0 \\ \theta^4 = \omega_2^1 - \kappa\omega = 0 \\ \omega = \omega^1 \neq 0 \end{cases} \qquad \text{(II.b.14)}$$

By (II.b.10) and (II.b.11) the integral manifolds of (I,ω) are given by $\sigma \to (x(\sigma), e_1(\sigma), e_2(\sigma), \kappa(\sigma))$ where $(x(\sigma), e_1(\sigma), e_2(\sigma))$ is the affine

Frenet frame of the curve $\gamma = \{x(\sigma)\} \subset \mathbb{A}^2$ and $\kappa(\sigma)$ is its affine curvature. Using $\equiv$ to denote congruence modulo terms $\theta^\alpha \wedge \theta^\beta$ (cf. (II.b.4)) we obtain from (i)-(iv) and (vi) in (0.c.3) the structure equations

$$\left\{\begin{aligned} d\omega &= \kappa\theta^1 \wedge \omega - \theta^3 \wedge \omega + \theta^1 \wedge \theta^4 \equiv \kappa\theta^1 \wedge \omega - \theta^3 \wedge \omega \\ d\theta^1 &= -\theta^2 \wedge \omega - \theta^1 \wedge \theta^3 \equiv -\theta^2 \wedge \omega \\ d\theta^2 &= 3\theta^3 \wedge \omega - \kappa\theta^1 \wedge \omega + 2\theta^3 \wedge \theta^2 - \theta^1 \wedge \theta^4 \equiv 3\theta^3 \wedge \omega - \kappa\theta^1 \wedge \omega \\ d\theta^3 &= \kappa\theta^2 \wedge \omega - \theta^4 \wedge \omega + \theta^2 \wedge \theta^4 \equiv \kappa\theta^2 \wedge \omega - \theta^4 \wedge \omega \\ d\theta^4 &= -d\kappa \wedge \omega + 3\kappa\theta^3 \wedge \omega - \kappa^2\theta^1 \wedge \omega + \kappa\theta^1 \wedge \theta^4 + 2\theta^3 \wedge \theta^4 \\ &\equiv -d\kappa \wedge \omega + 3\kappa\theta^3 \wedge \omega - \kappa^2\theta^1 \wedge \omega . \end{aligned}\right. \tag{II.b.15}$$

Remark. With our usual notation W_k^* for the kth derived system (cf. (I.c.17)), the derived flag of the Pfaffian differential system I is

$$\begin{array}{ccccccccc} W^* & \supset & W_1^* & \supset & W_2^* & \supset & W_3^* & \supset & W_4^* \\ \shortparallel & & \shortparallel & & \shortparallel & & \shortparallel & & \shortparallel \\ \mathrm{span}\{\theta^1,\theta^2,\theta^3,\theta^4\} & & \mathrm{span}\{\theta^1,\theta^2,\theta^3\} & & \mathrm{span}\{\theta^1,\theta^2\} & & \mathrm{span}\{\theta^1\} & & (0) \end{array}$$

This may be related to the fact that the affine curvature is a 3rd order invariant (cf. [34] for the *Cartan polygon method* of computing the order of contact necessary to determine a general Frenet frame).

On X we consider the variational problem $(I,\omega;\varphi)$ where

$$\varphi = \omega \qquad (= \text{affine arclength}).$$

The corresponding functional on $V(I,\omega)$ is (II.b.7). To compute the Euler-Lagrange system (J,ω) we follow the algorithm in Chapter I, Section e) and use (II.b.4). Thus we consider on $X \times \mathbb{R}^4$

$$\psi = \omega + \lambda_\alpha \theta^\alpha$$

$$\Psi = (\kappa\theta^1 - \theta^3) \wedge \omega + d\lambda_\alpha \wedge \theta^\alpha + \lambda_\alpha d\theta^\alpha$$

where the $d\theta^\alpha$ are given by (II.b.15), and where, by (II.b.4), to compute $C(\Psi)$ we need only consider them modulo the equivalence relation $\equiv$. By contracting with a basis for all tangent vectors to Z and using $\equiv$ to denote congruence modulo $\mathrm{span}\{\theta^\alpha\}$, we find that the Cartan system $C(\Psi)$ is generated by the Pfaffian equations

$$\begin{cases} \partial/\partial\lambda_\alpha \lrcorner \Psi = \theta^\alpha = 0 & \alpha = 1,2,3,4 \\ \text{(i)} \quad \partial/\partial\kappa \lrcorner \Psi \equiv \lambda_4\omega = 0 & \\ \text{(ii)} \quad \partial/\partial\theta^4 \lrcorner \Psi \equiv -d\lambda_4 - \lambda_3\omega = 0 & \\ \text{(iii)} \quad \partial/\partial\theta^3 \lrcorner \Psi \equiv -d\lambda_3 + (3\kappa\lambda_4 + 3\lambda_2 - 1)\omega = 0 & \\ \text{(iv)} \quad \partial/\partial\theta^2 \lrcorner \Psi \equiv -d\lambda_2 + (\kappa\lambda_3 - \lambda_1)\omega = 0 & \\ \text{(v)} \quad \partial/\partial\theta^1 \lrcorner \Psi \equiv -d\lambda_1 + (-\kappa^2\lambda_4 - \kappa\lambda_2 + \kappa)\omega = 0 \ . & \end{cases} \qquad \text{(II.b.16)}$$

By (i) we find that $Z_1 \subset Z$ is given by $\lambda_4 = 0$. Then by (ii) we see that $Z_2 \subset Z_1$ is given by $\lambda_3 = \lambda_4 = 0$. Then by (iii) it follows that $Z_3 \subset Z_2$ is given by (!)

$$\lambda_2 = 1/3 \ .$$

Next, since $d\lambda_2 = 0$ we use (iv) to find that $Z_4 \subset Z_3$ is given by $\lambda_1 = 0$. Finally, from (v) it follows that $Z_5 = Y$ is given (using $\lambda_2 = 1/3$) by $2/3\ \kappa = 0$; i.e.,

$$\kappa = 0 \ .$$

In other words, $Y \cong F(\mathbb{A}^2)$ and the Euler-Lagrange system (J,ω) is given by

$$\begin{cases} \omega^2 = \omega_1^1 = \omega_2^1 = 0 \\ \omega = \omega^1 = \omega_1^2 \neq 0 \end{cases} .$$

As noted above the integral manifolds of this system are the affine Frenet frames with zero affine curvature, and this establishes (II.b.13).

<u>Remark</u>. Since by (II.b.15) and our computation above

$$\begin{cases} \psi_Y = \omega + \frac{1}{3}\theta^2 \\ \Psi_Y = \frac{2}{3}(\theta^2 \wedge \theta^3 + \theta^1 \wedge \theta^4) \end{cases}$$

it follows that

$$\begin{cases} C(\Psi_Y) = \operatorname{span}\{\theta^1, \theta^2, \theta^3, \theta^4\} \\ \psi_Y \wedge (\Psi_Y)^2 \neq 0 \end{cases} .$$

Thus the variational problem in non-degenerate in the sense of Chapter I, Section e).

iii) We now study curves $\gamma \subset \mathbb{E}^3$. If we give γ parametrically by $s \to x(s) \in \mathbb{E}^3$ where s is arclength, then in a neighborhood where the curvature $\kappa(s) \neq 0$ we may determine a *Frenet frame* $s \to (x(s), e_1(s), e_2(s), e_3(s)) \in F(\mathbb{E}^3)$ such that the *Frenet-Serret* equations

$$\begin{cases} \dfrac{dx}{ds} = e_1 \\ \dfrac{de_1}{ds} = \kappa(s)e_2 \\ \dfrac{de_2}{ds} = -\kappa(s)e_1 + \tau(s)e_3 \\ \dfrac{de_3}{ds} = -\tau(s)e_2 \end{cases} \qquad \text{(II.b.17)}$$

are valid. Here the normal e_2, *binormal* e_3, curvature $\kappa(s)$, and *torsion* $\tau(s)$ are determined up to ± 1; the signs are fixed by specifying e_2, e_3 at one endpoint of γ. Moreover, as a consequence of (I.b.7) the functions $\kappa(s)$, $\tau(s)$ uniquely determine γ up to a rigid motion. This suggests that we set $X = F(\mathbb{E}^3) \times \mathbb{R}^2$ where $\mathbb{R}^2$ has coordinates (κ,τ), and on X we consider the Pfaffian differential system (I,ω) given by

$$\begin{cases} \theta^1 = \omega^2 = 0 \\ \theta^2 = \omega^3 = 0 \\ \theta^3 = \omega_1^3 = 0 \\ \theta^4 = \omega_1^2 - \kappa\omega = 0 \\ \theta^5 = \omega_2^3 - \tau\omega = 0 \\ \omega = \omega^1 \neq 0 \end{cases} \qquad \text{(II.b.18)}$$

The Frenet lifting of a curve $\gamma \subset \mathbb{E}^3$ with $\kappa \neq 0$ is an integral manifold of (I,ω).

On the other hand, it may happen that $\kappa = 0$ on a subset of an integral manifold of (I,ω). In fact, it may be shown that any piecewise C^4-curve $\gamma \subset \mathbb{E}^3$ is the projection of a piecewise C^1-integral manifold $N \subset X$ of (I,ω) (if $\kappa = 0$ on a subset then N is not generally uniquely determined by γ). We shall not prove this here, but shall simply consider functionals defined on $V(I,\omega)$.[13]

In this discussion we shall consider the functional

$$\Phi(N) = \frac{1}{2}\int_N \kappa^2\, ds \qquad \text{(II.b.19)}$$

defined on integral manifolds N of (I,ω). We shall prove that: *The Euler-Lagrange equations associated to (II.b.19) are quasi-integrable by quadratures.*(14)
In fact, it will turn out that this assertion is almost clear from our general theory together with Noether's theorem.

In order to study the Euler-Lagrange system associated to (II.b.19) we need to record the structure equations of (I,ω). Using (i)-(v) in (0.c.3) and the notation (II.b.5) they are

$$\text{(II.b.20)}\quad \begin{cases} \text{(i)} & d\omega = -\kappa\theta^1\wedge\omega - \theta^1\wedge\theta^4 - \theta^2\wedge\theta^5 \equiv -\kappa\theta^1\wedge\omega \\ \text{(ii)} & d\theta^1 = -(\tau\theta^2+\theta^4)\wedge\omega - \theta^2\wedge\theta^5 \equiv -(\tau\theta^2+\theta^4)\wedge\omega \\ \text{(iii)} & d\theta^2 = (\tau\theta^1-\theta^3)\wedge\omega + \theta^1\wedge\theta^5 \equiv (\tau\theta^1-\theta^3)\wedge\omega \\ \text{(iv)} & d\theta^3 = (\tau\theta^4-\kappa\theta^5)\wedge\omega + \theta^4\wedge\theta^5 \equiv (\tau\theta^4-\kappa\theta^5)\wedge\omega \\ \text{(v)} & d\theta^4 = -\pi^4\wedge\omega + \kappa(\theta^1\wedge\theta^4+\theta^2\wedge\theta^3) - \theta^3\wedge\theta^5 \equiv -\pi^4\wedge\omega \\ \text{(vi)} & d\theta^5 = -\pi^5\wedge\omega + \tau(\theta^1\wedge\theta^4+\theta^2\wedge\theta^3) + \theta^3\wedge\theta^4 \equiv -\pi^5\wedge\omega \\ & d\varphi = L'\pi^4\wedge\omega + ((L'\kappa^2-L\kappa)\theta^1 - L'\tau\theta^3)\wedge\omega - (\theta^1\wedge\theta^4+\theta^2\wedge\theta^3) \end{cases}$$

where

$$\begin{cases} \varphi = L(\kappa)\omega \qquad (15) \\ \pi^4 = d\kappa + \tau\theta^3 - \kappa^2\theta^1 \\ \pi^5 = d\tau - \kappa\tau\theta^1 - \kappa\theta^3 \end{cases} .$$

<u>Remarks</u>. We note that (I,ω) is a Pfaffian system in good form (Chapter I, Section c)) and its derived flag (loc. cit.) is (16)

$$\begin{array}{ccccccc} W^* & \supset & W_1^* & \supset & W_2^* & \supset & W_3^* \\ \| & & \| & & \| & & \| \\ \operatorname{span}\{\theta^1,\theta^2,\theta^3,\theta^4,\theta^5\} & & \operatorname{span}\{\theta^1,\theta^2,\theta^3\} & & \operatorname{span}\{\theta^2\} & & (0) \end{array}$$

Moreover, the equations (I.e.28), (I.e.29) are valid.

The construction of the Euler-Lagrange system (J,ω) is rather interesting. Following the algorithm in Chapter I, Section e) we consider the differential forms

$$\psi = \varphi + \lambda_\alpha \theta^\alpha$$

$$\Psi = d\varphi + d\lambda_\alpha \wedge \theta^\alpha + \lambda_\alpha d\theta^\alpha$$

on $Z = X \times \mathbb{R}^5$. Letting $\equiv$ denote congruence modulo $\mathrm{span}\{\theta^\alpha\}$, the Cartan system $C(\Psi)$ is generated by the Pfaffian equations

$$\begin{cases} & \partial/\partial\lambda_\alpha \lrcorner \Psi = \theta^\alpha = 0 \qquad \alpha = 1,\dots,5 \\ \text{(i)} & \partial/\partial\pi^5 \lrcorner \Psi \equiv -\lambda_5\omega = 0 \\ \text{(ii)} & \partial/\partial\pi^4 \lrcorner \Psi \equiv (L' - \lambda_4)\omega = 0 \\ \text{(iii)} & \partial/\partial\theta^5 \lrcorner \Psi \equiv -d\lambda_5 - \lambda_3\kappa\omega = 0 \\ \text{(iv)} & \partial/\partial\theta^4 \lrcorner \Psi \equiv -d\lambda_4 - (\lambda_1 + \lambda_3\tau)\omega = 0 \\ \text{(v)} & \partial/\partial\theta^3 \lrcorner \Psi \equiv -d\lambda_3 - (\lambda_2 + L'\tau)\omega = 0 \\ \text{(vi)} & \partial/\partial\theta^2 \lrcorner \Psi \equiv -d\lambda_2 - \tau\lambda_1\omega = 0 \\ \text{(vii)} & \partial/\partial\theta^1 \lrcorner \Psi \equiv -d\lambda_1 + (\tau\lambda_2 + L'\kappa^2 - L\kappa)\omega = 0 . \end{cases} \qquad \text{(II.b.21)}$$

In computing these equations we have used (II.b.20) and the remark (II.b.4). From equations (i), (ii) we find that $Z_1 \subset Z$ is given by (cf. the argument in part (i) of this section).

$$Z_1 = \{\lambda_5 = 0,\ \lambda_4 = L'\} .$$

Note that $\dim Z = 13$ and $\dim Z_1 = 11$. However, there are no integral elements of $(C(\Psi),\omega)$ tangent to Z_1 at a general point p of Z_1, since any such integral element will be spanned by $v \in T_p(Z)$ satisfying

$$\langle d\lambda_5, v\rangle = 0, \quad \langle\omega, v\rangle \neq 0 .$$

Using (iii) the 1st equation implies that $\lambda_3\kappa = 0$, and taking the general case when $\kappa \neq 0$ we must then have $\lambda_3 = 0$. Thus $Z_2 \subset Z_1$ is given by

$$\{\lambda_5 = \lambda_3 = 0,\ \lambda_4 = L'\} .$$

Again a straightforward computation shows that there are no integral elements of $(C_1(\Psi),\omega)$ that are tangent to Z_2 at a general point, since by (v) we must have $\lambda_2 + L'\tau = 0$. Thus $Z_3 \subset Z_2$ is given by

$$\{\lambda_5 = \lambda_3 = 0,\ \lambda_4 = L',\ \lambda_2 = -L'\tau\} . \qquad \text{(II.b.22)}$$

We will show that:

(II.b.23) *If* $L'' \neq 0$, *then* $Z_3 = Y$ *and the Euler-Lagrange system* (J,ω) *is non-degenerate on the 9-dimensional momentum space given by (II.b.22).*

<u>Proof</u>. On Z_3 we have

$$d\lambda_4 \equiv L''\pi_4 \mod\{\theta^\alpha,\omega\}$$

$$d\lambda_2 \equiv -L'\pi_5 \mod\{\pi_4,\theta^\alpha,\omega\} \quad .$$

Since $L'' \neq 0$ we also have $L' \neq 0$ and may solve for π_4, π_5 taking $\{\omega,\theta^1,..,\theta^5,d\lambda_1,d\lambda_2,d\lambda_4\}$ as a coframe on Z_3. The Cartan system $(C_3(\Psi),\omega)$ on Z_3 is given by

$$\begin{cases} \theta^\alpha = 0 & \alpha = 1,..,5 \\ -d\lambda_4 - \lambda_1\omega = 0 & \\ -d\lambda_2 - \tau\lambda_1\omega = 0 & \\ -d\lambda_1 + (\tau\lambda_2 + L'\kappa^2 - L\kappa)\omega = 0 \; . & \end{cases} \tag{II.b.24}$$

It is clear that $V(C_3(\Psi),\omega) \to Z$ is surjective, and thus $Z_3 = Y$. From

$$\begin{cases} \psi_Y = \varphi + \lambda_1\theta^1 + \lambda_2\theta^2 + \lambda_4\theta^4 \\ \Psi_Y = d\varphi + d\lambda_1 \wedge \theta^1 + d\lambda_2 \wedge \theta^2 + d\lambda_4 \wedge \theta^4 + \lambda_\alpha d\theta^\alpha \end{cases}$$

we find using (II.b.20) that

$$\begin{aligned} \psi_Y \wedge (\Psi_Y)^4 &= -6!\,L\lambda_4\omega \wedge d\lambda_1 \wedge \theta^1 \wedge d\lambda_2 \wedge \theta^2 \wedge d\lambda_4 \wedge \theta^4 \wedge \theta^3 \wedge \theta^5 \\ &\neq 0 \; . \end{aligned}$$

From this it follows that the variational problem is non-degenerate.

Q.E.D.

(II.b.25) <u>Remark</u>. It is *not* the case that the variational problem given by the functional $\int L(\kappa)\omega$, $L'' \neq 0$, is strongly non-degenerate in the sense of Chapter I, Section e). On the other hand, the variational problem associated to the functional

$$\Phi = \int L(\kappa,\tau)\omega, \qquad \det\begin{Vmatrix} L_{\kappa\kappa} & L_{\kappa\tau} \\ L_{\tau\kappa} & L_{\tau\tau} \end{Vmatrix} \neq 0 \tag{II.b.26}$$

is easily shown to be strongly non-degenerate. In fact, the equations (I.e.28), (I.e.30) (and hence (I.e.29)) are valid for the variational problem corresponding to (II.b.26). As a consequence of $L_{\kappa\kappa}L_{\tau\tau} - L^2_{\kappa\tau} \neq 0$ the associated quadratic form $\|A_{\lambda\mu}\|$ (cf. Chapter II, Section e)) is non-singular, and by definition the variational problem associated to (II.b.26) is strongly non-degenerate so that theorem (I.e.34) applies.

At this point, since $\dim Y = 9$ to determine an integral curve of the Euler-Lagrange system associated to (II.b.19) we must specify 8 constants. Noether's theorem furnishes $\dim E(3) = 6$, and presumably the global Hamiltonian character of (J,ω) furnishes one.[17] Moreover, since $\varphi = L(\kappa)\omega$ does not involve τ we may expect one more analogous to the presence of a cyclic coordinate.

In the case $L(\kappa) = \kappa^2/2$ we shall by a simple computation find the two 1st integrals V_1, V_2 of (II.b.24) beyond those provided by Noether's theorem. This will establish theorem (II.a.33) in the case $n = 3$, $R = 0$.

Using (II.b.22) the Euler-Lagrange system (II.b.24) is

$$\begin{cases} \theta^\alpha = 0 & \\ \text{(i)} \quad d\lambda_4 + \lambda_1\omega = 0 & \\ \text{(ii)} \quad \lambda_4 d\lambda_2 - \lambda_1\lambda_2\omega = 0 & \\ \text{(iii)} \quad d\lambda_1 - \left(\dfrac{\lambda_4^3}{2} - \dfrac{\lambda_2^2}{\lambda_4}\right)\omega = 0 & . \end{cases} \tag{II.b.27}$$

Multiplying (i) by λ_2 and adding to (ii) gives

$$d(\lambda_2\lambda_4) = 0 .$$

Using (II.b.22) this is

$$d(\kappa^2\tau) = 0 ,$$

which gives the algebraic 1st integral

$$V_1 = \lambda_2\lambda_4 = -\kappa^2\tau = c_1 \tag{II.b.28}$$

on solution curves to (J,ω). Plugging $\lambda_2 = c_1/\lambda_4$ into (iii) and using (i) gives

$$\lambda_4'' + \left(\frac{\lambda_4^3}{2} - \frac{c_1}{\lambda_4^3}\right) = 0$$

on solution curves to (J,ω). This equation has the algebraic 1st integral

$$V_2 = (\lambda_4')^2 + \frac{(\lambda_4)^4}{4} + \frac{c_1}{(\lambda_4)^2} \quad ;$$

i.e.,

$$V_2 = \lambda_1^2 + \frac{(\lambda_4)^4}{4} + \frac{c_1}{(\lambda_4)^2} = c_2 \; . \qquad \text{(II.b.29)}$$

This is equivalent to

$$\kappa'^2 + \frac{\kappa^4}{4} + \frac{c_1}{\kappa^2} = c_2$$

on integral curves of (J,ω).

For general values of c_1, c_2 the algebraic curve

$$y^2 + \frac{x^4}{4} + \frac{c_1}{x^2} = c_2$$

is a singular plane sextic whose desingularization has genus one (this is because $y = \sqrt{c_2 - x^4/4 - c_1/x^2}$ is single-valued near $x = \infty$). Consequently the Euler-Lagrange equations associated to the functional (II.b.19) have a phase portrait that may be described in (τ,κ,κ') space by the 2-parameter family of elliptic curves

$$\begin{cases} \tau\kappa^2 = c_1 \\ \kappa'^2 + \dfrac{\kappa^4}{4} - \tau = c_2 \; . \end{cases} \qquad \text{(II.b.30)}$$

iv) We shall now study a slightly new type of functional, defined for curves $\gamma \subset \mathbb{E}^3$ by

$$\begin{cases} \Phi = \displaystyle\int_\gamma ds \\ \text{with the constraint } \kappa \equiv 1. \end{cases} \qquad \text{(II.b.31)}$$

In the literature this is called the *Delauney problem* and we shall show that:

(II.b.32) *The Euler-Lagrange equations associated to the Delauney problem (II.b.31) are quasi-integrable by quadratures with phase portrait a rational algebraic curve.*

<u>Remarks</u>. The original statement of the problem (cf. [13], page 373) is:
"Among all space curves with given constant curvature, determine the shortest and longest curves which join two line elements of the space." There is an enormous classical literature concerning this question (loc. cit.).

Before turning to the functional (II.b.31) we shall investigate the Euler-Lagrange equations associated to the usual arclength functional

$$\Theta = \int_\gamma ds \tag{II.b.33}$$

This is a variational problem $(I,\omega;\varphi)$ where (I,ω) on X is the same as in (iii) above and where $\varphi = \omega$ (using the notations (II.b.18)). Now $d\varphi$ is given by the 1st equation in (II.b.20), and following the algorithm in Chapter I, Section e) and the same format as in Sections (i)-(iii) of this section the form Ψ is (cf. (I.d.16))

$$\Psi \equiv -\kappa\theta^1 \wedge \omega + d\lambda_\alpha \wedge \theta^\alpha + \lambda_\alpha d\theta^\alpha$$

where the $d\theta^\alpha$ are given by (II.b.20). The Cartan system $C(\Psi)$ is generated by the Pfaffian equations

$$\begin{cases} & \partial/\partial\lambda_\alpha \lrcorner \Psi = \theta^\alpha = 0 \\ \text{(i)} & \partial/\partial\pi^5 \lrcorner \Psi \equiv \lambda_5\omega = 0 \\ \text{(ii)} & \partial/\partial\pi^4 \lrcorner \Psi \equiv \lambda_4\omega = 0 \\ \text{(iii)} & \partial/\partial\theta^5 \lrcorner \Psi \equiv -d\lambda_5 - \lambda_3\kappa\omega = 0 \\ \text{(iv)} & \partial/\partial\theta^4 \lrcorner \Psi \equiv -d\lambda_4 + \lambda_3\tau\omega - \lambda_1\omega = 0 \\ \text{(v)} & \partial/\partial\theta^3 \lrcorner \Psi \equiv -d\lambda_3 - \lambda_2\omega = 0 \\ \text{(vi)} & \partial/\partial\theta^2 \lrcorner \Psi \equiv -d\lambda_2 - \lambda_1\tau\omega = 0 \\ \text{(vii)} & \partial/\partial\theta^1 \lrcorner \Psi \equiv -d\lambda_1 - \kappa\omega + \lambda_2\tau\omega = 0 \end{cases} , \tag{II.b.34}$$

where $\equiv$ denotes congruence modulo span $\{\theta^\alpha\}$. By (i) and (ii), $Z_1 \subset Z$ is $\{\lambda_5 = \lambda_4 = 0\}$. Then $Z_2 \subset Z_1$ is $\{\lambda_5 = \lambda_4 = \lambda_3\kappa = \lambda_3\tau - \lambda_1 = 0\}$. If $\kappa \neq 0$ then we must have $\lambda_3 = 0$ and $Z_3 \subset Z_2$ is $\{\lambda_5 = \lambda_4 = \lambda_3 = \lambda_1 = \lambda_2 = \kappa = 0\}$, where the last equation results from (vii). We have shown that:
If $\gamma \subset \mathbb{E}^3$ *is a solution to the Euler-Lagrange equations associated to (II.b.33), then* $\kappa \equiv 0$ *on* γ.

Of course this is not the most efficient proof that the geodesics in $\mathbb{E}^3$ are straight lines, but essentially the same computation will also yield (II.b.32).

Proof of (II.b.32). The variational problem (II.b.31) may be posed as an ordinary variational problem $(I_0,\omega;\varphi_0)$ on X_0 where $X_0 \subset X$ is defined by $\kappa \equiv 1$ and where $I_0 = I|X_0, \varphi_0 = ds|X_0$.(18) The structure equations (i)-(iv) and (vi) in (II.b.20) are still valid, while (v) becomes

$$d\theta^4 \equiv (\kappa^2\theta^1 - \tau\theta^3) \wedge \omega$$

We set $\pi = \pi^5$ so that X_0 has $\{\omega,\theta^1,\theta^2,\theta^3,\theta^4,\theta^5,\pi\}$ as coframe. For

$$\begin{cases} \varphi_0 = \omega \\ \psi = \varphi_0 + \lambda_\alpha\theta^\alpha \\ \Psi = d\psi \end{cases}$$

the Cartan system $C(\Psi)$ is generated by the Pfaffian equations (compare with (II.b.34))

$$\begin{cases} & \partial/\partial\lambda_\alpha \lrcorner \Psi = \theta^\alpha = 0 \\ \text{(i)} & \partial/\partial\pi \lrcorner \Psi \equiv \lambda_5\omega = 0 \\ \text{(ii)} & \partial/\partial\theta^5 \lrcorner \Psi \equiv -d\lambda_5 - \lambda_3\omega = 0 \\ \text{(iii)} & \partial/\partial\theta^4 \lrcorner \Psi \equiv -d\lambda_4 - (\lambda_1 - \lambda_3\tau)\omega = 0 \\ \text{(iv)} & \partial/\partial\theta^3 \lrcorner \Psi \equiv -d\lambda_3 - (\lambda_2 + \lambda_4\tau)\omega = 0 \\ \text{(v)} & \partial/\partial\theta^2 \lrcorner \Psi \equiv -d\lambda_2 - \lambda_1\tau\omega = 0 \\ \text{(vi)} & \partial/\partial\theta^1 \lrcorner \Psi \equiv -d\lambda_1 - (1 - \lambda_2\tau - \lambda_4)\omega = 0 \quad \text{(the 1 is from } \kappa = 1) \end{cases}$$

Following the by now familiar pattern, (i), (ii), and (iv) give

$$\begin{cases} \lambda_5 = \lambda_3 = 0 \\ \lambda_2 + \lambda_4\tau = 0 \ . \end{cases} \tag{II.b.35}$$

Then (i)-(vi) collapse to

$$\begin{cases} \text{(i)} & d\lambda_1 + (1 + \lambda_4\tau^2 - \lambda_4)\omega = 0 \\ \text{(ii)} & d\lambda_2 + \lambda_1\tau\omega = 0 \\ \text{(iii)} & d\lambda_4 + \lambda_1\omega = 0 \ . \end{cases} \tag{II.b.36}$$

The 2nd equation in (II.b.35) together with (ii), (iii) in (II.b.36) give

$$\frac{2d\lambda_4}{\lambda_4} + \frac{d\tau}{\tau} \equiv 0 \quad .$$

Using (II.b.35) this yields the 1st integral

$$V_1 = -\lambda_2\lambda_4 = c_1 \qquad (= \lambda_4^2\tau) \tag{II.b.37}$$

on solutions to the Euler-Lagrange system (J,ω). We set

$$\lambda = \lambda_4$$

so that the curve $\gamma \subset \mathbb{E}^3$ is uniquely determined, up to rigid motion, by knowing the constant c_1 and the function $\lambda(s)$ (this is because $\kappa(s) \equiv 1$ and $\tau(s) = c_1/\lambda(s)^2$). Taking arclength as parameter along solution curves to (J,ω) we obtain from (i), (iii) in (II.b.36) the O.D.E.

$$\lambda'' + \left(\lambda - \frac{c_1^2}{\lambda^3} - 1\right) = 0 \quad .$$

This equation has 1st integral

$$V_2 = \lambda'^2 + \left(\lambda^2 + \frac{c_1^2}{2\lambda^4} - 2\lambda\right) = c_2 \tag{II.b.38}$$

(we note that $V_1 = -\lambda_2\lambda_4$ and $V_2 = \lambda_1^2 + \left(\lambda_4^2 + \frac{c_1^2}{2\lambda_4^4} - 2\lambda_4\right)$ are both rational functions on Z).

In summary, the phase portrait of solution curves to the Euler-Lagrange system associated the functional (II.b.31) is given in the (λ,λ') plane by the 2-parameter family of algebraic curves

$$\lambda'^2 + \frac{c_1^2}{2\lambda^4} + \lambda^2 - 2\lambda = c_2 \quad . \tag{II.b.39}$$

For general values of c_1, c_2 the rational function

$$r(\lambda) = c_2 + 2\lambda - \lambda^2 - \frac{c_1^2}{2\lambda^4}$$

has degree six with a 4-tuple zero at $\lambda = \infty$ plus two other distinct zeroes. Thus the curve $\lambda'^2 = r(\lambda)$ is a 2-sheeted covering of the λ-sphere with $2 = 6 - 4$ branch points, and by the Riemann-Hurwitz formula ([36], [59]) is therefore a rational curve. This completes the proof of (II.b.32).

Remark. In this problem the momentum space $Y \subset X_0 \times \mathbb{R}^5$ is given by (II.b.35) and thus has coframe $\{\omega, \theta^1, \ldots, \theta^5, d\lambda_1, d\lambda_2, d\lambda_4\}$. In particular $\dim Y = 9$. Six algebraic 1st integrals of (J,ω) are given by $E(3)$ using Noether's theorem, and the two more by (II.b.37) and (II.b.38). Presumably these are accounted for by the Hamiltonian formalism plus the fact that φ_0 does not contain τ ("cyclic coordinate"), but we do not know how to make this precise. Of the total of *eight* 1st integrals we suspect that *seven* are independent and that the motion is linear on a cylinder $\mathbb{R} \times \mathbb{R}/\mathbb{Z}$.

FOOTNOTES FOR CHAPTER II

(1) Both of these are fairly artificial examples of higher order Lagrangians. We include them only to illustrate some phenomena that are true for a classical variational problem of any order; namely, there is always a Hamiltonian function (II.a.18) (defined "upstairs") and a notion of cyclic coordinate. We shall also use these examples to point out that our definition of 1st integral is probably too restrictive.

A much more substantial and natural example of a 2nd order variational problem is given in example (II.a.50) at the end of Chapter II, Section a).

(2) Recall Euler's homogeneity relation

$$\dot{y}^{\alpha} L_{\dot{y}^{\alpha}} = \mu L \, ;$$

it is proven by differentiating $L(y^{\alpha};\eta\dot{y}^{\alpha}) = \eta^{\mu} L(y^{\alpha};\dot{y}^{\alpha})$ with respect to η and setting $\eta = 1$.

(3) This is just a special case of the famous *Maupertuis principle*, which is explained very well in [13], pages 257-259.

(4) We do not feel that this is the correct notion, since for example it does not include the "1st integral" (II.a.22) in example (II.a.16). The proper formulation should probably use the theory of differential algebra (cf. [48]).

(5) We shall use some elementary concepts from (classical) algebraic geometry. For example we shall utilize the terms: *real algebraic variety; rational function and rational differential form; algebraic function; Zariski open set*. A suitable reference is [59].

(6) This means that (J,ω) is given by rational, regular (i.e., no poles or indeterminacies) data.

(7) We have already encountered one interesting example of this--cf. (I.d.37).

(8) A *symplectic structure* on a manifold P is given by a closed 2-form Ω such that Ω^m is a nowhere vanishing volume form on P. Symplectic structures are explained in [1] and [2].

(9) This is the cyclic coordinate alluded to earlier.

(10) In this regard see footnote (16) to Chapter I.

(11) In fact we suspect that *all* the non-degenerate examples we consider in this monograph are globally in Hamiltonian form, but we are not able to verify this due to the lack of a *global* Pfaff-Darboux theorem (0.d.9).

(12) In general, the affine geodesics in $\mathbb{A}^n$ are *rational normal curves*, given parametrically by $t \to (t, t^2, .., t^n)$ (here t is a constant multiple of the affine arclength).

(13) The point we are (somewhat muddily) trying to make is this: *Any* integral manifold $N \subset X$ projects to a curve $\gamma \subset \mathbb{E}^3$; it just may not be the case that N is the Frenet lifting of γ (e.g., γ may be a straight line). Essentially the condition that N be the Frenet lifting of γ is that along N the function κ vanish at isolated points. However, for our purposes this doesn't matter; we simply consider the functional as *defined on* $V(I,\omega)$. To tie this in with variational problems defined on curves $\gamma \subset \mathbb{E}^3$ we remark (without proof) that any sufficiently smooth curve in $\mathbb{E}^3$ is locally the projection of a (perhaps less smooth) integral manifold of $V(I,\omega)$.

(14) Again it will turn out that both τ and κ may be expressed by elliptic functions whenever N is a solution to the Euler-Lagrange equations.

(15) For a while it will be convenient to consider the variational problem associated to the functional

$$\Phi(N) = \int_N L(\kappa)\omega$$

where $L(\kappa)$ is an arbitrary smooth function of κ.

(16) This reflects the fact that in general a curve in $\mathbb{E}^3$ is determined by 3rd order invariants (its curvature (2nd order) and torsion (3rd order)).

(17) See footnote (11) to this chapter.

(18) This is one advantage of formulating the calculus of variations in the general setting of functionals defined on integral manifolds of exterior differential systems. Imposing constraints is accomplished simply by restricting the variational problem to a submanifold. This will also be true of integral constraints which are discussed in Section a of the Appendix.

[19]Using Theorem (III.a.25) below we may explain this terminology as follows: Let

$$g:[a,b] \to G ,$$

given by $t \to g(t) \in G$, be a solution curve to the Euler-Lagrange equations and set

$$g^*\omega^i = q^i(t)$$

where $q(t) = \{q^i(t)\}$ is a curve in the Lie algebra. Determine a curve $\lambda(t) = \{\lambda_i(t)\}$ in the dual Lie algebra by

$$\lambda_i(t) = L_{q^i}(q(t))$$

where $\varphi = L(q)dt$ gives the invariant functional on G. Then, as emerged from discussions with Robert Bryant, the 3rd equation in (III.a.26) means that

(*) $$\mathrm{Ad}^* g(t)\lambda(t) = \lambda_0$$

is a *constant* vector in $\mathfrak{g}^*$. Let $H_{\lambda_0} \subset G$ be the stabilizer of λ_0 under the co-adjoint representation. Then equation (*) says that $\lambda(t)$ uniquely determines $g(t)$ up to a curve in H_{λ_0}.

In many important cases G will be a compact reductive group and λ_0 will be a regular element. Then H_{λ_0} is a compact torus $\mathbb{R}^{m-1}/\mathbb{Z}^{m-1}$, and *if* the Euler-Lagrange equations are quasi-integrable by quadratures then $\lambda(t)$ travels on a closed curve in $\mathfrak{g}^*$. It follows that $g(t)$ lies on a manifold diffeomorphic to a torus $\mathbb{R}^m/\mathbb{Z}^m$, and moreover it seems likely that the curve $\{g(t)\}$ in $\mathbb{R}^m/\mathbb{Z}^m$ is the projection of a straight line (in many examples). This is the reason for our choice of terminology.

[20]Referring to footnote [28] in Chapter I we may think of 1st integrals as functions on $V(J,\omega)$. In case $(I,\omega;\varphi)$ is non-degenerate we have remarked that $V(J,\omega)$ is a sympletic manifold. In this case the usual Poisson bracket of functions on $V(J,\omega)$ is defined (cf. below) and gives our modified Poisson bracket.

[21]Added in proof: Using the invariance of the Wilmore integrand under inversions in spheres, the complete integrability of (II.a.53) may be established.

III. EULER EQUATIONS FOR VARIATIONAL PROBLEMS IN HOMOGENEOUS SPACES

a) Derivation of the Equations.

i) It is well-known that a general curve $\gamma \subset \mathbb{E}^n$ has a Frenet frame. We may express this as follows: If $N = \{a \leq s \leq b\}$ and we give γ by a map $x: N \to \mathbb{E}^n$ where s is the arclength, there is an essentially canonical lifting

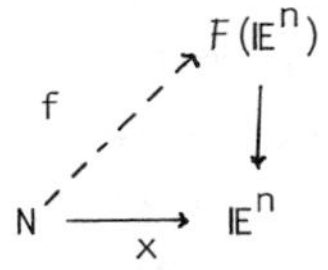

$$
\begin{array}{ccc}
 & & F(\mathbb{E}^n) \\
 & \overset{f}{\nearrow} & \downarrow \\
N & \underset{x}{\longrightarrow} & \mathbb{E}^n
\end{array}
\qquad\qquad (\text{III.a.1})
$$

of x to the manifold $F(\mathbb{E}^n)$ of frames $(x; e_1, .., e_n)$ on $\mathbb{E}^n$. By definition

$$f(s) = (x(s); e_1(s), .., e_n(s))$$

where for $1 \leq k \leq n$

$$\mathrm{span}\{e_1(s), .., e_k(s)\} = \mathrm{span}\{x'(s), .., x^{(k)}(s)\} \subset T_{x(s)}(\mathbb{E}^n).$$

$$(\text{III.a.2})$$

Here "general" should mean that the right-hand side of (III.a.2) always has dimension k. To say that the lifting is "essentially canonical" means that $e_2(s), \ldots, e_n(s)$ are determined up to ± 1; in particular, once they are chosen for $s = a$ then they are uniquely determined.

Now upon choice of a reference frame we may identify $F(\mathbb{E}^n)$ with the group $E(n)$ of Euclidean motions. In general, given a Lie group G and closed subgroup H we may consider curves γ in the homogeneous manifold G/H. If we give γ by a map $g: N \to G/H$, then again there is in general an essentially unique lifting

$$
\begin{array}{ccc}
 & & G \\
 & \overset{f}{\nearrow} & \downarrow \pi \\
N & \underset{g}{\longrightarrow} & G/H
\end{array}
\qquad\qquad (\text{III.a.3})
$$

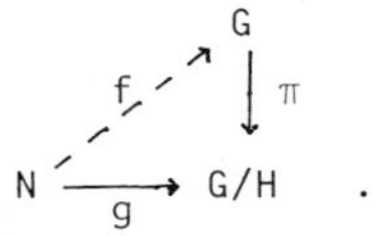

More precisely, f always exists in case the curve γ has *constant type* (cf. [34] and [44] for definitions and proofs). Moreover the "Frenet images" f(N) are integral manifolds of a Pfaffian differential system I on G generated by left invariant 1-forms (Maurer-Cartan forms). Finally, under very mild smoothness assumptions every curve $\gamma \subset G/H$ is $\pi(f(N))$ for some integral manifold $f: N \to G$ of I; the point is that just as for a curve $\gamma \subset \mathbb{E}^k \subset \mathbb{E}^n$ we may not have an essentially unique Frenet lifting. *This suggests that we study variational problems for curves in homogeneous spaces by studying variational problems defined for integral manifolds of invariant Pfaffian systems on Lie groups.*

It is well-known (cf. [1],[2]) that the behavior of a rigid body in $\mathbb{R}^3$ rotating about its center of gravity under motion free of external forces is described by the Euler-Lagrange equations associated to a left-invariant kinetic energy T defined on the tangent bundle of SO(3). The particular left invariant metric on SO(3) corresponding to T depends on the 3 principal moments of inertia of the rigid body (loc. cit). If we write

$$T(SO(3)) \cong SO(3) \times \mathfrak{so}(3)$$

where $\mathfrak{so}(3)$ is the Lie algebra of SO(3), then the projection to $\mathfrak{so}(3)$ of the solution curves to the Euler-Lagrange equations are themselves solutions to what are called the Euler equations. Moreover, the essential qualitative properties of the rigid body motion (stable and unstable equilibria, periodic motion, etc.) can be inferred from these Euler equations).

This theory of rigid body motion has been generalized to the study of the motion associated to left-invariant kinetic energies on an arbitrary Lie group G (cf. [1],[2],[50],[61] and the references cited there). Using the isomorphisms

$$\begin{cases} T(G) \cong G \times \mathfrak{g} \\ \mathfrak{g} \cong \mathfrak{g}^* \end{cases},$$

the second being given by the metric on $\mathfrak{g}$, the solutions to the Euler-Lagrange equations again project to curves $p(t) \in \mathfrak{g}^*$ which are solutions to the generalized Euler equations. A basic feature of these equations is the theorem of Kostant-Souriau: *The solution curves lie in coadjoint orbits.*

In this section we will generalize the essential aspects of this theory to the Euler-Lagrange differential system associated to any left-invariant variational problem $(I,\omega;\varphi)$ on a Lie group, and shall then apply these results to several examples of Frenet liftings.

ii) We shall briefly derive the classical theory. Let G be a connected Lie group with Lie algebra $\mathfrak{g}$ having basis $\{e_i\}$ with

$$[e_i,e_j] = c^k_{ij}e_k\ , \qquad c^k_{ij}+c^k_{ji} = 0\ . \tag{III.a.4}$$

If $\{\omega^i\}$ is the dual basis for the space $\mathfrak{g}^*$ of left-invariant 1-forms then the Maurer-Cartan equation is

$$d\omega^i + \frac{1}{2}c^i_{jk}\omega^j\wedge\omega^k = 0\ . \tag{III.a.5}$$

On $X = T(G)\times\mathbb{R}$ we write points as (v,t) where $v\in T(G)$ and we set

$$\omega^i(v) = p^i\ .^{(1)}$$

Then

$$X \cong G\times\mathfrak{g}\times\mathbb{R}$$

and (p^i,t) give coordinates on $\mathfrak{g}\times\mathbb{R}$.

Observe now that $X\cong J^1(\mathbb{R},G)$ and consider on X the standard Pfaffian system (0.e.2). Using our present coframing (which is generally *not* a coordinate coframing, but this doesn't matter since everything is intrinsic), (I,ω) is given in the form (I.a.1) by

$$\begin{cases}\theta^i = \omega^i - p^i dt = 0\\ \omega = dt \neq 0\end{cases}\ . \tag{III.a.6}$$

The integral manifolds are 1-jets of curves $t\to g(t)\in G$.

We consider the Lagrangian given by the left-invariant metric on G having $\{\omega^i\}$ as an orthonormal basis. Thus, as a function on $J^1(\mathbb{R},G)\cong\mathbb{R}\times T(G)$,

$$L = \frac{1}{2}\left(\sum(p^i)^2\right)\ .$$

The corresponding standard variational problem (example (I.a.5)) has $\varphi = Ldt$. Then

$$d\varphi = \sum p^i dp^i\wedge dt\ . \tag{III.a.7}$$

To find the Euler-Lagrange system we follow our prescription as given in Chapter I, Section e). Thus, let $Z = X\times\mathbb{R}^n$ where $\mathbb{R}^n$ has coordinates $\lambda = (\lambda_1,\ldots,\lambda_n)$ $(n = \dim G)$, and on Z consider the differential forms

$$\begin{cases} \psi = \varphi + \lambda_i \theta^i \\ \Psi = d\psi \end{cases} .$$

Using (III.a.5) and (III.a.7) we obtain

$$\Psi = \sum (p^i - \lambda_i) dp^i \wedge dt + d\lambda_i \wedge \theta^i - \frac{1}{2} \lambda_i c^i_{jk} \omega^j \wedge \omega^k \quad . \tag{III.a.8}$$

Accordingly the Cartan system $C(\Psi)$ is generated by the Pfaffian equations

$$\begin{cases} \partial/\partial p^i \lrcorner \Psi = (p^i - \lambda_i) dt = 0 \\ \partial/\partial \lambda_i \lrcorner \Psi = \theta^i = 0 \\ \partial/\partial \omega^i \lrcorner \Psi = -d\lambda_i - \lambda_j c^j_{ik} \omega^k = 0 \end{cases} . \tag{III.a.9}$$

We define $Z_1 \subset Z$ by $p^i = \lambda_i$ and use $\{\omega;\omega^i;d\lambda_i\}$ as a coframe on Z_1. By example (I.e.18) the variational problem $(I,\omega;\varphi)$ is non-degenerate with $Y = Z_1$, and by example (II.a.23) the solution curves to the Euler-Lagrange system (J,ω) project to the geodesics for the corresponding left-invariant metric on G.

To derive and explain the Euler equations we consider $\mathfrak{g}^*$ as an *abstract* vector space with basis $\{e^*_i\}$ dual to the basis $\{e_i\}$ of $\mathfrak{g}$ (we do *not* consider the e^*_i as 1-forms on G). The metric on $\mathfrak{g}$ gives an intrinsic identification

$$\eta: \mathbb{R}^n \tilde{\to} \mathfrak{g}^*$$

with

$$\eta(\lambda) = \lambda_i e^*_i \ .$$

The momentum space

$$Y \subset J^1(\mathbb{R},G) \times \mathbb{R}^n \cong J^1(\mathbb{R},G) \times \mathfrak{g}^*$$

and we consider the projection to $\mathfrak{g}^*$ of a solution curve N to the Euler-Lagrange system (J,ω) on Y. By the last equation in (III.a.9) the motion of the projection $\lambda \in \mathfrak{g}^*$ satisfies

$$d\lambda_i + \lambda_j p^k c^j_{ik} = 0 \quad . \tag{III.a.10}$$

<u>Explanation</u>. Taking t as a parameter we may give N by

$$t \to \underbrace{(t;g(t);p^i(t)}_{\substack{\shortparallel \\ j^1(g)(t)}};\lambda_i(t)) \in J^1(\mathbb{R},G) \times \mathbb{R}^n ,$$

and then (III.a.10) means that

$$\frac{d\lambda_i(t)}{dt} + \lambda_j(t)p^k(t)c^j_{ik} = 0 .$$

At this juncture, classically one uses the isomorphism $\mathfrak{g} \tilde{=} \mathfrak{g}^*$ given by the metric to write (III.a.10) as

$$dp^i + \sum_{j,k} p^j p^k c^j_{ik} = 0 . \tag{III.a.11}$$

These are the *Euler equations*; there are $n = \dim G = \frac{1}{2} \dim T(G)$ of them.(2)

(III.a.12) <u>Remark</u>. As a portent of things to come we consider a variational problem $(I,\omega;\varphi)$ where

$$\begin{cases} \varphi = L(p)dt \\ \det \left\| L_{p^i p^j} \right\| \neq 0 . \end{cases}$$

This is a general non-degenerate left-invariant Lagrangian, and we note that the above discussion up through (III.a.10) remains valid where

$$\lambda_i = L_{p^i} .$$

In particular this is true if

$$L(p) = \frac{1}{2}\left(\sum \mu_i (p^i)^2\right) , \qquad \mu_i > 0 . \tag{III.a.13}$$

Returning to the general discussion we consider the *adjoint representation*

$$Ad: G \to Aut(\mathfrak{g}) . \tag{III.a.14}$$

We recall that by definition

$$Ad(g)(v) = (L_g)_*(R_{g^{-1}})_*(v)$$

where $g \in G$ and $v \in \mathfrak{g} = T_e(G)$. (Since left and right translations commute we may write this as $(R_{g^{-1}})_*(L_g)_*(v)$.) It is well-known that the differential

$$\mathrm{ad}: \mathfrak{g} \to \mathrm{Hom}(\mathfrak{g},\mathfrak{g}) \qquad (\mathrm{ad} = \mathrm{Ad}_*)$$

of the map (III.a.14) is given by

$$\mathrm{ad}(e_i)(e_j) = [e_i,e_j] = c_{ij}^k e_k \quad .$$

The *coadjoint representation* is by definition the dual to (III.a.15). Its differential

$$\mathrm{ad}^*: \mathfrak{g} \to \mathrm{Hom}(\mathfrak{g}^*,\mathfrak{g}^*)$$

is given by

$$\langle \mathrm{ad}^*(e_i)e_k^*, e_j\rangle = -\langle e_k^*, \mathrm{ad}(e_i)e_j\rangle = -c_{ij}^k \quad .$$

Consequently, for $\lambda = \lambda_i e_i^* \in \mathfrak{g}^*$

$$\mathrm{ad}^*(e_i)\lambda = -\lambda_k c_{ij}^k e_j^* \quad . \qquad \text{(III.a.15)}$$

Recall that the *coadjoint orbit* $\mathcal{O}_{\mathrm{Ad}^*}(\lambda) \subset \mathfrak{g}^*$ of λ is the image of the map

$$\begin{cases} G \to \mathfrak{g}^* \\ g \mapsto \mathrm{Ad}^*(g)\lambda \quad . \end{cases}$$

By (III.a.15) the tangent space

$$T_\lambda(\mathcal{O}_{\mathrm{Ad}^*}(\lambda)) \subset \mathfrak{g}^*$$

is the image of the map

$$\begin{cases} \mathfrak{g} \to \mathfrak{g}^* \\ e_i \to -\lambda_k c_{ij}^k e_j^* \quad ; \end{cases} \qquad \text{(III.a.16)}$$

i.e.

$$T_\lambda(\mathcal{O}_{\mathrm{Ad}^*}(\lambda)) = \{p^i \lambda_k c_{ij}^k e_j^* \text{ where } (p^1,\dots,p^n) \in \mathbb{R}^n\} \qquad \text{(III.a.17)}$$

Comparing (III.a.10) with (III.a.17) gives the

KOSTANT-SOURIAU THEOREM (Partial Statement): *The solution curves* $\lambda(t) \in \mathfrak{g}^*$ *to Euler's equations lie in a coadjoint orbit.*

Actually, much more is known (cf. [2] and the references cited therein). Namely, the coadjoint orbits have natural symplectic structures (Kirilov) and Euler's equations are in Hamiltonian form relative to this symplectic structure. Very briefly, the surjective map

$$\pi_\lambda : \mathfrak{g} \to T_\lambda(\mathcal{O}_{Ad^*}(\lambda))$$

given by (III.a.16) has kernel given by

$$\ker \pi_\lambda = \{p \in \mathfrak{g} : [p, e_i] \in \lambda^\perp \text{ for all } e_i\} \qquad \text{(III.a.18)}$$

where $\lambda^\perp \subset \mathfrak{g}$ is the annihilator of λ. For $p_1, p_2 \in \mathfrak{g}$ we define

$$\Omega(p_1, p_2) = \langle \lambda, [p_1, p_2] \rangle \qquad \text{(III.a.19)}$$

It follows from (III.a.18) that Ω induces a non-degenerate alternating bilinear form

$$\Omega : \mathfrak{g}/\ker \pi_\lambda \times \mathfrak{g}/\ker \pi_\lambda \to \mathbb{R} \, ,$$

and since

$$T_\lambda(\mathcal{O}_{Ad^*}(\lambda)) \cong \mathfrak{g}/\ker \pi_\lambda$$

this induces a canonical non-degenerate 2-form on $\mathcal{O}_{Ad^*}(\lambda)$ that is invariant under the action of G. Under the surjective map $G \to \mathcal{O}_{Ad^*}(\lambda)$, the Maurer-Cartan equations (III.a.5) and definition (III.a.19) imply that Ω pulls back to $-d\lambda$ (where we now view λ as a 1-form on G). Consequently, $d\Omega = 0$ and this establishes the Kirilov symplectic structure on the coadjoint orbits.

Finally, for the function $H(\lambda) = \frac{1}{2}(\sum \lambda_i^2)$ on $\mathfrak{g}^*$ it is straightforward to verify that the vector field on $\mathcal{O}_{Ad^*}(\lambda)$ corresponding to $dH_{\mathcal{O}_{Ad^*}(\lambda)}$ under the symplectic isomorphism $\Omega : T(\mathcal{O}_{Ad^*}(\lambda)) \xrightarrow{\sim} T^*(\mathcal{O}_{Ad^*}(\lambda))$ has integral curves given by Euler's equations (III.a.11).

(III.a.20) Example. In $\mathbb{R}^3$ we consider a *rigid body* B rotating about a fixed point $0 \in B$.

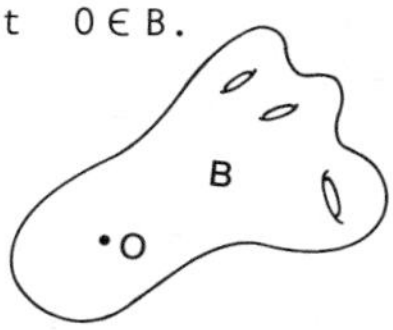

For the purposes of this discussion we may think of B as a large number particles with position vectors $\vec{x}_i \in \mathbb{R}^3$ and masses m_i. Then the rigidity is expressed by the condition

$$\|\vec{x}_i - \vec{x}_j\| = \text{constant}$$

that the particles be at a fixed distance from one another. It is clear that the *configuration space* of B; i.e., the space of all possible positions of B, is equal to the group $G = SO(3)$ of proper rigid motions leaving 0 fixed (this assumes that not all $\vec{x}^i$ lie in a plane). In a moment we will make a *natural* identification of the configuration space with the frame manifold $F_0(\mathbb{E}^3) \cong G$ (cf. Chapter 0, Section c); here we agree to only use oriented frames).

We consider the motion of B under Newton's laws in the absence of external forces. This motion is given by the solution curves of the Euler-Lagrange equations associated to a mechanical system on G where the potential energy is zero (cf. (I.d.18)). Equivalently, on $J^1(\mathbb{R},G)$ we have a Lagrangian given by a kinetic energy function. Using the identification

$$J^1(\mathbb{R},G) \cong \mathbb{R} \times G \times \mathfrak{g},$$

where the trivialization

$$T(G) \cong G \times \mathfrak{g}$$

is given by isomorphisms

$$(L_g)_*: \underset{\substack{\| \\ \mathfrak{g}}}{T_e(G)} \tilde{\to} T_g(G) ,$$

it is clear that T is independent of $t \in \mathbb{R}$ and of $g \in G$. This latter property merely formulates mathematically the condition that the kinetic energy is preserved by a fixed rigid motion of $\mathbb{R}^3$. Thus, T gives a left-invariant metric G, and first we want to interpret this metric in terms of the geometry of the body.

For this we use a special feature of $\mathbb{R}^3$. Namely, there is an isomorphism

$$\eta: \mathbb{R}^3 \tilde{\to} \mathfrak{g}$$

where by definition

$$\eta(v) = \left\{ \begin{array}{l} \text{tangent vector to rotation of B} \\ \text{about } \frac{v}{\|v\|} \text{ with angular velocity } \|v\| \end{array} \right\}$$

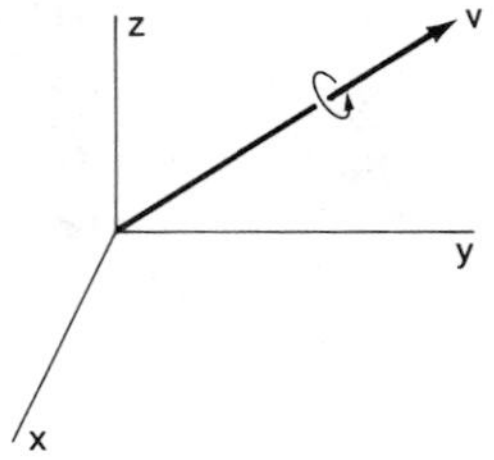

For example

$$\eta(1,0,0) = x^2\,\partial/\partial x^3 - x^3\,\partial/\partial x^2 ,$$

where on the right hand side we are interpreting $\mathfrak{g}$ as a Lie algebra of vector fields on $\mathbb{R}^3$ (cf. remark (III.a.29) below). Under this isomorphism

$$\eta(v \times w) = [\eta(v),\eta(w)]$$

where $v \times w$ is the vector cross-product of $v,w \in \mathbb{R}^3$.

Suppose that $e \in \mathbb{R}^3$ is a unit vector and $\eta(pe)$ is the tangent vector to rotating about the axis $\mathbb{R}e$ with angular velocity p. By definition of the kinetic energy

$$T(\eta(pe)) = \frac{1}{2}\left(\sum_i m_i v_i^2\right)$$

where v_i is the velocity of $\vec{x}_i$. But by the obvious formula for the angular velocity of $\vec{x}_i$,

$$v_i = r_i(e)\cdot p$$

where $r_i(e)$ is the distance from $\vec{x}_i$ to the line $\mathbb{R}e$. Thus

$$T(\eta(pe)) = \frac{1}{2}\left(\sum_i m_i r_i(e)^2\right)p^2 .$$

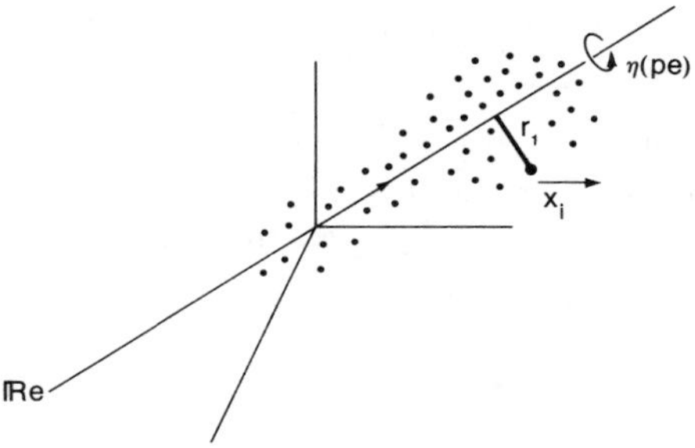

Definition. For $e \in \mathbb{R}^3$ a unit vector

$$I_e(B) = \sum_i m_i r_i(e)^2$$

is the *moment of inertia* of B with respect to the axis $\mathbb{R}e$.

It follows that

$$T(\eta(pe)) = \frac{p^2}{2} I_e(B) . \qquad \text{(III.a.21)}$$

Using the isomorphism η we consider T as a quadratic form on $\mathbb{R}^3$. By a well-known linear algebra result there exists an orthonormal basis $e_1(B)$, $e_2(B)$, $e_3(B)$ for $\mathbb{R}^3$ relative to which T is diagonalized.

Definitions: i) $e_1(B)$, $e_2(B)$, $e_3(B)$ are called the *principal axes of inertia* of B; ii) $I_i = I_{e_i(B)}(B)$ are its *principal moments of inertia*; and iii) the ellipsoid in $\mathbb{R}^3$ with principal axes $e_i(B)$ and with equation

$$(y^1)^2 I_1 + (y^2)^2 I_2 + (y^3)^2 I_3 = 1$$

in the coordinate system with basis $e_1(B)$, $e_2(B)$, $e_3(B)$ is called the *inertia ellipsoid* $E(B)$ associated to the body.
It is intuitively clear that $E(B)$ is the ellipsoid whose shape most closely resembles that of B.

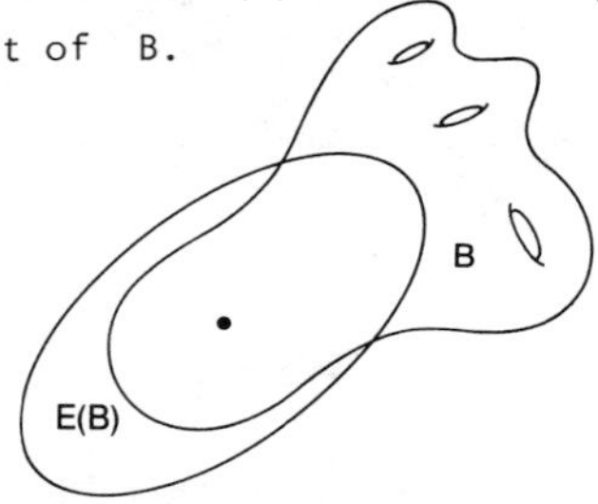

As will be seen below, the motion of B is the same as that of its inertia ellipsoid, so that in considering rigid body we may reduce to this case.

We now consider the motion of B as being described by the position of the frame given by the principal axes of inertia. Using the moving frame notations from Chapter 0, Section c), on the basis of our discussion it is clear that the value of the kinetic energy on $v \in \mathfrak{g}$ is

$$T(v) = \frac{1}{2}\left(I_1\langle\omega_2^3, v\rangle^2 + I_2\langle\omega_1^3, v\rangle^2 + I_3\langle\omega_1^2, v\rangle^2\right) .$$

For example, if v represents infinitesimal rotation about $e_1(B)$ with velocity p^1

$$T(v) = I_1(p^1)^2 = I_1\langle\omega_2^3, v\rangle^2 ,$$

since from

$$de_2 = \omega_2^1 e_1 + \omega_2^3 e_3$$

we see that ω_2^3 measures infinitesimal rotation about e_1. In summary, T *is given by the kinetic energy corresponding to the left-invariant metric*

$$ds^2 = I_1(\omega_2^3)^2 + I_2(\omega_1^3)^2 + I_3(\omega_1^2)^2$$

on $0(3)$. If we set

$$\begin{cases} \tilde{\omega}^1 = \omega_2^3 \\ \tilde{\omega}^2 = \omega_1^3 \\ \tilde{\omega}^3 = \omega_1^2 \end{cases} \qquad (3)$$

then using $\tilde{\omega}^i$ in place of ω^i in the general discussion above, the kinetic energy is

$$T = \frac{1}{2}\left(I_1(p^1)^2 + I_2(p^2)^2 + I_3(p^3)^2\right) .$$

<u>Note</u>: The case when $I_1 = I_2 = I_3$ corresponds to a "spherical" body, and then ds^2 is the bi-invariant metric on $0(3)$. The case when $I_1 = I_2 \neq I_3$ corresponds to a body symmetric about an axis, and then the ds^2 is invariant under a 1-parameter subgroup acting on the right.

The structure equations (cf. (0.c.4))

$$\begin{cases} d\tilde{\omega}^1 = \tilde{\omega}^2 \wedge \tilde{\omega}^3 \\ d\tilde{\omega}^2 = \tilde{\omega}^1 \wedge \tilde{\omega}^3 \\ d\tilde{\omega}^3 = \tilde{\omega}^1 \wedge \tilde{\omega}^2 \end{cases}$$

give

$$c_{23}^1 = 1, \qquad c_{13}^2 = -1, \qquad c_{12}^3 = 1$$

and all other $c_{jk}^i = 0$ $(j < k)$. Taking into account the remark (III.a.12) the Euler equations (III.a.10) give the familiar equations

$$\begin{cases} d\lambda_1 + \dfrac{(I_3 - I_2)}{I_2 I_3} \lambda_2\lambda_3 = 0 \\[2ex] d\lambda_2 + \dfrac{(I_1 - I_3)}{I_1 I_3} \lambda_1\lambda_3 = 0 \\[2ex] d\lambda_3 + \dfrac{(I_2 - I_1)}{I_1 I_2} \lambda_1\lambda_2 = 0 \end{cases}$$

found in any book on mechanics. The coadjoint orbits are the spheres

$$V = \lambda_1^2 + \lambda_2^2 + \lambda_3^2 = c$$

(this is in $\mathfrak{g}^* \cong \mathbb{R}^3$; in $\mathfrak{g} \cong \mathbb{R}^3$ the coadjoint orbits are ellipsoids since the isomorphism $\mathfrak{g}^* \xrightarrow{\sim} \mathfrak{g}$ given by the metric is not the usual one unless

$I_1 = I_2 = I_3 = 1$ (cf. footnote (3))). The Kirilov symplectic form is

$$\Omega = \lambda_3 d\lambda_1 \wedge d\lambda_2 - \lambda_2 d\lambda_1 \wedge d\lambda_3 + \lambda_1 d\lambda_2 \wedge d\lambda_3$$

and the Hamiltonian

$$\begin{aligned} H &= \frac{1}{2}\,(I_1 (p^1)^2 + I_2 (p^2)^2 + I_3 (p^3)^2) \\ &= \frac{1}{2}\left(\frac{\lambda_1^2}{I_1} + \frac{\lambda_2^2}{I_2} + \frac{\lambda_3^2}{I_3}\right) \end{aligned}$$

since $\lambda_i = L_{p_i}$. The motion of the point $\lambda(t) \in \mathfrak{g}^*$ is described by the curves of intersection

$$\begin{cases} V = c \\ H = c^1 \end{cases}.$$

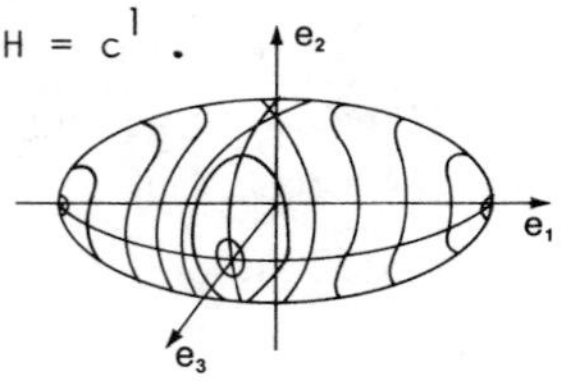

Determining the motion of the point $\lambda(t) \in \mathfrak{g}^*$ is equivalent to determining the motion of the axis of infinitesimal rotation of the body in space (as remarked above this axis traverses a closed curve on an ellipsoid in $\mathbb{R}^3$). The remaining step of describing how the body rotates about this moving axis is perhaps most conveniently done by Poinsot's description given in mechanics books (cf. [2], pages 145-148).

iii) We retain the preceding notations and assume given a subspace $W^* \subset \mathfrak{g}^*$. Using the ranges of indices

$$1 \leqq i,j \leqq n = \dim G;\quad 1 \leqq \alpha,\beta \leqq s = \dim W^*;\quad s+1 \leqq \mu,\nu \leqq n$$

we choose a basis $\{\omega^i\}$ for $\mathfrak{g}^*$ so that $\{\omega^\alpha\}$ gives a basis for W^*. The equations

$$\omega^\alpha = 0$$

generate a left-invariant Pfaffian differential system on G. To express this in our usual form we set

$$\begin{cases} V = (W^*)^\perp \subset \mathfrak{g} \\ X = G \times V \times \mathbb{R} \subset T(G) \times \mathbb{R} \end{cases}.$$

Note that $\{\omega^\mu\}$ gives a basis for V^* and we let $(p^\mu;t)$ be the corresponding coordinates on $V \times \mathbb{R}$. We then define (I,ω) on X to be the Pfaffian system generated by

$$\begin{cases} \theta^{\alpha} = \omega^{\alpha} = 0 \\ \theta^{\mu} = \omega^{\mu} - p^{\mu}dt = 0 \\ \omega = dt \neq 0.^{(4)} \end{cases} \qquad \text{(III.a.22)}$$

Integral manifolds of (I,ω) are simply curves in G with parameter t and everywhere tangent to the left invariant distribution $V \subset \mathfrak{g}$.

(III.a.23) Example. Let $H \subset G$ be a closed subgroup with Lie algebra $\mathfrak{h} \subset \mathfrak{g}$. We consider a curve $\gamma \subset G/H$ with parameter t as given by a map

$$\gamma : N \to G/H, \qquad N = [a,b],$$

and denote by $(j^k\gamma)(t) \in J^k(\mathbb{R}, G/H)$ the k-jet of γ at t. Now G acts on $J^k(\mathbb{R}, G/H)$. We set

$$G^k_{\gamma}(t) = \text{stability group of } (j^k\gamma)(t)$$

and say that γ has *constant type* in case the subgroups $G^k_{\gamma}(t) \subset G$ are all conjugate (cf. [34],[44]). For instance, if G is compact then this is a generic condition. For curves of constant type satisfying the condition (also generic) that $\dim G^k_{\gamma}(t) = 0$ for $k \gg 0$ there is an essentially unique Frenet lifting

$$\begin{array}{ccc} & \overset{f}{\nearrow} & G \\ & & \downarrow \\ N & \xrightarrow{\gamma} & G/H \end{array} .$$

Moreover, $f: N \to G$ determines an integral manifold of (I,ω) for a suitable $W^* \subset \mathfrak{g}^*$. We refer to the above references for numerous examples of this construction.

Returning to the general discussion we consider a variational problem $(I,\omega;\varphi)$ where

$$\varphi = L(p)dt .$$

Associated to this variational problem is the Euler-Lagrange differential system (J,ω) on $Y \subset X \times \mathfrak{g}^*$, and integral manifolds of (J,ω) project onto curves $\lambda(t) \in \mathfrak{g}^*$. Moreover, in $\mathfrak{g}^*$ we have the coadjoint orbit $\mathcal{O}_{Ad^*}(\lambda)$ whose tangent space is the image of the map

$$\pi_{\lambda} : \mathfrak{g} \to T_{\lambda}(\mathcal{O}_{Ad^*}(\lambda)) \subset \mathfrak{g}^*$$

given by (III.a.16). We set

$$V(\lambda) = \pi_\lambda(V) \subset T_\lambda(\mathcal{O}_{Ad^*}(\lambda)) ,$$

so that the $V(\lambda)$ give a distribution on the coadjoint orbit $\mathcal{O}_{Ad^*}(\lambda)$.

(III.a.25) THEOREM. *The solution curves to the Euler-Lagrange system* (J,ω) *on* Y *project in* $\mathfrak{g}^*$ *to curves that lie in coadjoint orbits* $\mathcal{O}_{Ad^*}(\lambda)$. *Moreover, these curves are integral curves of the distribution* $V(\lambda)$.

Proof. We follow the prescription in Chapter I, Section e) and consider $Z = X \times \mathbb{R}^n$ where $\mathbb{R}^n$ has coordinates $\lambda = (\lambda_1,..,\lambda_n)$. Intrinsically, $\mathbb{R}^n \cong \mathfrak{g}^*$ with $\lambda = \lambda_i e_i^*$ where $\{e_i\} \subset \mathfrak{g}$ is the dual basis to $\{\omega^i\}$. On Z we consider the differential forms

$$\begin{cases} \psi = \varphi + \lambda_i \theta^i \\ \Psi = d\psi \end{cases} .$$

Clearly by (III.a.5)

$$\Psi = (L_{p^\mu} - \lambda_\mu) dp^\mu \wedge dt + d\lambda_i \wedge \theta^i - \frac{1}{2} \lambda_j c^j_{ik} \omega \wedge \omega^k ,$$

so that the Cartan system $C(\Psi)$ is generated by the Pfaffian equations

$$\begin{cases} \partial/\partial p^\mu \lrcorner \Psi = (L_{p^\mu} - \lambda_\mu) dt = 0 \\ \partial/\partial \lambda_i \lrcorner \Psi = \theta^i = 0 \\ \partial/\partial \omega^i \lrcorner \Psi \equiv d\lambda_i + \frac{1}{2} \lambda_j p^\mu c^j_{i\mu} dt \mod\{\theta^i\} \end{cases} . \tag{III.a.26}$$

Using (III.a.17) the last equations imply the theorem. Q.E.D.

Remark. The last equations (III.a.26) are not, at least on the face of it, sufficient to uniquely determine $\lambda(t)$ with given initial conditions. The point is that the 1st equations $L_{p^\mu} = \lambda_\mu$ may not uniquely determine the $p^\mu(t)$ in terms of $\lambda_\mu(t)$. However, if the variational problem is non-degenerate (cf. Chapter I, Section e)), then it follows that at least locally the equations

$$d\lambda_i + \frac{1}{2} c^j_{i\mu} \lambda_j p^\mu dt = 0 \tag{III.a.27}$$

are expressible in terms of the λ_i's, and so we shall call (III.a.27) the *Euler equations associated to* $(I,\omega;\varphi)$.

(III.a.28) Example. We reconsider the functional (cf. (II.b.1))

$$\Phi(\gamma) = \frac{1}{2} \int_\gamma \kappa^2 \, ds , \qquad \gamma \subset \mathbb{E}^2 ,$$

and shall derive (II.b.3) *without computation.* In this case the Frenet frame gives a lifting of γ to the Euclidean group $E(2)$ as an integral curve of the Pfaffian system $\omega^2 = 0$. Before going on we pause to make a general

(III.a.29) Remark. The Lie algebra of the Euclidean group $E(n)$ acting on $\mathbb{R}^n$ is generated by the vector fields

$$e_i = \partial/\partial x^i$$

$$e_i^j = x^j \, \partial/\partial x^i - x^i \, \partial/\partial x^j = -e_j^i$$

with the bracket relations

$$\begin{cases} [e_i, e_j] = 0 \\ [e_i, e_j^k] = \delta_i^k e_j - \delta_i^j e_k \\ [e_i^j, e_k^\ell] = \delta_i^\ell e_k^j - \delta_j^k e_\ell^i \end{cases} . ^{(5)} \qquad \text{(III.a.30)}$$

Thus the coadjoint orbits are

$$\begin{array}{c} \mathcal{O}_{\mathrm{Ad}^*}(\lambda) \\ \cap \\ 0 \to \mathfrak{so}(n)^* \to \mathfrak{e}(n)^* \xrightarrow{\pi} \mathbb{R}^{n*} \to 0 \end{array}$$

where $\pi(\mathcal{O}_{\mathrm{Ad}^*}(\lambda))$ is a sphere in $\mathbb{R}^{n*} \cong \mathbb{R}^n$.

When $n = 2$ the last equation in (III.a.30) is zero and the coadjoint orbits are surfaces in $\mathfrak{e}(2)^*$ that project onto circles in $\mathbb{R}^2$, and this observation is sufficient to establish (II.b.3).

It is, however, instructive to explicitly integrate the Euler equations in this case. On $X = E(2) \times \mathbb{R}^3$ where $\mathbb{R}^3$ has coordinates (t,q,r) we consider the differential system (I,ω) given by

$$\begin{cases} \theta^1 = \omega^1 - q\,dt = 0 \\ \theta^2 = \omega^2 = 0 \\ \theta^3 = \omega_1^2 - r\,dt = 0 \\ \omega = dt \neq 0 \end{cases} . ^{(6)} \qquad \text{(III.a.31)}$$

Each integral of (III.a.31), taken with t as a parameter, gives a curve in X projecting to a curve $t \to x(t) \in \mathbb{E}^2$ along which

$$\begin{cases} ds = q(t)dt \\ \kappa(s) = r(t)/q(t) \ . \end{cases} \tag{III.a.32}$$

Thus (II.b.1) corresponds to a variational problem $(I,\omega;\varphi)$ where

$$\varphi = \frac{r^2}{2q} dt \ .$$

To compute the Euler-Lagrange equations we have, using the structure equations (0.c.3),

$$\psi = \frac{r^2}{2q} dt + \lambda_1(\omega^1 - qdt) + \lambda_2\omega^2 + \lambda_3(\omega_1^2 - rdt)$$

$$\Psi = \frac{r}{q} dr \wedge dt - \frac{1}{2}\left(\frac{r}{q}\right)^2 dq \wedge dt + d\lambda_1 \wedge (\omega^1 - qdt) + d\lambda_2 \wedge \omega^2$$

$$+ d\lambda_3 \wedge (\omega_1^2 - rdt) - \lambda_1(\omega^2 \wedge \omega_1^2 + dq \wedge dt) + \lambda_2\omega^1 \wedge \omega_1^2 - \lambda_3 dr \wedge dt \ .$$

The Cartan system is then generated by

$$\begin{cases} \partial/\partial r \lrcorner \Psi = \left(\frac{r}{q} - \lambda_3\right) dt = 0 \\ \partial/\partial q \lrcorner \Psi = \left(-\frac{1}{2}\left(\frac{r}{q}\right)^2 - \lambda_1\right) dt = 0 \\ \partial/\partial\lambda_i \lrcorner \Psi = \theta^i = 0 \\ \text{(i)} \quad \partial/\partial\omega^1 \lrcorner \Psi \equiv -d\lambda_1 + \lambda_2 r dt \mod\{\theta^i\} \\ \text{(ii)} \quad \partial/\partial\omega^2 \lrcorner \Psi \equiv -d\lambda_2 + \lambda_1 r dt \mod\{\theta^i\} \\ \text{(iii)} \quad \partial/\partial\omega_1^2 \lrcorner \Psi \equiv -d\lambda_3 - \lambda_2 q dt \mod\{\theta^i\} \ . \end{cases} \tag{III.a.33}$$

The 1st two equations give

$$\begin{aligned} \lambda_3 &= r/q & &(= \kappa) \\ \lambda_1 &= -\frac{1}{2}\left(\frac{r}{q}\right)^2 & &\left(= -\frac{\kappa^2}{2}\right) . \end{aligned} \tag{III.a.34}$$

Equations (i)-(iii) in (III.a.33) are the Euler equations. The linear combination

$$-\lambda_1(\text{i}) + \lambda_2(\text{ii}) = 0$$

gives

$$\lambda_1^2 + \lambda_2^2 = c \ . \tag{III.a.35}$$

Using (iii) and the 1st equation in (III.a.34) gives

$$\lambda_2 = -\kappa' \quad ,$$

where by (III.a.32) the derivative is with respect to arclength. Then (III.a.34) and (III.a.35) give

$$\kappa'^2 + \frac{\kappa^4}{4} = c \; ,$$

which is just the 1st integral $V_4 = c$ in i) in Chapter II, Section b). The solution curves to Euler's equations are in this case the closed curves

$$\begin{cases} \lambda_1 + \frac{1}{2}\lambda_3^2 = 0 \\ \lambda_1^2 + \lambda_2^2 = c \end{cases}$$

that we may picture as follows: The coadjoint orbits are surfaces E_c lying in the space of variables $(\lambda_1,\lambda_2,\lambda_3) \in \mathbb{R}^3 \cong e(2)^*$. Under the projection $(\lambda_1,\lambda_2,\lambda_3) \overset{\pi}{\to} (\lambda_1,\lambda_2)$ discussed above, E_c projects onto the circle $\lambda_1^2 + \lambda_2^2 = c$. In this case the distribution $V(\lambda) \subset T(E_c)$ is for $c \neq 0$ the whole tangent bundle, so that the condition to be an integral curve of $V(\lambda)$ imposes no restriction. (This will change radically when we consider the functional $\frac{1}{2}\int \kappa^2\, ds$ for curves in $\mathbb{E}^n$, $n \geqq 3$.) Nevertheless the solution curves to the Euler equations are the above closed curves on E_c.

b) Investigation of the Euler Equations in Some Differential-Geometric Examples.

i) We shall apply theorem (III.a.25) to study variational problems for curves $\gamma \subset \mathbb{E}^n$, among other things completing the proof of theorem (II.a.32) in the case $R = 0$.

First we establish our notations. To emphasize the group-theoretic character of what is occurring we identify $F(\mathbb{E}^n)$ with the Euclidean group $E(n)$ and set $X = E(n) \times \mathbb{R}^{n+1}$ where $\mathbb{R}^{n+1}$ has coordinates $(t,q,r_1,\ldots,r_{n-1}) = (t,q;r_j)$. On X we consider the differential system (I,ω) generated by the Pfaffian equations

$$\begin{cases} \theta^1 = \omega^1 - q dt = 0 & \\ \theta^\rho = \omega^\rho = 0 & 2 \leqq \rho \leqq n \\ \theta_j^{j+1} = \omega_j^{j+1} - r_j dt = 0 & 1 \leqq j \leqq n-1 \\ \theta_j^k = \omega_j^k = 0 & k \geqq j+2 \\ \omega = dt \neq 0 . & \end{cases} \tag{III.b.1}$$

Integrals of (I,ω) have parameter t and project to curves $t \to x(t) \in \mathbb{E}^n$, denoted by γ, along which $e_i(t)$ is a frame satisfying the general Frenet-Serret formulas

$$\begin{cases} \dfrac{dx}{dt} = q(t) e_1(t) \\ \dfrac{de_j}{dt} = r_j(t) e_{j+1}(t) - r_{j-1}(t) e_{j-1}(t) \end{cases} \tag{III.b.2}$$

where we set $r_0(t) = r_n(t) = 0$. The *curvatures* of the curve γ are defined by

$$\kappa_1(s) = \frac{r_1(t)}{q(t)}, \ldots, \kappa_{n-1}(s) = \frac{r_{n-1}(t)}{q(t)} \tag{III.b.3}$$

where $s(t)$ is the arclength (thus $ds(t)/dt = q(t)$). In the non-degenerate case when $\kappa_1 \cdots \kappa_{n-1} \neq 0$, the curvatures $\kappa_2, \ldots, \kappa_{n-1}$ are uniquely determined up to ± 1 by γ and, as a consequence of (I.b.7), uniquely determine the curve up to a rigid motion. However, on integral manifolds of (I,ω) some of the κ_j may vanish at isolated points or even be identically zero; for example, the condition that γ lie in an $\mathbb{E}^\ell \subset \mathbb{E}^n$ is

$$\kappa_\ell(s) \equiv 0 \quad .^{(7)} \tag{III.b.4}$$

We consider a variational problem $(I,\omega;\varphi)$ where

$$\varphi = \ell(q, r_1, \ldots, r_{n-1}) dt .$$

Since

$$\begin{cases} ds = q dt \\ \kappa_j = r_j / q \end{cases}$$

there is an obvious condition that

$$\ell(q, r_1, \ldots, r_{n-1}) dt = L(\kappa_1, \ldots, \kappa_{n-1}) ds \tag{III.b.5}$$

for a function $L(\kappa_1, \ldots, \kappa_{n-1})$, and we shall consider only variational problems for which (III.b.5) holds. An integral manifold $N \subset X$ of (I,ω) determines a curve $\gamma \subset \mathbb{E}^n$ and we shall write the functional as

$$\Phi(\gamma) = \int_\gamma L(\kappa_1,\ldots,\kappa_{n-1})\, ds \quad . \tag{III.b.6}$$

(III.b.7) Remark. There are a couple of technical points that should be mentioned. The first is that as a consequence of (I.b.9) there are integral manifolds N of (I,ω) along which $\kappa_1 \equiv \cdots = \kappa_j \equiv 0$ but $\kappa_{j+1} \neq 0$. Then N projects to a line γ in $\mathbb{E}^n$ and is therefore a "strange" Frenet frame for γ; in particular, the functional (III.b.6) is *not* an integral of the curvatures of γ. Moreover, although the domain of Φ includes all curves $\gamma \subset \mathbb{E}^n$ for which $\kappa_1 \cdots \kappa_{n-1} \neq 0$ it is not clear that it includes all C^∞ curves. In other words, minimizing (III.b.6) over integral manifolds N of (I,ω) may not be the same as minimizing the functional over all smooth curves in $\mathbb{E}^n$. For the particular Lagrangians $L(\kappa_1,\ldots,\kappa_{n-1})$ that we shall be concerned with, it will be a consequence of the particular form of the Euler-Lagrange equations that this difficulty does not arise.

As noted in remark (III.a.29) there is a natural $\mathrm{Ad}\, O(n)$-invariant decomposition $e(n) = \mathbb{R}^n \oplus \mathfrak{so}(n)$ of the Lie algebra of $E(n)$, and we write the $\mathrm{Ad}^*\, O(n)$-invariant decomposition of $e(n)^*$ as (cf. (III.a.30))

$$e(n)^* = \operatorname{span}\{e_i^*\} \oplus \operatorname{span}\{e_i^{j*}\} \quad . \tag{III.b.8}$$

Corresponding to this decomposition it will be convenient to write $\lambda \in e(n)^*$ as

$$\begin{cases} \lambda = \mu \oplus \zeta & \text{where} \\ \mu = (\mu_i) \in \mathbb{R}^{n*} & \text{and} \quad \zeta = (\|\zeta_i^j\| \ \text{ with } \ \zeta_i^j + \zeta_j^i = 0) \in \mathfrak{so}(n)^*. \end{cases} \tag{III.b.9}$$

As a consequence of theorem (III.a.25) we have the

(III.b.10) THEOREM. *The projection* $\mu(t)$ *to* $\mathbb{R}^{n*}$ *of any solution to the Euler-Lagrange equations associated to (III.b.6) satisfies*

$$\begin{cases} \|\mu(t)\| = \text{constant}, & \textit{and} \\ \dfrac{d\mu}{ds} = \kappa \cdot \mu \end{cases}$$

where

$$\kappa = \begin{pmatrix} 0 & \kappa_1 & & & \\ -\kappa_1 & 0 & & & \\ & & \ddots & & \\ & & & 0 & \kappa_{n-1} \\ & & & -\kappa_{n-1} & 0 \end{pmatrix} .$$

Proof. This is an immediate consequence of (III.a.25) and (III.a.30). However, for use in the proof of theorem (II.a.32) it is worthwhile to give the computation explicitly. For this we use the notation (III.b.9) and follow the prescription in Chapter I, Section e) for computing the Euler-Lagrange differential system. Thus the 1-form $\psi = \varphi + \lambda_\alpha \theta^\alpha$ is in this case

$$\psi = \ell(q;r_j)dt + \mu_i \omega^i + \frac{1}{2}\zeta^i_j \omega^j_i - \mu_i q dt - \zeta^j_{j+1} r_j dt$$

where $r_n = 0$ and the summation is over all indices and not just those with $i < j$ (whence the factor of 1/2). For $\Psi = d\psi$ we have

$$\Psi = \ell_q dq \wedge dt + \ell_{r_j} dr_j \wedge dt - \mu_i dq \wedge dt - \zeta^j_{j+1} dr_j \wedge dt$$
$$+ (\text{terms not involving } dq, dr_j) .$$

The Cartan system $C(\Psi)$ *contains* the original differential system (III.b.1) plus the following Pfaffian equations

$$\begin{cases} \partial/\partial q \;\lrcorner\; \Psi = (\ell_q - \mu_i)dt = 0 \\ \partial/\partial r_j \;\lrcorner\; \Psi = (\ell_{r_j} - \zeta^j_{j+1})dt = 0 \end{cases} .$$

It follows that $Z_1 \subset Z$ is defined by

$$\begin{cases} \ell_q = \mu_i \\ \ell_{r_j} = \zeta^j_{j+1} \end{cases} \tag{III.b.11}$$

Next, we use the structure equations (0.c.3) to infer that as a bilinear form on the tangent spaces $T_p(Z)$ *at points* $p \in Z_1$

$$\Psi = d\mu_i \wedge \omega^i - d\mu_i \wedge q dt + \frac{1}{2} d\zeta^i_j \wedge \omega^j_i - d\zeta^j_{j+1} \wedge r_j dt$$
$$+ \mu_j \omega^i \wedge \omega^j_i + \frac{1}{2}\zeta^i_j \omega^k_i \wedge \omega^j_k \quad . \tag{III.b.12}$$

Thus, using (III.b.1)

$$\partial/\partial\omega^i \lrcorner \Psi = -d\mu_i + \mu_j\omega^j_i$$

$$\equiv -d\mu_i + (\mu_{i+1}r_i - \mu_{i-1}r_{i-1})dt \mod\{\theta^i\}. \qquad \text{(III.b.13)}$$

Using $ds = q\,dt$ and $\kappa_j = r_j/q$ this gives for the $\mathbb{R}^{n*}$ part of the Euler equations that

$$\frac{d\mu_i}{ds} = \mu_{i+1}\kappa_i - \mu_{i-1}\kappa_{i-1} \ ,$$

which implies theorem (III.b.10). Q.E.D.

What we have done so far is valid for any functional (III.b.6). Moreover, we have only used the "easy" or $\mathbb{R}^{n*}$ part of the Euler equations. To go further we must use all the Euler equations, and for this we first write (III.b.13) as

$$d\mu \equiv R\mu\, dt \qquad \text{(III.b.14)}$$

where

$$R = \begin{pmatrix} 0 & r_1 & & & \\ -r_1 & 0 & & & \\ & & \ddots & & \\ & & & 0 & r_{n-1} \\ & & & -r_{n-1} & 0 \end{pmatrix} \qquad \text{(III.b.15)}$$

and where "$\equiv$" means congruent modulo $\{\theta^i\}$. Next, for δ^i_j the usual Kronecker symbol we define the skew symmetric matrix $\mu \wedge \delta$ by

$$(\mu \wedge \delta)^i_j = \mu_j\delta^i_1 - \mu_i\delta^j_1 \ .$$

Considering $\partial/\partial\omega^i_j \lrcorner \Psi$ in (III.b.12) gives

$$d\zeta \equiv ([\zeta,R] + \mu \wedge \delta)dt \qquad \text{(III.b.16)}$$

where $[\zeta,R]$ is the usual matrix commutator. This equation is the $\mathfrak{so}(n)^*$ part of Theorem (III.a.25); the special form (III.b.15) of R is the "$V(\lambda)$ part" of that result. Using (III.b.14) and (III.b.16) we shall prove the

(III.b.17) THEOREM. *For the variational problem (III.b.6) with Lagrangian* $L = L(\kappa_1)$ *depending only on the* 1st *curvature, on any solution to the Euler equations we have*

$$r_3 \equiv \cdot\cdot \equiv r_{n-1} \equiv 0 \ .$$

Proof. Here we shall only establish the result under the additional assumption:

if some $\kappa_j \equiv 0$, then $\kappa_{j+1} \equiv \cdot\cdot \equiv \kappa_{n-1} \equiv 0$.

This will be sufficient to prove theorem (II.a.32) and will avoid the necessity of considering strange integral manifolds of (I,ω).[8] It is worth noting that if $r_{j+1} \equiv \cdot\cdot \equiv r_{n-1} \equiv 0$ then equations (III.b.14) and (III.b.16) reduce to the same equations for $\mathbb{E}^j$. Thus we may assume that $r_{n-1} \not\equiv 0, r_{n-2} \not\equiv 0, .., r_1 \not\equiv 0$ and shall arrive at a contradiction unless $n = 3$. By restricting to a neighborhood we may assume that

$$r_{n-2} \cdots r_1 \neq 0 \quad . \tag{III.b.18}$$

From (III.b.11) we have

$$\zeta_1^2 \neq 0, \quad \zeta_2^3 = \cdot\cdot = \zeta_{n-1}^n = 0 \quad . \tag{III.b.19}$$

This analogue of having a "cyclic coordinate" in a classical variational is the ultimate reason for theorem (III.b.17). When written out and noting that $(\mu \wedge \delta)^i_j = 0$ unless $i = 1$ or $j = 1$, (III.b.16) is *for* $2 \leqq j < i$

$$d\zeta^i_j \equiv (\zeta^i_{j+1} r_j - \zeta^i_{j-1} r_{j-1} - \zeta^{i-1}_j r_{i-1} + \zeta^{i+1}_j r_i)dt \tag{III.b.20}$$

If $n \geqq 3$ then by (III.b.19)

$$0 = d\zeta^n_{n-1} \equiv -\zeta^n_{n-2} r_{n-2} dt \quad ,$$

and since $r_{n-2} \not\equiv 0$ we must have

$$\zeta^n_{n-2} = 0 \quad . \tag{III.b.21}$$

If $n \geqq 4$ then by (III.b.19)-(III.b.21)

$$\begin{cases} 0 = d\zeta^{n-1}_{n-2} \equiv -\zeta^{n-1}_{n-3} r_{n-3} dt \\ 0 = d\zeta^n_{n-2} \equiv -\zeta^n_{n-3} r_{n-3} dt \ , \end{cases}$$

and these imply that

$$\zeta^{n-1}_{n-3} = \zeta^n_{n-3} = 0 \quad . \tag{III.b.22}$$

If $n \geqq 5$ then from (III.b.19)-(III.b.22)

$$\begin{cases} 0 = d\zeta_{n-3}^{n-2} \equiv -\zeta_{n-4}^{n-2} r_{n-4} dt \\ 0 = d\zeta_{n-3}^{n-1} = -\zeta_{n-4}^{n-1} r_{n-4} dt \\ 0 = d\zeta_{n-3}^{n} = -\zeta_{n-4}^{n} r_{n-4} dt \end{cases} .$$

Continuing in this way we conclude:

<u>For $j < i$</u>

$$\zeta_j^i = 0 \quad \text{except that} \quad \zeta_1^2 \neq 0 . \qquad \text{(III.b.23)}$$

We shall show that (III.b.23) leads to a contradiction if $n \geqq 4$.

For this we use the remaining case of the Euler equations (III.b.16)

$$d\zeta_1^i \equiv (\zeta_2^i r_1 - \zeta_1^{i-1} r_{i-1} + \zeta_1^{i+1} r_i + \mu_i) dt .$$

For $i \geqq 3$ (III.b.23) gives

$$0 = d\zeta_1^i \equiv (-\zeta_1^{i-1} r_{i-1} \delta_3^i + \mu_i) dt . \qquad \text{(III.b.24)}$$

In particular, taking $i \geqq 4$ we have

$$\mu_4 = \cdots = \mu_n = 0 .$$

Taking $i = 4$ in $d\mu_i \equiv (\mu_{i+1} r_i - \mu_{i-1} r_{i-1}) dt$ (cf. (III.b.14)) gives

$$\mu_3 r_3 = 0 \Rightarrow \mu_3 = 0 .$$

Repeating the argument gives

$$\mu_2 = \mu_1 = 0 .$$

Finally, taking $i = 3$ in (III.b.24) we obtain

$$0 = d\zeta_1^3 \equiv (-\zeta_1^2 r_2) dt$$

which implies that

$$r_2 = 0 .$$

This is a contradiction arising from the assumption $r_{n-2} \neq 0$, $n \geqq 4$.

Q.E.D.

(III.b.25) Remarks. If $n = 3$ then (III.b.23) becomes

$$\zeta_1^3 = 0 .$$

The relation $d\zeta_1^3 = 0$ may then be used to integrate the Euler equations associated to $\frac{1}{2}\int_\gamma \kappa^2\, ds$ for $\gamma \subset \mathbb{E}^3$, as was done in Chapter II, Section b). In any case, if we combine the discussion in iii) of Chapter II, Section b) with Theorem (III.b.17) then we have completed the proof of Theorem (II.a.32) in the case $R = 0$. In fact, much more has been proved as (II.b.30) gives explicitly the phase portrait of the Euler equations in this case.

As a generalization of Theorem (III.b.17), it seems likely that if $L = L(\kappa_1,\ldots,\kappa_{j-1})$ then any solution curve to the Euler-Lagrange equations lies in an $\mathbb{E}^{2j-1}$. When $j = 1$ this is just the property that geodesics are straight lines.

iii) We shall now use the method of Euler equations and coadjoint orbits to determine the solutions to the Euler-Lagrange equations associated to the functional

$$\frac{1}{2}\int_\gamma \kappa^2\, ds , \qquad \gamma \subset S^n .$$

As for curves in $\mathbb{E}^n$, a general curve $\gamma \subset S^n$ has an essentially unique Frenet frame with curvatures $\tau_1(s),\ldots,\tau_{n-1}(s)$ that determine the curve up to a rigid motion. We shall prove the

(III.b.27) THEOREM. *The Euler equations associated to (III.b.26) are quasi-integrable by quadratures. On any solution curve* γ *we have* $\tau_3 \equiv \cdots \equiv \tau_{n-1} \equiv 0$, *and therefore* γ *lies in an* S^3. *Setting* $\kappa = \tau_1$ *and* $\sigma = \tau_2$ *the "phase portrait" of* γ *is the elliptic curve*

$$\begin{cases} \sigma\kappa^2 = c_1 \\ (\kappa')^2 = \dfrac{\kappa^4}{4} - \kappa^2 - \dfrac{c_1}{\kappa^2} + c_2 \end{cases} .$$

Remarks. i) The rational function $p(x) = \frac{x^4}{4} - x^2 - \frac{c_1}{x^2} + c_2$ has degree six with a double zero at $x = \infty$. Therefore, for general constants c_1, c_2 the algebraic curve $y^2 = p(x)$ is branched at four points over the x-sphere and consequently has genus one (cf. [36], [59]).

ii) Once we establish theorem (III.b.27) the proof of theorem (II.a.32) will be complete.

Using the range of indices $0 \leqq i,j \leqq n$, the structure equations for the manifold $F_0(\mathbb{R}^{n+1}) \cong O(n+1)$ of frames $(e_0,e_1,\ldots,e_n)$ at the origin in $\mathbb{R}^{n+1}$ are

$$\begin{cases} de_i = \omega_i^j e_j \ , & \omega_i^j + \omega_j^i = 0 \\ d\omega_i^j = \omega_i^k \wedge \omega_k^j \ . \end{cases} \tag{III.b.28}$$

On $X = O(n+1) \times \mathbb{R}^{n+1}$ where $\mathbb{R}^{n+1}$ has coordinates $(t,q_0,\ldots,q_{n-1}) = (t;q_i)$ we consider the Pfaffian system (I,ω) given by

$$\begin{cases} \theta_i^{i+1} = \omega_i^{i+1} - q_i dt = 0 & i = 0,\ldots,n-1 \\ \theta_i^{i+k} = \omega_i^{i+k} = 0 & k \geqq 2 \\ \omega = dt \neq 0 \ . \end{cases} \tag{III.b.29}$$

Via the map $F_0(\mathbb{R}^{n+1}) \to S^n$ given by

$$(e_0,e_1,\ldots,e_n) \to e_0$$

the integral manifolds N of (I,ω) project to curves $\gamma \subset S^n$ on which the arclength is $ds = q_0 dt$. If $q_1 \cdots q_{n-1} \neq 0$, then N is a Frenet lifting of γ with curvatures given by

$$\tau_1 = q_1/q_0, \ldots, \tau_{n-1} = q_{n-1}/q_0 \ .$$

The condition that γ lie in an Euclidean subsphere $S^k \subset S^n$ is

$$\tau_k \equiv 0 \ . \tag{III.b.30}$$

As before we consider a variational problem $(I,\omega;\varphi)$ where

$$\varphi = \ell(q_0,\ldots,q_{n-1})dt = L(\tau_1,\ldots\tau_{n-1})ds \ . \tag{III.b.31}$$

For example, (III.b.26) is given by

$$\ell(q) = \frac{(q_1)^2}{2q_0} \ .$$

In $\mathbb{R}^{n(n+1)/2}$ we use coordinates $\lambda = (\lambda_j^i)$ where $\lambda_j^i + \lambda_i^j = 0$, and following the algorithm in Chapter 1, Section e) we set

$$\psi = \ell(q)dt + \frac{1}{2}(\lambda_j^i \theta_i^j)$$

on $X \times \mathbb{R}^{n(n+1)/2}$, the summation being over all indices (whence the 1/2). By (III.b.28) and (III.b.29), $\Psi = d\psi$ is given by

$$\Psi = (\ell_{q_i} - \lambda^i_{i+1})dq_i \wedge dt + \frac{1}{2} d\lambda^i_j \wedge \theta^j_i + \frac{1}{2} \lambda^i_j \omega^k_i \wedge \omega^j_k , \qquad \text{(III.b.32)}$$

and then the Cartan system is generated by the Pfaffian equations

$$\begin{cases} \text{(i)} & \partial/\partial\lambda^i_j \lrcorner \Psi = \theta^j_i = 0 \\ \text{(ii)} & \partial/\partial q_i \lrcorner \Psi = (\ell_{q_i} - \lambda^i_{i+1})dt = 0 \\ \text{(iii)} & \partial/\partial\omega^j_i \lrcorner \Psi = -d\lambda^i_j + [\omega,\lambda]^i_j = 0 \end{cases} \qquad \text{(III.b.33)}$$

where $[\omega,\lambda]$ is the matrix commutator. Without indices, (iii) is the Euler equation

$$d\lambda - [\omega,\lambda] = 0 .$$

Letting "$\equiv$" denote congruence modulo $\{\theta^i\}$, the Euler equations are

$$\begin{cases} \text{(i)} & d\lambda^i_{i+1} \equiv (q_{i+1}\lambda^i_{i+2} - q_{i-1}\lambda^{i-1}_{i+1})dt \\ \text{(ii)} & d\lambda^i_j \equiv (q_j\lambda^i_{j+1} - q_{i-1}\lambda^{i-1}_j + q_i\lambda^{i+1}_j - q_{j-1}\lambda^i_{j-1})dt, \quad j \geqq i+2 . \end{cases} \qquad \text{(III.b.34)}$$

Note that (ii) reduces to (i) when $j = i+1$ if we use $\lambda^{i+1}_{i+1} = \lambda^i_i = 0$.

For a Lagrangian $L = L(\tau_1)$ not depending on $\tau_2,\dots,\tau_{n-1}$ we infer from (III.b.31) and (ii) in (III.b.33) that

$$\lambda^2_3 = \dots = \lambda^{n-1}_n = 0 . \qquad \text{(III.b.35)}$$

If we assume that γ does not lie in an $S^k \subset S^n$ for any $k < n$, then as in the proof of Theorem (III.b.17) we may assume that

$$q_0 \cdot q_1 \cdots q_{n-1} \neq 0 , \qquad \text{(III.b.36)}$$

The argument now breaks into several steps.

Step One: We will show that

$$\lambda^i_{i+2} = 0 \quad \text{for} \quad i \geqq 1 . \qquad \text{(III.b.37)}$$

Proof. Equation (i) in (III.b.34) together with (III.b.35), (III.b.36) give

$$0 = d\lambda^{n-1}_n \equiv -q_{n-1}\lambda^{n-2}_n dt$$

$$\Rightarrow \lambda^{n-2}_n = 0 .$$

Repeating the calculation when $i = n-2$ and using what we just proved gives

$$0 = d\lambda^{n-2}_{n-1} \equiv (q_{n-1}\lambda^{n-2}_{n} - q_{n-2}\lambda^{n-3}_{n-1})dt$$

$$\equiv - q_{n-2}\lambda^{n-3}_{n-1}\ dt$$

$$\Rightarrow \lambda^{n-3}_{n-1} = 0 .$$

The inductive statement is:

$$\lambda^{j}_{j+1} = 0 \quad \text{for} \quad j \geqq i \Rightarrow \lambda^{k}_{k+2} = 0 \quad \text{for} \quad k \geqq i-1 .$$

Using (III.b.35) this establishes (III.b.37).

Step Two: We next show that

$$\lambda^{i}_{i+3} = 0 \qquad \text{for} \quad i \geqq 1 . \qquad \text{(III.b.38)}$$

Proof. Equations (ii) in (III.b.34) together with (III.b.35)-(III.b.37) give

$$0 = d\lambda^{i}_{i+2} \equiv (q_{i+2}\lambda^{i}_{i+3} - q_{i-1}\lambda^{i-1}_{i+2})dt, \quad i \geqq 2 .$$

As in the proof of (III.b.37), beginning with $i = n-2$ and working downward we obtain

$$0 = d\lambda^{i}_{i+2} \equiv -q_{i-1}\lambda^{i-1}_{i+2}dt \quad , \qquad i \geqq 2 ,$$

which implies (III.b.38).

Step Three: Now we show that

$$\lambda^{i}_{i+4} = 0 \qquad \text{for} \quad i \geqq 0 . \qquad \text{(III.b.39)}$$

Proof. Equations (ii) in (III.b.34) together with (III.b.35)-(III.b.38) give

$$0 = d\lambda^{i}_{i+3} \equiv (q_{i+3}\lambda^{i}_{i+4} - q_{i-1}\lambda^{i-1}_{i+3})dt, \quad i \geqq 1 .$$

As before this gives inductively

$$0 = d\lambda^{i}_{i+3} \equiv -q_{i-1}\lambda^{i-1}_{i+3}dt \quad , \qquad i \geqq 1 ,$$

and this implies (III.b.39).

Continuing in this manner we arrive at

$$\lambda^{i}_{i+j} = 0 \qquad \text{for} \quad i \geqq 0,\ j \geqq 4 . \qquad \text{(III.b.40)}$$

Taking $L = \frac{1}{2}(\tau_1)^2$ and using (III.b.31), (ii) in (III.b.33), (III.b.35), and (III.b.37)-(III.b.40) we infer that the only non-zero λ^i_j $(i<j)$ are

$$\begin{cases} \lambda^0_1 = \ell_{q_0} = -\frac{1}{2}(q_1/q_0)^2 = -\kappa^2/2 \\ \lambda^1_2 = \ell_{q_1} = q_1/q_0 = \kappa \\ \lambda^0_2 = ? \\ \lambda^0_3 = ? \end{cases} \tag{III.b.41}$$

where we set

$$\kappa = \tau_1, \quad \sigma = \tau_2 .$$

By (ii) in (III.b.34)

$$0 = d\lambda^1_3 \equiv (-q_0\lambda^0_3 - q_2\lambda^1_2)\ dt \quad ,$$

which by (III.b.41) gives

$$\lambda^0_3 = -\sigma\kappa \quad . \tag{III.b.42}$$

Similarly, by (i) in (III.b.34) and (III.b.37)

$$d\lambda^1_2 \equiv -q_0\lambda^0_2 dt \quad .$$

Letting ' denote the derivative with respect to arclength $ds = q_0 dt$ we obtain from (III.b.41) that

$$\lambda^0_2 = -\kappa' \quad . \tag{III.b.43}$$

At this juncture all the λ^i_j have been determined.

By (ii) in (III.b.34)

$$d\lambda^0_3 \equiv -q_2\lambda^0_2 dt \quad ,$$

which by (III.b.42), (III.b.43) yields

$$\begin{aligned} & (-\sigma\kappa)' = \sigma\kappa' \\ \Rightarrow\quad & 2\sigma\kappa' = -\sigma'\kappa \\ \Rightarrow\quad & (\sigma\kappa^2)' = 0 \quad . \end{aligned}$$

This gives the 1st integral

$$V_1 = \sigma\kappa^2 = c_1 \tag{III.b.44}$$

on solution curves to Euler's equations.

By (ii) in (III.b.34)

$$d\lambda_2^0 \equiv (q_2\lambda_3^0 + q_0\lambda_2^1 - q_1\lambda_1^0)dt \quad .$$

Using (III.b.41)-(III.b.44) we obtain

$$-\kappa'' = -\sigma^2\kappa + \kappa - \frac{\kappa^3}{2}$$

$$\Rightarrow \kappa'' = \frac{\kappa^3}{2} - \kappa + \frac{c_1}{\kappa^3} \quad ,$$

and we find another 1st integral

$$V_2 = (\kappa')^2 - \frac{\kappa^4}{4} + \kappa^2 + \frac{c_1}{\kappa^2} = c_2 \quad . \qquad \text{(III.b.45)}$$

To complete the proof of theorem (III.b.27) we shall show that

$$q_3 \equiv 0 \quad . \qquad \text{(III.b.46)}$$

This implies that $\tau_3 \equiv \cdots \equiv \tau_{n-1} \equiv 0$, which by (III.b.30) and (III.b.44), (III.b.45) finishes the argument for the theorem.

By (III.b.39), (ii) in (III.b.34), and (III.b.42)

$$0 = d\lambda_4^0 \equiv -q_3\lambda_3^0 dt$$

$$\equiv q_3\sigma\kappa dt$$

$$\Rightarrow q_3 \equiv 0$$

since if $\sigma\kappa \equiv 0$ then either $\sigma \equiv 0$ or $\kappa \equiv 0$, and either of these gives $q_3 \equiv 0$. Q.E.D.

It seems that the proofs of (III.b.17) and (III.b.27) establish the point that theorem (III.a.25) is useful in practice as well as being of theoretical interest.

iii) For a curve $\gamma \subset \mathbb{E}^n$ given by $t \to x(t)$, any 1st order functional

$$\Phi(\gamma) = \int L(t, x(t), x'(t))\, dt$$

invariant under time shift and rigid motions is of the form

$$\Phi(\gamma) = \int_\gamma L(\|x'(t)\|)\, dt \quad .$$

If $L(r)$ is homogeneous then the solution curves to the associated Euler-Lagrange equations are just the geodesics (= straight lines; cf.

example (II.a.23)). The same result is true in any constant curvature space.

However, it is also interesting to consider functionals defined on curves lying in other homogeneous spaces. For example we may consider the Grassmannian $G(k,n)$ of k-planes Λ through the origin in $\mathbb{R}^n$. It is well-known that there is a canonical identification

$$T_\Lambda(G(k,n)) \cong \mathrm{Hom}(\Lambda,\Lambda^\perp) \ , \tag{III.b.47}$$

and that the isotropy subgroup at $\Lambda_0 = \mathbb{R}^k \subset \mathbb{R}^n$ of $O(n)$ acting on $G(k,n)$ is $O(n-k)\times O(k)$ acting in $T_{\Lambda_0}(G(k,n)) \cong \mathrm{Hom}(\mathbb{R}^k,\mathbb{R}^{n-k})$ by

$$(A\times B)\cdot\xi \ = \ A\circ\xi\circ B^{-1} \tag{III.b.48}$$

where $A\in O(n-k)$, $B\in O(k)$, and $\xi\in \mathrm{Hom}(\mathbb{R}^k,\mathbb{R}^{n-k})$.

<u>Proof</u>. We represent points of $G(k,n)$ by their *Plücker coordinates*

$$p(\Lambda) \ = \ e_1\wedge\cdots\wedge e_k \in \Lambda^k\mathbb{R}^n$$

where $e_1,\ldots,e_k$ is an orthonormal basis for Λ (thus $p(\Lambda)$ is a unit vector defined up to ± 1; i.e., $p(\Lambda)\in \mathbb{RP}^{\binom{n}{k}}$). For a curve $\Lambda(t) = \mathrm{span}\{e_1(t),\ldots,e_k(t)\}$ we define $\xi e_i \in \Lambda^\perp$ by

$$(\xi e_i, v) \ = \ \left(\frac{de_i}{dt}(0), v\right), \qquad v\in\Lambda^\perp. \tag{III.b.49}$$

Since $\left(\frac{de_i}{dt}(0), e_i\right) = 0$ it follows that

$$\left.\frac{dp(\Lambda(t))}{dt}\right|_{t=0} \ = \ \sum_\alpha (-1)^{\alpha-1} e_1\wedge\cdots\wedge\xi e_\alpha\wedge\cdots\wedge e_n \ ,$$

and therefore the identification (III.b.47) is given by the map (here we set $\Lambda'(t) = dp(\Lambda(t))/dt$)

$$\Lambda'(0) \to \xi$$

where $\xi\in\mathrm{Hom}(\Lambda,\Lambda^\perp)$ is defined by (III.b.49).

In case $n = 2m$ and $k = m$ we are considering the "middle" Grassmannian $G(m,2m)$, and a tangent vector is a linear map

$$\xi\colon \mathbb{R}^m \to \mathbb{R}^m$$

(keep in mind that these are "different $\mathbb{R}^m$'s"; i.e., ξ is determined only up to a transformation (III.b.48)). To determine the invariants of

ξ under the action of the isotropy group we consider the image E of the unit sphere in $\mathbb{R}^m$. Then E is an ellipsoid (possibly degenerate if ξ is singular), and by choosing the principal axes of E as part of a basis for the image $\mathbb{R}^m$ we may put ξ in the normal form

$$\xi = \begin{pmatrix} \xi_1 & & 0 \\ & \ddots & \\ 0 & & \xi_m \end{pmatrix} \qquad \text{(III.b.50)}$$

Let $L(\xi) = L(\xi_1,\ldots,\xi_m)$ be any function of $\xi_1,\ldots,\xi_m$ invariant under permutations and sign changes $\xi_i \to -\xi_i$. Then, using the identification (III.b.47) and denoting by $\Lambda'(t) \in \mathrm{Hom}(\Lambda(t),\Lambda(t)^{\perp})$ the tangent vector to a curve $\gamma = \{\Lambda(t)\} \subset G(m,2m)$, the functional

$$\Phi(\gamma) = \int L(\Lambda'(t))\, dt$$

gives a well-defined, invariant 1st order variational problem on $G(m,2m)$. For example, when

$$L(\xi) = \sqrt{\xi_1^2 + \cdots + \xi_m^2}$$

we have arclength. At the other extreme we consider the functional

$$\Phi(\gamma) = \int |\det\|\Lambda'(t)\||\, dt \qquad \text{(III.b.51)}$$

corresponding to $L(\xi) = |\xi_1 \cdots \xi_m|$. We shall say that γ is *non-degenerate* in case $\det\|\Lambda'(t)\| \neq 0$.

<u>Special Case</u>: When $m = 3$ we may picture γ as a *ruled surface* in $\mathbb{RP}^3$

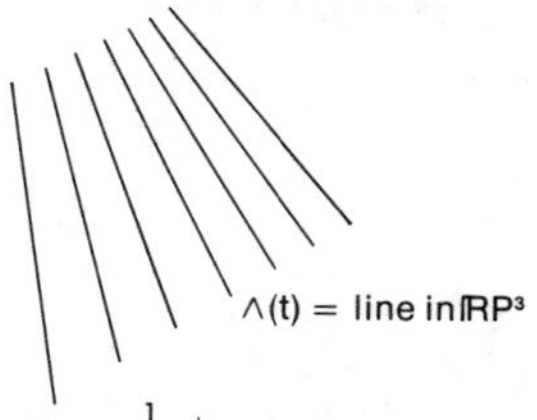

and then

$$|\det\|\Lambda'(t)\|| = \lim_{h\to 0} \frac{1}{h} |\text{distance from } \Lambda(t+h) \text{ to } \Lambda(t)| .$$

For example, if we imagine the motion of a rod *but where no work is done when the rod pivots about a fixed point*, then $\frac{1}{2}(\det\|\Lambda'(t)\|)^2$ is a form of *kinetic energy*. We shall prove the following

(III.b.52) THEOREM. *The solution curves to the Euler-Lagrange equations corresponding to the functional (III.b.51) in the case* $m = 2$ *are given by the orbits in* $G(2,4)$ *of* 1-*parameter subgroups of* $O(4)$.

It is well-known that the geodesics in any Riemannian symmetric space are the orbits of 1-parameter subgroups (cf. [53]), and so the above result suggests the following

(III.b.53) Question. Are the solution curves to the Euler-Lagrange equations associated to *any* homogeneous invariant 1st order functional defined on a symmetric space always given by geodesics?

Of course, we must assume that the competing curves are non-degenerate in the sense that the functional is non-zero.

Proof of Theorem (III.b.52). We begin by noting that non-degenerate curves in $G(2,4)$ have Frenet liftings to the frame manifold $F_0(\mathbb{R}^4) \cong O(4)$ whose Maurer-Cartan matrix looks like (cf. [34], [35], [44])

$$\begin{pmatrix} 0 & \omega_1^2 & \omega_1^3 & 0 \\ -\omega_1^2 & 0 & 0 & \omega_2^4 \\ -\omega_1^3 & 0 & 0 & \omega_3^4 \\ 0 & -\omega_2^4 & -\omega_3^4 & 0 \end{pmatrix} . \qquad \text{(III.b.54)}$$

In other words, the Frenet liftings are given by integral manifolds of the invariant differential system

$$\omega_1^4 = \omega_2^3 = 0$$

on $O(4)$, which are just the conditions expressing that the frames for $\Lambda(t)$ and $\Lambda(t)^{\perp}$ have been chosen so as to diagonalize the differential $\Lambda'(t) \in \mathrm{Hom}(\Lambda(t), \Lambda(t)^{\perp})$.

On $X = O(4) \times \mathfrak{so}(4) \times \mathbb{R}$ we consider the differential system

$$\begin{cases} \theta_i^j = \omega_i^j - q_i^j dt = 0 \\ q_1^4 = q_2^3 = 0 \\ \omega = dt \neq 0 \end{cases} . \qquad \text{(III.b.55)}$$

The functional (III.b.51) corresponds to the variational problem $(I, \omega; \varphi)$ where

$$\varphi = L dt$$

$$L = \frac{|q_1^3 q_2^4|}{\ell} \tag{III.b.56}$$

$$\ell = \sqrt{(q_1^3)^2 + (q_2^4)^2} = ds/dt \ .$$

Following the algorithm in Chapter I, Section e) we consider

$$\psi = L dt + \frac{1}{2} \lambda_j^i \theta_i^j$$

on $Z = X \times \mathbb{R}^6$ where $\mathbb{R}^6$ is the space of skew-symmetric matrices $\lambda = \|\lambda_j^i\|$. Then $\Psi = d\psi$ is given by

$$\Psi = (L_{q_i^j} - \lambda_j^i) dq_i^j \wedge dt + \frac{1}{2} d\lambda_j^i \wedge \theta_i^j + \lambda_j^i \omega_i^k \wedge \omega_k^j \ , \tag{III.b.57}$$

where it is understood that we must formally set $q_1^4 = q_2^3 = dq_1^4 = dq_2^3 = 0$. The Cartan system is

$$\text{(i)} \quad \partial/\partial q_i^j \lrcorner \Psi = (L_{q_i^j} - \lambda_j^i) dt = 0 \qquad i < j \text{ and } (i,j) \neq (1,4) \text{ or } (2,3)$$

$$\text{(ii)} \quad \partial/\partial \lambda_i^j \lrcorner \Psi = \theta_j^i = 0 \tag{III.b.58}$$

$$\text{(iii)} \quad -d\lambda + [\omega, \lambda] = 0 \ .$$

Equations (iii) are the Euler equations.

From (III.b.56) and (i) we have

$$\lambda_2^1 = \lambda_4^3 = 0 \ . \tag{III.b.59}$$

Using "$\equiv$" to denote congruence modulo span $\{\theta_i^j\}$, we obtain from (III.b.55) and the Euler equations (iii) in (III.b.58) that

$$0 = d\lambda_2^1 \equiv (\lambda_4^1 q_2^4 - \lambda_3^2 q_1^3) dt \ .$$

This gives

$$\lambda_4^1 q_2^4 = \lambda_3^2 q_1^3 \ . \tag{III.b.60}$$

Similarly, from $d\lambda_4^3 = 0$ we obtain

$$\lambda_4^1 q_1^3 = \lambda_3^2 q_2^4 \ . \tag{III.b.61}$$

From (III.b.60), (III.b.61) we have

$$\lambda_3^2 (q_1^3)^2 = \lambda_3^2 (q_2^4)^2 \ .$$

We now examine cases.

Case (i). $q_1^3 \neq \pm q_3^4$. (Note: if $q_1^3 = -q_2^4$ then we may reverse e_4 to have $q_1^3 = q_2^4$. Curves in $G(2,4)$ satisfying

$$q_1^3 = q_2^4 \neq 0 \qquad \text{(III.b.62)}$$

may be called umbilics; they play a special role on the theory.) We then have

$$\lambda_3^2 = 0 = \lambda_4^1 , \qquad \text{(III.b.63)}$$

where the second equation follows also from (III.b.60), (III.b.61) and $q_1^3 \neq q_2^4$. From (iii) in (III.b.58) and (III.b.59) we have

$$0 = d\lambda_3^2 \equiv (\lambda_4^2 q_3^4 - \lambda_3^1 q_1^2)\,dt.$$

This together with the same equation for $0 = d\lambda_4^1$ give

$$\begin{cases} \lambda_4^2 q_3^4 = \lambda_3^1 q_1^2 \\ \lambda_4^2 q_1^2 = \lambda_3^1 q_3^4 \end{cases} . \qquad \text{(III.b.64)}$$

As before this gives

$$\begin{cases} \lambda_4^2 (q_3^4)^2 = \lambda_4^2 (q_1^2)^2 \\ \lambda_3^1 (q_3^4)^2 = \lambda_3^1 (q_1^2)^2 \end{cases} .$$

This leads to

Subcase (i). $q_1^2 \neq \pm q_3^4$ (again we may take $+$ in case of equality). Then

$$\lambda_4^2 = 0 = \lambda_3^1 , \qquad \text{(III.b.65)}$$

and (III.b.59), (III.b.63), (III.b.65) give

$$\lambda_j^i = 0 \qquad \text{for all } i,j . \qquad \text{(III.b.66)}$$

The solution to the Euler equations is thus $\lambda = 0$.

All of this is valid for any $L = L(q_1^3, q_2^4)$. In the case at hand, (III.b.56) and (i) in (III.b.58) give

$$\frac{q_2^4}{\ell} - \frac{(q_1^3)^2 q_2^4}{\ell^3} = 0$$

$$\frac{q_1^3}{\ell} - \frac{(q_2^4)^2 q_1^3}{\ell^3} = 0 \quad .$$

These equations imply that

$$q_1^3 = 0 = q_2^4 \quad , \tag{III.b.67}$$

and this is a contradiction. Hence Subcase (i) cannot hold and we must have

$$q_1^2 = q_3^4 \quad .$$

Then (III.b.64) gives

$$\lambda_3^1 = \lambda_4^2 \quad ; \tag{III.b.68}$$

note that all other $\lambda_j^i = 0$ by (III.b.59) and (III.b.63). By (iii) in (III.b.58) we have

$$d\lambda_3^1 = 0 = d\lambda_4^2 \ .$$

Equation (i) in (III.b.58) now gives

$$\begin{cases} \left(\dfrac{q_2^4}{\ell}\right)^3 = \lambda_4^2 = \text{constant} \\ \left(\dfrac{q_1^3}{\ell}\right)^3 = \lambda_3^1 = \text{constant} \quad . \end{cases}$$

If we normalize our parameter ℓ so that $\ell = 1$ this is

$$q_2^4 = c_2^4 \ , \qquad q_1^3 = c_1^3$$

where c_i^j are constants.

Finally, we may rotate the Frenet frames $e_1(t), e_2(t)$ and $e_3(t), e_4(t)$ through the *same* angle $\theta(t)$ with (cf. (III.b.67))

$$\frac{d\theta(t)}{dt} = -q_1^2 = -q_3^4$$

Then the curve $\Lambda(t) \subset G(2,4)$ remains the same with

$$q_1^2 = q_3^4 = 0 \quad .$$

The mapping $t \to g(t) \in O(4)$ thus has Maurer-Cartan matrix

$$\begin{pmatrix} 0 & 0 & c_1^3 & 0 \\ 0 & 0 & 0 & c_2^4 \\ -c_1^3 & 0 & 0 & 0 \\ 0 & -c_2^4 & 0 & 0 \end{pmatrix} dt \quad ,$$

and is therefore a 1-parameter subgroup.[9]

<u>Case (ii)</u>. $q_1^3 \equiv q_2^4$. This is handled by similar methods, and so we will not give the details.

FOOTNOTES TO CHAPTER III

(1) Our usual notation has been to write points in $T(G)$ as (g,v) where $g \in G$ and $v \in T_g(G)$. In this chapter we shall drop reference to g and use the isomorphisms $(L_g)_*: T_e(G) \tilde{\to} T_g(G)$ to identify $T(G)$ with $G \times \mathfrak{g}$. Finally, we shall use t (for time) rather than x to denote the independent variable.

(2) The left action of G on the variational problem gives, by Noether's theorem, $\dim G = n$ 1st integrals. In principle then this allows us to reduce the number of Euler-Lagrange equations from $2n$ to n, by fixing n constants. In practice this reduction is in some sense accomplished by Euler's equations; to understand the precise mechanism one needs the Marsden-Weinstein reduction (cf. [1]).

(3) Above we have used the isomorphism

$$\eta: \mathbb{R}^3 \to \mathfrak{g} \quad .$$

On the other hand, the bi-invariant metric (Cartan-Killing form) gives an isomorphism

$$\mathfrak{g} \tilde{\to} \mathfrak{g}^* \ .$$

Under the resulting isomorphism

$$\mathbb{R}^3 \tilde{\to} \mathfrak{g}^*$$

we see that

$$\begin{cases} (1,0,0) \to \tilde{\omega}^1 \\ (0,1,0) \to \tilde{\omega}^2 \\ (0,0,1) \to \tilde{\omega}^3 \end{cases} .$$

(4) The notation is slightly at variance with the previous chapters. The Pfaffian system is now

$$\theta^i = 0, \quad \omega \neq 0$$

on $G \times V \times \mathbb{R}$. Its integral manifolds correspond to curves in G whose tangent lines lie in the sub-bundle $\omega^\alpha = 0$ of $T(G)$.

(5) These may be verified directly; alternatively they follow from (iii), (iv) in (0.c.3) and the Maurer-Cartan equation.

(6) It may be useful to compare this setup with that in (i) of Chapter II, Section b). In general there is no unique way to set up a variational problem in the form $(I,\omega;\varphi)$. In the situation at hand, if one tries to do the Euler equations for $\omega^{\alpha}=0$, $\omega\neq 0$ without introducing a new variable t and writing the independence condition as $\omega - q\,dt = 0$ where $q\neq 0$, then a considerable mess ensues.

(7) It is not necessary that $\kappa_{\ell+1}\equiv .. \equiv\kappa_{n-1}\equiv 0$. For example, an integral manifold of (III.b.1) is given by a line in $\mathbb{E}^3$ along which the normal frame spins arbitrarily.

(8) Observe that if γ lies in an $\mathbb{E}^{\ell}\subset\mathbb{E}^n$, then we may arbitrarily spin the $(e_{\ell+1},..,e_n)$-part of the generalized Frenet frame without affecting $L(\kappa_1)$. What is really going on here is that the variational problem for $L(\kappa_1)$ *descends* to one on the Stiefel manifold of all "partial frames" $(x;e_1,e_2)$, but for computational purposes it is more convenient to work up on the group.

(9) It is a well-known consequence of (I.b.7) that the condition for a map

$$f: \mathbb{R}\to G$$

from the t-line to a Lie group G to be a (left translate of) a 1-parameter subgroup is that

$$f^*(\omega) = c\,dt \qquad (c = \textit{constant})$$

for any left-invariant 1-form ω on G. For example, the curves in $\mathbf{E}^3$ with constant curvature and torsion are exactly the orbits of 1-parameter subgroups of $E(3)$.

IV. ENDPOINT CONDITIONS; JACOBI EQUATIONS AND THE 2ND VARIATION; THE HAMILTON-JACOBI EQUATION

a) Endpoint Conditions.

We consider a variational problem $(I,\omega;\varphi)$ with one independent variable on a manifold X. In Chapter I, Section e) we canonically associated to $(I,\omega;\varphi)$ a manifold Y and Pfaffian system (J,ω) on Y which we *defined* to be the Euler-Lagrange system. However, (J,ω) was arrived at only by *heuristically* expressing the condition

$$\frac{d}{dt}\left(\int_{N_t} \varphi\right)_{t=0} = 0 \qquad \text{(IV.a.1)}$$

where N_t is a family of integral manifolds of (I,ω) with $N_0 = N$. It was *not* proved that N being a solution to (J,ω) implies (IV.a.1); indeed, this doesn't even make sense without specifying the endpoint conditions imposed on the variation N_t. Of course, there is no trouble when N is compact; in this case the Euler-Lagrange equations imply (IV.a.1). However, the endpoint conditions (cf. (I.d.3))

$$v_{\partial N} = 0$$

imposed on normal vector fields to N satisfying (I.b.15) in order to arrive at the Euler-Lagrange equations (I.d.14) are much too stringent for two reasons.

Before illustrating these, we remark that an analysis of the derivation of (I.d.14) shows that the weaker conditions

$$\langle\varphi, v\rangle = \langle\theta^{\alpha}, v\rangle = 0 \quad \text{on } \partial N, \qquad \text{(IV.a.2)}$$

rather than (I.d.3), are all that was really used. The precise statement is this: *If* $N \in V(I,\omega)$ *is a solution to the Euler-Lagrange equations (I.d.14), then (IV.a.1) holds for all families* N_t *whose infinitesimal variation satisfies (IV.a.2).*

But this is still not good enough.

(IV.a.3) Example. We consider the analysis in Chapter II, Section b) of the functional (II.b.19) for curves $\gamma \subset \mathbb{E}^3$. Using (II.b.18) we see that the conditions (IV.a.2) imposed on a variation $\gamma_t \subset \mathbb{E}^3$ are that the curves γ_t should all have a common Frenet frame at the end-points of γ (more precisely, this should hold up to 1st order in t). But clearly the correct conditions are that the γ_t should have the same endpoints A,B and tangent lines L_A, L_B; the osculating 2-planes need *not* coincide.

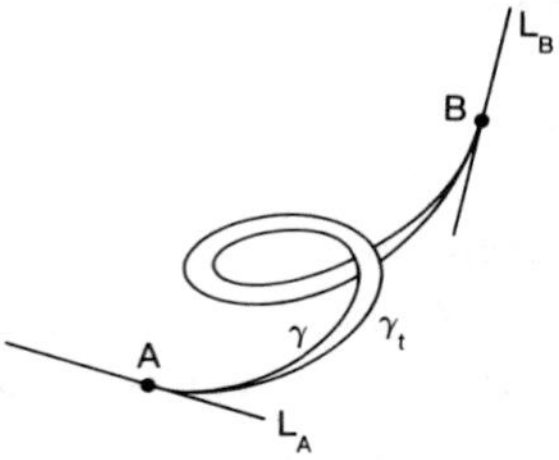

In this case it is clear that there are "a lot" of compactly supported infinitesimal variations $[v] \in T_N(V(I,\omega))$, so that the Euler-Lagrange equations (I.d.14) should in any case be a *necessary* condition that (IV.a.1) hold for variations satisfying the correct endpoint conditions, but it is by no means clear that they are sufficient.

(IV.a.4) Example. We now consider the Delauney problem given by the functional (II.b.31) on curves $\gamma \subset \mathbb{E}^3$ with the constant curvature constraint $\kappa \equiv 1$. Here the endpoint conditions (IV.a.2) are the same as those in the preceding example, and the same objections are also valid in this case (cf. the remark following (II.b.32)).

More seriously, due to the constraint $\kappa \equiv 1$ it is certainly doubtful that there exist "a lot" of compactly supported infinitesimal variations $[v] \in T_N(V(I_0,\omega))$ (we are using the notations from Chapter II, Section b)). Hence, it is not clear that the Euler-Lagrange equations (I.d.14) are even necessary for (IV.a.1); as beautiful a system as (II.b.34) (with the constraint $\kappa \equiv 1$) is, it may be irrelevant to the problem of minimizing (II.b.31).

As remarked several times, $(I,\omega;\varphi)$ may be a disguised version of a complicated constrained higher order problem. Finding the correct endpoint conditions and Euler-Lagrange equations appears in general to necessitate working through the entire derived flag (cf. Chapter I, Section c)). Hence, a better goal of the theory is to formulate a set of *sufficient* conditions on $(I,\omega;\varphi)$ implying that our Euler-Lagrange system gives the right answer. These conditions should be reasonably simple to state and, perhaps above all, should lend themselves to computation of interesting special cases. It is not expected that these conditions should cover all variational problems, but they should be valid in reasonable generality as measured by application to examples. Formulating and studying one such set of conditions is the goal of this section.

We consider a variational problem $(I,\omega;\varphi)$ on a manifold X with associated Euler-Lagrange differential system (J,ω) on the momentum manifold Y (cf. Chapter I, Section e) for this construction). As will now be explained, the point is to *formulate the endpoint conditions on* Y *instead of on* X. Without essential loss of generality we may assume that

$$\varphi = L\omega \tag{IV.a.5}$$

where L is a function on X; this has been the case in all our examples.

<u>Definition</u>. The variational problem $(I,\omega;\varphi)$ is said to be *well-posed* if the following two conditions are satisfied:

i) $(I,\omega;\varphi)$ is non-degenerate with

$$\dim Y = 2m+1$$

(cf. Chapter I, Section e); this implies in particular that the characteristic direction field $\Psi_Y^{\perp} \subset T(Y)$ gives a line sub-bundle); and

ii) we may determine a sub-bundle $U^* = \text{span}\{\theta^1,\ldots,\theta^m\}$ of $W^* = \text{span}\{\theta^1,\ldots,\theta^s\}$ such that *on* Y

$$K^* = \text{span}\{\omega,\theta^1,\ldots,\theta^m\} \subset T^*(Y) \tag{IV.a.6}$$

is a completely integrable sub-bundle of rank $m+1$ transverse to the line sub-bundle $\Psi_Y^{\perp} \subset T(Y)$, and where (most importantly)

$$\lambda_{m+1} = \cdots = \lambda_s = 0 \tag{IV.a.7}$$

on Y.

<u>Remarks</u>. i) Of course the second part of this definition may be phrased intrinsically by the requirements that K^* be a Frobenius system and that $Y \subset U^*$ (cf. remark (I.e.8)). We have given the definition in the above form for computational purposes.

ii) We note that, as a consequence of (IV.a.5),

$$\psi_Y \in K^* .$$

iii) It is crucial that the definition takes place up on Y.

Returning to the general discussion, the completely integrable Pfaffian system K^* defines a foliation of Y with m-dimensional leaves. In examples the quotient manifold of Y by this foliation always seems to exist, and so we simply assume this to be the case and denote the quotient by

$$Q = Y/K^* .$$

<u>Definition</u>. We call Q the *reduced momentum space*.

Note that

$$K^* = \tilde{\omega}^* T^*(Q)$$

where $\tilde{\omega}: Y \to Q$ is the projection; in particular, $\dim Q = m+1$. The cast of characters may be summarized by the diagram

$$\begin{array}{ccc} & Y \subset W^* \subset T^*(X) & \\ \tilde{\omega} \swarrow & & \searrow \pi \\ Q & & X \end{array} \quad . \qquad \text{(IV.a.8)}$$

<u>Definition</u>. Assume that the variational problem $(I,\omega;\varphi)$ is well-posed. Then the *endpoint conditions of admissable variations* are defined by

$$\omega = \theta^1 = \cdots = \theta^m = 0. \qquad \text{(IV.a.9)}$$

<u>Explanation</u>. Let $N = \{a \leqq s \leqq b\}$ and

$$f: N \to X$$

be an integral manifold of (I,ω). An *admissable variation* is by definition given by a map

$$F: N \times [0,\varepsilon] \to X$$

such that, if we set $F(s,t) = f_t(s)$, then

$$f_t: N \to X$$

is an integral manifold of (I,ω). That the *admissable variation should satisfy the endpoint conditions* (IV.a.9) means by definition that

$$F^*\omega = F^*\theta^1 = \cdots = F^*\theta^m = 0 \qquad \text{on} \quad \partial N \times [0,\varepsilon]. \qquad \text{(IV.a.10)}$$

As usual we will not need to use the whole variation $\{f_t\}$ but only the part up to 2nd order in t (i.e., f_0 and $\left.\frac{\partial f_t}{\partial t}\right|_{t=0}$). Then (IV.a.10) is

$$F^*\omega = F^*\theta^1 = \cdots = F^*\theta^m = 0 \qquad \text{in } \left. T^*(\partial N \times [0,\varepsilon])\right|_{\partial N \times \{0\}} .$$

For clarity of exposition we will continue to work with the whole variation.

If we set $N_t = f_t(N) \subset X$, then in general there will (non-uniquely) exist curves $\Gamma_t \subset Y$ with $\pi(\Gamma_t) = N_t$. Setting $\gamma_t = \tilde{\omega}(\Gamma_t) \subset Q$ in the basic diagram (IV.a.8), the endpoint conditions mean that *the* γ_t *all have same endpoints*. We may visualize the situation as looking something like

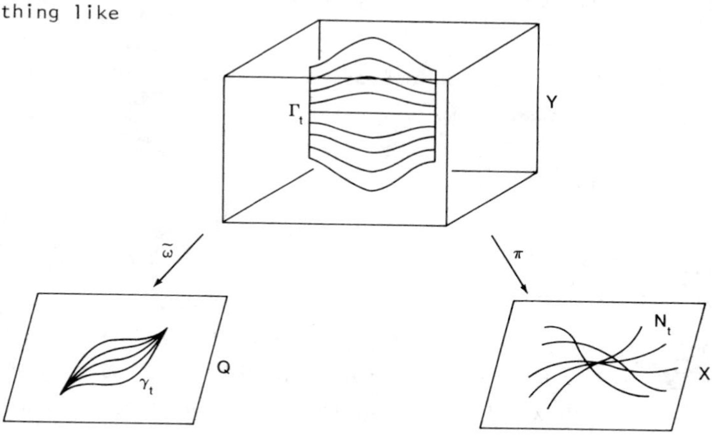

where the vertical lines are the fibres of $\tilde{\omega}$. All this should become clearer when we do examples below.

(IV.a.11) Remark. The picture to have in mind of a solution to the Euler-Lagrange equations (I.d.14) "joining two endpoints" is the following: On the momentum space Y the foliation defined by K^* has leaves W_q that locally look the standard linear foliation of $\mathbb{R}^{2m+1}$ by the 1st m-coordinate $\mathbb{R}^m$'s. Given two leaves W_{q_1}, W_{q_2}, a family of integral manifolds Γ_t of the differential system $\pi^*\theta^1 = \cdot\cdot = \pi^*\theta^s = 0$, $\pi^*\omega \neq 0$ on Y whose two endpoints lie in W_{q_1}, W_{q_2} respectively projects in X to an admissable variation $N_t = \pi(\Gamma_t)$ satisfying the endpoint conditions (IV.a.9). (In the language of algebraic geometry we may think of the leaves W_q as "blown up" endpoint conditions.) The characteristic directions of the 2-form Ψ_Y (i.e., the direction field $\Psi_Y^\perp \subset T(Y)$) are transverse to the W_q's. If we fix W_{q_1} then the characteristic curves leaving W_{q_1} fill out a manifold $C(W_{q_1})$ of dimension $m+1$, and by dimension considerations in general we expect $C(W_{q_1})$ to meet a W_{q_2} in one point. In this way, given q_1 and q_2 we may expect to find unique points $y_1 \in W_{q_1}$, $y_2 \in W_{q_2}$ and solution curve to the Euler-Lagrange system (J,ω) joining y_1 to y_2. This would then be *the* extremal for the functional in question corresponding to the endpoint conditions q_1, q_2.

Of course this nice picture will not always hold; sufficient conditions analogous to the classical Legendre and Jacobi conditions will be given in theorems (IV.c.7) and (IV.d.1) below. Here we shall prove the following

(IV.a.12) THEOREM. *If* N *satisfies the Euler-Lagrange equations of a well-posed variational problem, then*

$$\frac{d}{dt}\left(\int_{N_t} \varphi\right)_{t=0} = 0$$

for any admissable variation N_t *of* N *that satisfies the endpoint conditions (IV.a.9).*

<u>Proof</u>. We follow the notations of Chapter I, Section d) and use the Euler-Lagrange equations in the form (I.d.14). Thus we have an integral manifold $N \subset X$ of (I,ω) with N being diffeomorphic to $\{a \leqq s \leqq b\}$, and there are functions λ_α defined on X such that

$$v \lrcorner\, d(\varphi + \lambda_\alpha \theta^\alpha) \equiv 0 \quad \text{mod } N \tag{IV.a.13}$$

for all tangent vectors v to the ambient manifold X at points of N. (For later use we remark that since $\lambda_{m+1} = \cdots = \lambda_s = 0$ on N, if we use the additional index range $1 \leqq i,j \leqq m$ and fact that $\theta^\alpha \equiv 0$ mod N, (IV.a.13) is the same as

$$v \lrcorner\, d(\varphi + \lambda_i \theta^i) \equiv 0 \quad \text{mod } N \quad .) \tag{IV.a.14}$$

Suppose now that $v \in C^\infty(N, T(X))$ is tangent to an admissable variation $\{N_t\}$ of N satisfying the endpoint conditions (IV.a.9). Then by (I.b.15), (IV.a.5), and (IV.a.9)

$$\begin{cases} \text{(i)} & v \lrcorner\, d\theta^\alpha + d(v \lrcorner\, \theta^\alpha) \equiv 0 \quad \text{mod } N \\ \text{(ii)} & \langle\varphi, v\rangle = \langle\theta^1, v\rangle = \cdots = \langle\theta^m, v\rangle = 0 \quad \text{on } \partial N \ . \end{cases} \tag{IV.a.15}$$

By proposition (I.b.5) and Stokes' theorem

$$\frac{d}{dt}\left(\int_{N_t} \varphi\right)_{t=0} = \int_N v \lrcorner\, d\varphi + \int_{\partial N} v \lrcorner\, \varphi$$
$$= \int_N v \lrcorner\, d\varphi$$

by (ii) in (IV.a.15)

$$= -\int_N v \lrcorner\, d(\lambda_\alpha \theta^\alpha)$$

by (IV.a.13)

$$= \int_N d(v \lrcorner\, \lambda_\alpha \theta^\alpha)$$

by remark (iii) following (I.b.15)

$$= \int_{\partial N} v \lrcorner\, (\lambda_i \theta^i)$$

by Stokes' theorem and (IV.a.7)

$$= 0$$

by (ii) in (IV.a.15). Q.E.D.

Before giving some examples of well-posed variational problems, it is convenient to give one general condition that insures well-posedness.

(IV.a.16) PROPOSITION. *Suppose that the variational problem* $(I,\omega;\varphi)$ *is strongly non-degenerate (as defined in Chapter I, Section e)). Then it is well-posed.*

Proof. This is an immediate consequence of theorem (I.e.34) and the condition that $L^* = \text{span}\{\omega,\theta^1,..,\theta^s\}$ be completely integrable.
Q.E.D.

Remark. We note that in this case $L^* = K^*$.

(IV.a.17) Example. We consider a non-degenerate classical variational problem as discussed in example (I.e.18). Either directly or by applying proposition (IV.a.16), this problem is well-posed with

$$K^* = \text{span}\{dx;\theta^1,..,\theta^m\}$$
$$= \text{span}\{dx;dy^1,..,dy^m\} \ .$$

The endpoint conditions (IV.a.9) thus have the following geometric meaning. An admissable variation is given by 1-jets

$$x \to \left(x, y(x,t), \frac{\partial y(x,t)}{\partial x}\right) \quad , \qquad a \leqq x \leqq b \quad ,$$

of curves $x \to y(x,t)$ in $\mathbb{R}^m$. The endpoint condition (cf. (IV.a.10)) means that the variation vector field

$$v = \frac{\partial y^\alpha(x,0)}{\partial t} \frac{\partial}{\partial y^\alpha} + \frac{\partial^2 y^\alpha(x,0)}{\partial x \partial t} \frac{\partial}{\partial \dot{y}^\alpha}$$

satisfies

$$\langle dx,v\rangle = \langle dy^1,v\rangle = \cdot\cdot = \langle dy^m,v\rangle = 0$$

on ∂N. This is equivalent to

$$\frac{\partial y}{\partial t}(x,a) = \frac{\partial y}{\partial t}(x,b) = 0 \quad ;$$

i.e., the endpoint conditions are just as expected: *We consider paths joining two fixed points in* $\mathbb{R}^m$.

(Note: The condition $\langle dx,v\rangle = 0$ on ∂N is vacuous due to our particular parametrization of integral elements.) The basic diagram (IV.a.8) is

$$\begin{array}{ccc} & Y \subset J^1(\mathbb{R},\mathbb{R}^m) \times \mathbb{R}^m & \\ \tilde{\omega} \swarrow & & \searrow \pi \\ \mathbb{R} \times \mathbb{R}^m & & J^1(\mathbb{R},\mathbb{R}^m) \end{array} ,$$

and from this the interpretation of the endpoint conditions is transparent

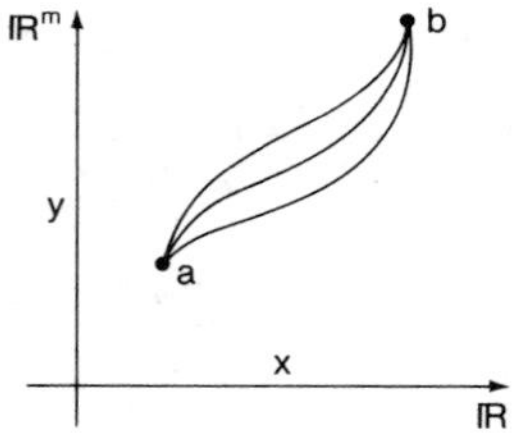

(IV.a.18) Example. We consider a non-degenerate 2nd order variational problem as in example (I.e.23). Again, either directly or by applying proposition (IV.a.16) this problem is well-posed with

$$K^* = \text{span}\{dx;\theta^1,..,\theta^m;\dot{\theta}^1,..,\dot{\theta}^m\}$$

$$= \text{span}\{dx;dy^1,..,dy^m;d\dot{y}^1,..,d\dot{y}^m\} \quad .$$

An admissable variation is given by 2-jets

$$x \to \left(x; y(x,t); \frac{\partial y(x,t)}{\partial x}; \frac{\partial^2 y(x,t)}{\partial x^2} \right), \qquad a \leqq x \leqq b \ ,$$

of curves $x \to y(x,t)$ in $\mathbb{R}^m$. The variation vector field is

$$v = \frac{\partial y^\alpha(x,0)}{\partial t} \frac{\partial}{\partial y^\alpha} + \frac{\partial^2 y^\alpha(x,0)}{\partial t \partial x} \frac{\partial}{\partial \dot{y}^\alpha} + \frac{\partial^3 y^\alpha(x,0)}{\partial t \partial x^2} \frac{\partial}{\partial \ddot{y}^\alpha} \ ,$$

and again as expected the endpoint conditions (IV.a.9) turn out to be

$$\begin{cases} \dfrac{\partial y(a,0)}{\partial t} = \dfrac{\partial y(b,0)}{\partial t} = 0 \\ \dfrac{\partial^2 y(a,0)}{\partial t \partial x} = \dfrac{\partial^2 y(b,0)}{\partial t \partial x} = 0 \ . \end{cases}$$

Thus admissable variations are given by families of curves joining two points and having the same tangent vectors there. The basic diagram (IV.a.8) is

$$\begin{array}{ccc} & Y \subset J^2(\mathbb{R},\mathbb{R}^m) \times \mathbb{R}^{2m} & \\ \tilde{\omega} \swarrow & & \searrow \pi \\ J^1(\mathbb{R},\mathbb{R}^m) & & J^2(\mathbb{R},\mathbb{R}^m) \ , \end{array}$$

and again the interpretation of the endpoint conditions is transparent:

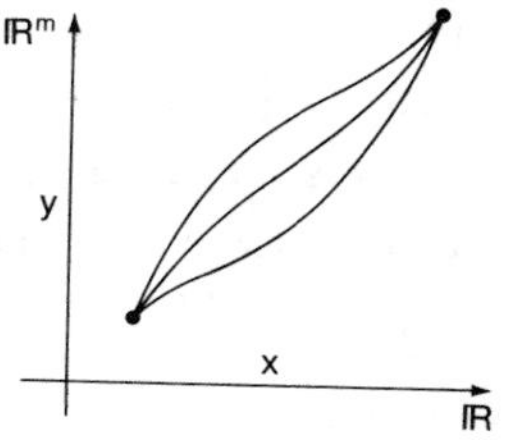

(family of curves $y_t(x)$ with fixed tangent line at each endpoint).

(IV.a.19) Example. Referring to remark (II.b.25) we consider the variational problem associated to the functional (II.b.26). Again by proposition (IV.a.16) this is well-posed with, using the notation (II.b.18),

$$K^* = \mathrm{span}\{\omega, \theta^1, \ldots, \theta^5\}$$

$$= \mathrm{span}\{\omega^1, \omega^2, \omega^3, \omega_1^2, \omega_1^2, \omega_2^3\} \ .$$

The basic diagram (IV.a.8) is

$$\begin{array}{ccc} & Y\subset X\times\mathbb{R}^5 & \\ {\scriptstyle\tilde{\omega}}\swarrow & & \searrow{\scriptstyle\pi} \\ F(\mathbb{E}^3)=Q & & X=F(\mathbb{E}^3)\times\mathbb{R}^2 \end{array} ;$$

i.e., the reduced momentum space is just the frame manifold. An admissable variation then consists of a family of curves $\gamma_t\subset\mathbb{E}^3$ joining two points A,B and having the same Frenet frames (i.e., same tangent lines and osculating 2-planes) at these points:

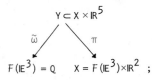

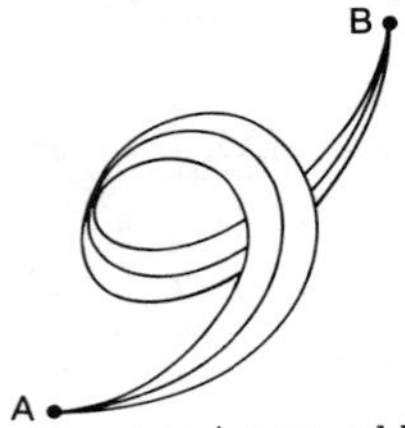

We note that the curves γ_t need not all have the same length, and the curvature at the endpoints need not be constant in t.

(IV.a.20) Example. We consider the variational problem associated to the functional (II.b.19) (similar considerations will also apply to this functional defined on curves $\gamma\subset\mathbb{E}^n$). In this case proposition (IV.a.16) does not apply (unless $n=2$) since the variational problem associated to (II.b.19) fails to be strongly non-degenerate. However, as proved below (II.b.23) the variational problem is non-degenerate with Y being given by (II.b.22). We will show that the variational problem is also well-posed with endpoint conditions (IV.a.9) where, using the notations (II.b.18),

$$\begin{aligned} K^* &= \operatorname{span}\{\omega,\theta^1,\theta^2,\theta^3,\theta^4\} \\ &= \operatorname{span}\{\omega^1,\omega^2,\omega^3,\omega_1^2,\omega_1^3\} . \end{aligned}$$

In fact, by (II.b.23) $\dim Y=9$ and so $m=4$ and $\operatorname{rank} K^*=5=m+1$. Moreover, from the structure equations (0.c.3) it follows that K^* is completely integrable, and finally the crucial condition (IV.a.7) is a consequence of (II.b.22) (actually all we need is that $\lambda_5=0$ on Y). The basic diagram (IV.a.8) is

$$\begin{array}{ccc} & Y\subset X\times\mathbb{R}^5 & \\ {\scriptstyle\tilde{\omega}}\swarrow & & \searrow{\scriptstyle\pi} \\ Q & & X=F(\mathbb{E}^3)\times\mathbb{R}^2 \end{array}$$

where the reduced momentum space is the *Stiefel manifold*

$$Q = \{(x,e_1): x \in \mathbb{E}^3 \text{ and } e_1 \text{ is a unit vector}\}.$$

Thus admissable variations satisfying the endpoint conditions are given by families of curves $\gamma_t \subset \mathbb{E}^3$ joining two points and having the same tangent lines there; this is exactly the right condition.

(IV.a.21) Example. We consider the Delauney problem associated to the constrained functional (II.b.31). This turns out to be an instructive example:

(IV.a.22) *The variational problem associated to (II.b.31) is well-posed in* ∞^1 *different ways. One among these corresponds to the original intent of the problem as explained in the remark following (II.b.32).*

Proof. We must first show that the variational problem is non-degenerate. For this we follow the notations in the previous discussion of (II.b.31), and consider the momentum space $Y \subset X_0 \times \mathbb{R}^5$ defined by the equations (II.b.35). On Y we take as coframe

$$\{\omega, \theta^1, \theta^2, \theta^3, \theta^4, \theta^5, \eta = -\lambda_4 d\tau, d\lambda_1, d\lambda_4\}$$

(thus $d\lambda_2 = \eta - \tau d\lambda_4$). By (II.b.35)

$$\begin{cases} \psi_Y = \omega + \lambda_1\theta^1 + \lambda_2\theta^2 + \lambda_4\theta^4 \\ \Psi_Y = d\omega + d\lambda_1 \wedge \theta^1 + \eta \wedge \theta^2 + d\lambda_4 \wedge (\theta^4 + \tau\theta^2) \\ \qquad + \lambda_1 d\theta^1 + \lambda_2 d\theta^2 + \lambda_4 d\theta^4 \end{cases},$$

where all forms are understood to be restricted to Y. Now in any case

$$\psi_Y \wedge (\Psi_Y)^4 = C\Theta$$

for some constant C where

$$\Theta = \omega \wedge d\lambda_1 \wedge \theta^1 \wedge \eta \wedge \theta^2 \wedge d\lambda_4 \wedge \theta^4 \wedge \theta^3 \wedge \theta^5$$

is a non-zero volume form on Y. Writing

$$\Psi_Y = d\lambda_1 \wedge \theta^1 + \eta \wedge \theta^2 + d\lambda_4 \wedge (\theta^4 + \tau\theta^2) + \sigma$$

where σ is a sum of terms not containing $d\lambda_1, \eta$, or $d\lambda_4$, it follows by exterior algebra that

$$\begin{aligned}\psi_Y \wedge (\Psi_Y)^4 &= 6\psi_Y \wedge d\lambda_1 \wedge \theta^1 \wedge \eta \wedge \theta^2 \wedge d\lambda_4 \wedge (\theta^4 + \tau\theta^2) \wedge \sigma \\ &= 6\psi_Y \wedge d\lambda_1 \wedge \theta^1 \wedge \eta \wedge \theta^2 \wedge d\lambda_4 \wedge \theta^4 \wedge \sigma \\ &= 6\omega \wedge d\lambda_1 \wedge \theta^1 \wedge \eta \wedge \theta^2 \wedge d\lambda_4 \wedge \theta^4 \wedge \sigma\end{aligned}$$

where the last step follows from the expression for ψ_Y. Now

$$\sigma = d\omega + \lambda_1 d\theta^1 + \lambda_2 d\theta^2 + \lambda_4 d\theta^4 ,$$

and in order to have $C \neq 0$ it is necessary and sufficient that the coefficient of $\theta^3 \wedge \theta^5$ in σ be non-zero. Referring to (II.b.20) this coefficient is $-\lambda_4$, and this shows that $\psi_Y \wedge (\Psi_Y)^4 \neq 0$.

We now consider the following ∞^1 choices for K^* with parameter μ:

$$\begin{aligned}K^*_\mu &= \operatorname{span}\{\omega, \theta^1, \theta^2, \theta^4, (1-\mu)\theta^3 + \mu\theta^5\} \\ &= \operatorname{span}\{\omega^1, \omega^2, \omega^3, \omega_1^2, (1-\mu)\omega_1^3 + \mu\omega_2^3\} \quad .\end{aligned}$$

Each of these is completely integrable and defines a set of endpoint conditions relative to which the variational problem is well-posed. The diagrams (IV.a.8) are

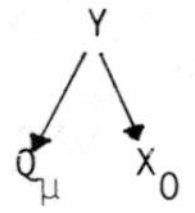

.

For the cases $\mu = 0,1$ it is immediate that

$$\begin{aligned}Q_0 &= \{(x, e_1) : x \in \mathbb{E}^3 \text{ and } e_1 \text{ is a unit vector}\} \\ Q_1 &= \{(x, e_3) : x \in \mathbb{E}^3 \text{ and } e_3 \text{ is a unit vector}\}.\end{aligned}$$

Consequently, the respective endpoint conditions on a family of curves $\gamma_t \subset \mathbb{E}^3$ are: Fix the endpoints and tangent lines (corresponding to K_0^*), and fix the endpoints and osculating 2-plane; i.e., the binomial (corresponding to K_1^*). For general μ the endpoint condition is: Fix the endpoints together with a linear combination of the tangent and binormal.

Clearly the desired endpoint conditions correspond to $\mu = 0$.

(IV.a.23) Example. An instructive example of endpoint conditions is provided by the affine geodesics discussed in part ii) of Chapter II, Section b). For ordinary Euclidean geodesics, since two points determine a line the endpoint conditions are obvious. However, if we think of a conic $C \subset \mathbb{A}^2$ as given by

$$ax^2 + 2bxy + cy^2 + dx + ey + f = 0 \tag{IV.a.24}$$

and take $[a,b,c,d,e,f] \in \mathbb{P}^5$ as the "coordinates" of C, then we see that in general **5** points are needed to determine a conic ("in general" means: no 3 points are colinear--cf. the Steiner constructions in [36]). Thus, for example there are ∞^1 conics passing through two points $a,b \in \mathbb{A}^2$ and having given tangent lines ℓ_0, ℓ_b

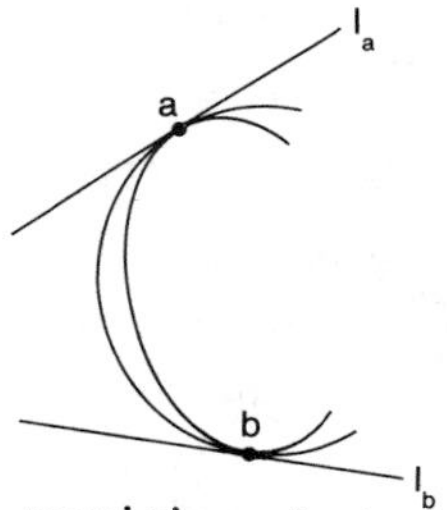

In the equation (IV.c.24) requiring C to pass through the origin with tangent line $y = 0$ imposes the two linear conditions $d = f = 0$, so there is in fact a $\mathbb{P}^1$ of conics satisfying the endpoint conditions (a,ℓ_a), (b,ℓ_b).

We may also see from (II.b.14), (II.b.15) that these are not the correct endpoint conditions. They correspond to the subspace

$$\begin{aligned} K_1^* &= \operatorname{span}\{\omega^1, \omega^2, \omega_1^2\} \\ &= \operatorname{span}\{\omega, \theta^1, \theta^2\} \end{aligned}$$

of $L^* = \operatorname{span}\{\omega, \theta^1, \theta^2, \theta^3, \theta^4\}$. In fact, setting $\omega^1 = \omega^2 = \omega_1^2 = 0$ exactly corresponds to fixing the point x and line $\mathbb{R}e_1$ in an affine frame (x, e_1, e_2). Certainly K_1^* is completely integrable, but the condition $Y \subset K_1^*$ is violated due to the fact that $\lambda_2 = 1/3$ on Y!

Another possible choice of endpoint conditions is given by the subspace

$$K_2^* = \text{span}\{\omega^1, \omega^2, \omega_2^1\}$$
$$= \text{span}\{\omega^1, \theta^1, \theta^4\}$$

of L^*. Again K_2^* is completely integrable, since setting $\omega^1 = \omega^2 = \omega_2^1 = 0$ corresponds to fixing x and the line $\mathbb{R}e_2$ in an affine frame (x, e_1, e_2). Moreover, the crucial condition $Y \subset K_2^*$ in the definition of well-posed endpoint conditions is satisfied since $\lambda_1 = \lambda_4 = 0$ on Y. Now K_2^* corresponds to the endpoint conditions (a, ν_a) where $a \in \mathbb{A}^2$ and ν_a is the line determined by the affine normal, and at first sight these may seem to be a good candidate to give a well-posed problem. But this is still not correct due to the nice geometric fact that:

(IV.a.25) *For a conic the affine normal remains fixed.*

In fact, the affine curvature is just the rate of charge of the affine normal.

To see geometrically what (IV.a.25) means we recall from Section ii) in Chapter II, Section b) that the affine normal ν_a at a point $a \in \gamma$, where γ is any curve in $\mathbb{A}^2$, is constructed as follows: For each $\varepsilon > 0$ draw the line ℓ_ε parallel to the tangent line at ε-distance away and denote by p_ε the center of gravity of the region included between γ and ℓ_ε

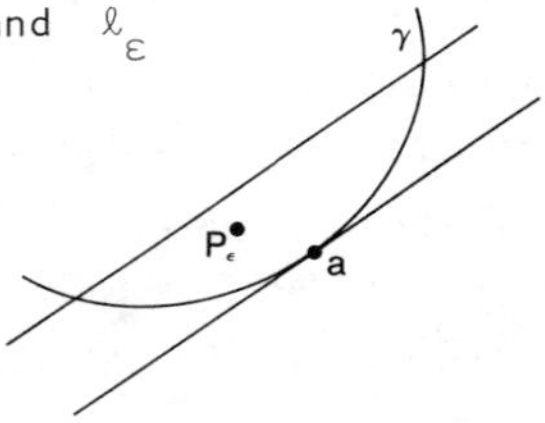

Since the notion of *area* is an affine invariant, the limiting position ν_a of the line $\overline{p_\varepsilon a}$ as $\varepsilon \to 0$ is well-defined and gives the affine normal line. For a conic we have a picture like

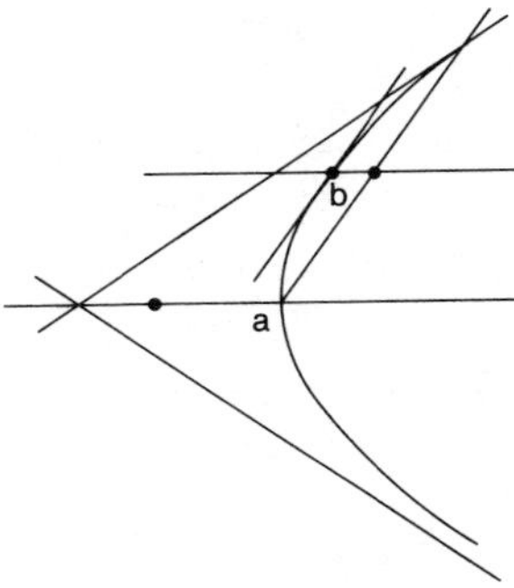

The analytic explanation for the failure of the endpoint conditions (a,ν_a) to give a well-posed variational problem is that *the transversality requirement in the definition is not satisfied.*

To illustrate a somewhat subtle aspect of the endpoint conditions in constrained variational problems (cf. Chapter IV, Section e) we shall conclude this section with the following

(IVa.26) Example. In the space $\tilde{X} = \mathbb{R}^{2k+1}$ with variables $(x;v^1,..,v^k;\dot{v}^1,..,\dot{v}^k) = (x;v^i;\dot{v}^i)$ we consider a classical variational problem $(\tilde{I},\omega;\tilde{\varphi})$ where $(\tilde{I},\omega)$ is the differential system

$$\begin{cases} \tilde{\theta}^i = dv^i - \dot{v}^i dx = 0 & i = 1,..,k \\ \omega = dx \neq 0 \end{cases}$$

and

$$\tilde{\varphi} = L(x,v,\dot{v})dx .$$

We assume that

$$\left\| L_{\dot{v}^i \dot{v}^j} \right\| > 0 \qquad ;$$

then the variational problem $(\tilde{I},\omega;\tilde{\varphi})$ is strongly non-degenerate and well-posed. The Euler-Lagrange system is generated by the Pfaffian equations

$$\begin{cases} \tilde{\theta}^i = 0 \\ dL_{\dot{v}^i} - L_{v^i} dx = 0 \\ dx \neq 0 \end{cases} \qquad \text{(IV.a.27)}$$

As will be proved below in Chapter IV, Sections c) and d), given v_0 there is a neighborhood V of v_0 such that any $v_1 \in V$ is joined to v_0 by a unique solution curve to (IV.a.27) remaining in V. In this sense the O.D.E. system with *endpoint conditions*

$$\begin{cases} \dfrac{d}{dx}\left(L_{\dot{v}^i}\left(x,v(x),\dfrac{dv(x)}{dx}\right)\right) = L_{v^i}\left(x,v(x),\dfrac{dv(x)}{dx}\right) \\ v(x_0) = v_0, \quad v(x_1) = v_1 \end{cases} \qquad \text{(IV.a.28)}$$

has a unique solution.

Moreover, the O.D.E. system with *initial conditions*

$$\begin{cases} \dfrac{d}{dx}\left(L_{\dot{v}^i}\left(x,v(x),\dfrac{dv(x)}{dx}\right)\right) = L_{v^i}\left(x,v(x),\dfrac{dv(x)}{dx}\right) \\ v(x_0) = v_0 , \quad \dfrac{dv}{dx}(x_0) = \dot{v}_0 \end{cases} \qquad \text{(IV.a.29)}$$

also has a unique solution. For example, if $(\tilde{I},\omega;\tilde{\varphi})$ is a mechanical system then these initial conditions represent initial position and velocity, and therefore they are in some sense more natural physically than the endpoint conditions in (IV.a.28).

Now suppose in the space X with variables $(x;v^1,\dots,v^k;\dot{v}^1,\dots,\dot{v}^k;z^1,\dots,z^\ell) = (x;v^i;\dot{v}^i;z^\rho)$ we consider the variational problem $(I,\omega;\varphi)$ where (I,ω) is the Pfaffian differential system

$$\begin{cases} \text{(i)} \quad \tilde{\theta}^i = dv^i - \dot{v}^i dx = 0 \\ \text{(ii)} \quad \tilde{\theta}^\rho = dz^\rho - f^\rho(x,v,\dot{v})dx = 0 \\ \qquad \omega = dx \neq 0 \end{cases} \qquad \text{(IV.a.30)}$$

and

$$\varphi = L(x,v,\dot{v})dx \quad .$$

For a concrete special case we may take the example (II.a.40) of a wheel rolling in the plane without slipping $(v^1 = \alpha,\ v^2 = \beta,\ z^1 = x,\ z^2 = y,\ L = (1-\cos v^2)(\dot{v}^2)^2 + \frac{\sin^2 v^2}{2}(\dot{v}^1)^2$, and the equations (ii) above are (cf. (II.a.42))

$$\begin{cases} dz^1 - \dot{v}^2 \cos v^1 dx = 0 \\ dz^2 - \dot{v}^2 \sin v^1 dx = 0 \end{cases} \quad) .$$

We may think of $(I,\omega;\varphi)$ as a classical variational problem with the constraints

$$\dot{z}^\rho = f^\rho(x,v,z,\dot{v}) \qquad \text{(IV.a.31)}$$

in which *the functions* f^ρ *and Lagrangian* L *do not depend on* z (any non-holonomic constraints are expressed locally by equations of the form (IV.a.31), where in general f^ρ depends on z).

To each solution curve $\tilde{\gamma}\subset\tilde{X}$, given by $x \to (x,v(x),\frac{dv(x)}{dx})$, of the Euler-Lagrange equations (IV.a.27) associated to $(\tilde{I},\omega;\tilde{\varphi})$ there corresponds a unique curve $\gamma\subset X$, given by $x \to (x,v(x),z(x),\frac{dv(x)}{dx})$, with given initial point $z(x_0) = z_0$ and where

$$\frac{dz^\rho(x)}{dx} = f^\rho\left(x, v(x), \frac{dv(x)}{dx}\right) . \qquad \text{(IV.a.32)}$$

We shall show that:

(IV.a.33) *The variational problem* $(I,\omega;\varphi)$ *is strongly non-degenerate (in particular, it is well-posed).*

(IV.a.34) *Each curve* γ *obtained from* $\tilde{\gamma}$ *by solving (IV.a.32) is a solution curve to the Euler-Lagrange equations for* $(I,\omega;\varphi)$. *However, these curves* γ *do not form a general solution.*

Following the proofs of these two statements we shall discuss their meaning *vis à vis* endpoint and initial conditions.

Proof of (IV.a.33). In the space with variables $(x;v^i;z^\rho;\dot{v}^i;\dot{z}^\rho)$ we consider the submanifold X defined by

$$\dot{z}^\rho = f^\rho(x,v,\dot{v}) .$$

On X we consider the classical variational problem with constraints given by the data

$$\begin{cases} \theta^i = dv^i - \dot{v}^i dx = 0 \\ \theta^\rho = dz^\rho - \dot{z}^\rho dx = 0 \\ \omega = dx \neq 0 \\ \varphi = L(x,v,\dot{v})dx . \end{cases} \qquad \text{(IV.a.35)}$$

Clearly the variational problem $(I,\omega;\varphi)$ is equivalent to one given by this data, and so we shall use $(I,\omega;\varphi)$ to denote (IV.a.35). Writing

$$\begin{aligned} df^\rho &\equiv f^\rho_i dv^i + f^\rho_{\dot{i}} d\dot{v}^i \mod dx \\ &\equiv f^\rho_i \theta^i + f^\rho_{\dot{i}} d\dot{v}^i \mod dx , \end{aligned}$$

the structure equations of (IV.a.35) are

$$\begin{cases} d\theta^i \equiv -d\dot{v}^i \wedge dx & \mod\{\theta^i,\theta^\rho\} \\ d\theta^\rho \equiv -f^\rho_{\dot{i}} d\dot{v}^i \wedge dx & \mod\{\theta^i,\theta^\rho\} \end{cases}$$

Setting

$$\eta^\rho = \theta^\rho - f^\rho_{\dot{i}}\theta^i$$

it follows that the 1st derived system I_1 of I is generated by the sub-bundle

$$W_1^* = \mathrm{span}\{\eta^1,..,\eta^\ell\}$$

of $T^*(X)$. Accordingly we take

$$\{dx;\theta^i;\eta^\rho;d\dot{v}^i\}$$

as coframe on X. Then

$$\begin{cases} dL \equiv L_{\dot{v}^i}d\dot{v}^i \mod \mathrm{span}\ \{dx;\theta^i;\eta^\rho\}; \\ dL_{\dot{v}^i} \equiv L_{\dot{v}^i\dot{v}^j}d\dot{v}^j \mod \mathrm{span}\ \{dx;\theta^i;\eta^\rho\}\ . \end{cases}$$

It follows that the quadratic form $\mathcal{Q}$ is represented by the matrix $\|L_{\dot{v}^i\dot{v}^j}\|$, which by assumption is positive definite.

In Chapter I, Section e) the definition of strong non-degeneracy requires that there be an admissable coframe so that (I.e.28) is satisfied. In the present situation we may take this coframe to be $\{dx;\theta^i;\eta^\rho;d\dot{v}^i\}$ *in case* $f^\rho(x,v,\dot{v})$ *is linear in* $\dot{v}^i$. In general it is not always possible to find the admissable coframe, and so for the purposes of this illustrative discussion the general case will be omitted (the linear case covers our rolling wheel example).

The strong non-degeneracy then follows from these remarks together with $\|L_{\dot{v}^i\dot{v}^j}\| > 0$. Q.E.D.

<u>Proof of (IV.a.34)</u>. Following our usual algorithm (cf. Chapter I, Section e)) for computing the Euler-Lagrange system associated to $(I,\omega;\varphi)$, we consider on $Z = X \times \mathbb{R}^{k+\ell}$ the 1-form

$$\psi = \varphi + \lambda_i\theta^i + \lambda_\rho\theta^\rho$$

whose exterior derivative $\Psi = d\psi$ is given by

$$\Psi = (L_{\dot{v}^i} - (\lambda_i + \lambda_\rho \dot{f}_i^\rho))d\dot{y}^i \wedge dx + d\lambda_\rho \wedge \theta^\rho$$
$$+ (d\lambda_i - (L_{v^i} + \lambda_\rho f_i^\rho)dx) \wedge \theta^i\ .$$

The Cartan system $C(\Psi)$ is generated by the Pfaffian equations

$$\begin{cases} \text{(i)} & \partial/\partial\dot{y}^i \lrcorner \Psi = \left(L_{\dot{v}^i} - (\lambda_i + \lambda_\rho \dot{f}_i^\rho)\right)dx = 0 \\ \text{(ii)} & \partial/\partial\theta^i \lrcorner \Psi = -d\lambda_i + (L_{v^i} + \lambda_\rho f_i^\rho)dx = 0 \\ \text{(iii)} & \partial/\partial\theta^\rho \lrcorner \Psi = -d\lambda_\rho = 0 \end{cases} \qquad \text{(IV.a.36)}$$

together with the original system $\theta^i = \theta^\rho = 0$. From (i) it follows that the momentum space $Y \subset Z$ is given by

$$L_{\dot{v}^i} = \lambda_i + \lambda_\rho \dot{f}^\rho_i .$$

On Y the Euler-Lagrange system is given by (ii), (iii) plus the original system $\theta^i = \theta^\rho = 0$.

Now let $\tilde{\gamma}$ given by $x \to (x; v^i(x); \frac{dv^i(x)}{dx})$ be a solution curve to (IV.a.27), and consider the curve γ in Y given by

$$x \to \left(x; v^i(x); z^\rho(x); \frac{dv^i(x)}{dx}; \frac{dz^\rho(x)}{dx}; \lambda_i(x); \lambda_\rho(x)\right)$$

where

$$\begin{cases} \lambda_i(x) = L_{\dot{v}^i}\left(x, v(x), \dfrac{dv(x)}{dx}\right) \\ \lambda_\rho(x) \equiv 0 \end{cases}$$

and where (IV.a.32) is satisfied. By comparing (IV.a.36) with (IV.a.27) we see that γ is a solution curve to the Euler-Lagrange system for $(I, \omega; \varphi)$.

However, it is clear that γ is not the *general* solution curve since along a general solution curve the constants λ_ρ need not be zero. Q.E.D.

Discussion. What is going on may be explained as follows:

Suppose we are interested in solution curves to the Euler-Lagrange equations for $(I, \omega; \varphi)$ having given *initial conditions*

$$\begin{cases} v(x_0) = v_0 \\ \dfrac{dv}{dx}(x_0) = \dot{v}_0 \\ z(x_0) = z_0 \end{cases} .$$

Then there is a unique such curve along which all $\lambda_\rho \equiv 0$, and it is obtained from the solution curve to the problem (IV.a.29) by solving the equations (IV.a.32) with the initial conditions $z(x_0) = z_0$. This initial value problem for the Euler-Lagrange equations associated to $(I, \omega; \varphi)$ is what has physical meaning as in the example (II.a.40) of the rolling wheel.

On the other hand, by (IV.a.33) and (IV.a.16) the variational problem $(I, \omega; \varphi)$ is well-posed with *endpoint conditions* given by

$$dx = \theta^i = \theta^\rho = 0 \ .$$

These endpoint conditions mean: consider all curves $x \to (x, v(x), z(x))$ connecting (x_0, v_0, z_0) to (x_1, v_1, z_1). In the case of the rolling wheel we should roll the wheel in a given time so that Q_0 goes to Q_1.

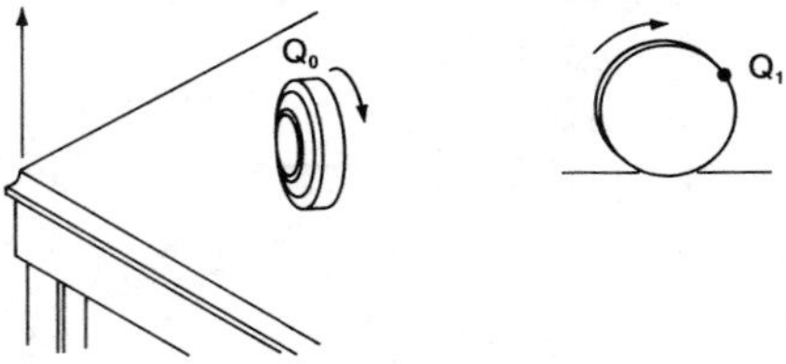

As will be proved below, if the endpoint conditions are sufficiently close there will be a unique solution curve to the Euler-Lagrange equations for $(I, \omega; \varphi)$ connecting (x_0, v_0, z_0) to (x_1, v_1, z_1) and remaining in a neighborhood of (x_0, y_0, z_0). However, in general the λ_ρ must be non-zero constants along this curve. Put another way, there are not enough solution curves to (II.a.36) along which $\lambda_\rho \equiv 0$ in order to connect two general points (x_0, v_0, z_0) and (x_1, v_1, z_1).

In summary the point is this: *For classical unconstrained variational problems the solution curves to the initial value problem and to the endpoint problem coincide. However, for constrained problems the latter set of curves may be strictly larger than the former.*

b) Jacobi Vector Fields and Conjugate Points; Examples.

As we shall now see, the general formulation of the calculus of variations in the setting of exterior differential systems leads naturally to the Jacobi equations associated to a variational problem $(I, \omega; \varphi)$ on a manifold X, and also to relatively efficient computation of examples. Again the point turns out to be that *the Jacobi equations live naturally on the momentum space* Y, which we recall is the manifold where the Euler-Lagrange differential system (J, ω) associated to $(I, \omega; \varphi)$ is defined (cf. Chapter I, Section e)).

Although the definition of Jacobi equations given below is valid in complete generality, the variational problem $(I, \omega; \varphi)$ must be well-posed (in the sense of the preceding section) in order that we may also define conjugate points. Since in any case our main examples satisfy this condition it seems reasonable to make this assumption throughout. Therefore in this section we will use the following

<u>Notations</u>. $(I,\omega;\varphi)$ will be a well-posed variational problem on X with associated Euler-Lagrange differential system (J,ω) on Y. The endpoint conditions will be given by the completely integrable subsystem $K^* \subset L^*$, and written in the form (IV.a.9) as

$$\omega = \theta^1 = \cdots = \theta^m = 0 \quad . \tag{IV.b.1}$$

The reduced momentum space is denoted by Q and the basic diagram (IV.a.8) is

$$\begin{array}{ccc} & Y \subset W^* & \\ {\scriptstyle\tilde{\omega}} \swarrow & & \searrow {\scriptstyle\pi} \\ Q & & X \end{array} \tag{IV.b.2}$$

Integral curves of (J,ω) will be denoted by $\Gamma \subset Y$ with respective projections

$$\begin{cases} \pi(\Gamma) = N \subset X \\ \tilde{\omega}(\Gamma) = \gamma \subset Q \quad . \end{cases}$$

Thus $N \subset X$ is an integral manifold of (I,ω) that is also a solution of the Euler-Lagrange equations (I.d.14).

From Chapter I, Section b) we recall that to an integral manifold of any Pfaffian differential system, in our case the the integral manifold $\Gamma \subset Y$ of (J,ω), there is associated the space of vector fields V to Y defined along Γ and that give the infinitesimal deformations of Γ as an integral manifold of (J,ω) (cf. (I.b.16))

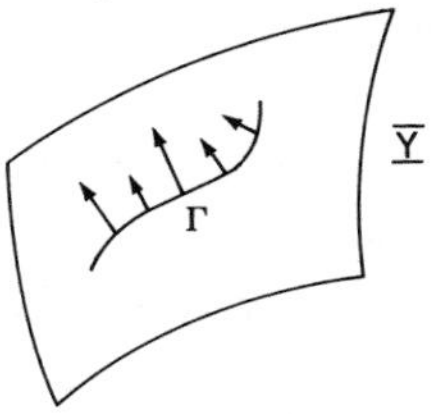

Recalling our notation $V(J,\omega)$ for the space of integral manifolds of (J,ω), we denote the above space of vector fields V by $T_\Gamma(V(J,\omega))$, where as before the notation is meant to suggest the tangent space to $V(J,\omega)$ at Γ. The linear differential equations that define $T_\Gamma(V(J,\omega))$ among all normal vector fields to $\Gamma \subset Y$ are (I.b.15).

Definitions: i) Let $N \subset X$ be an integral manifold of (I,ω) that satisfies the Euler-Lagrange equations associated to the variational problem $(I,\omega;\varphi)$. Then the vector space of *Jacobi vector fields* is $T_\Gamma(V(J,\omega))$ where $\pi(\Gamma) = N$ in (IV.b.2).(1)

ii) The two endpoint conditions of N are said to be *conjugate* if there exists a Jacobi vector field V satisfying

$$\langle\omega,V\rangle = \langle\theta^1,V\rangle = \cdots = \langle\theta^m,V\rangle = 0 \quad \text{on } \partial\Gamma \quad .$$

Remarks. i) Thus, as in the classical case, *the Jacobi equations are the variational equations to solution curves of the Euler-Lagrange equations*. However, somewhat in contrast to the traditional definition it is essential that they be defined as normal vector fields to Γ *up on* Y *and not down on* X (actually, in the classical unconstrained case this point of view is taken in [13]).

ii) Associated to each Jacobi field V defined along $\Gamma \subset Y$ is its projection $w = \pi_* V$ defined along $N \subset X$ and which gives the actual variation of our solution manifold to the Euler-Lagrange equations

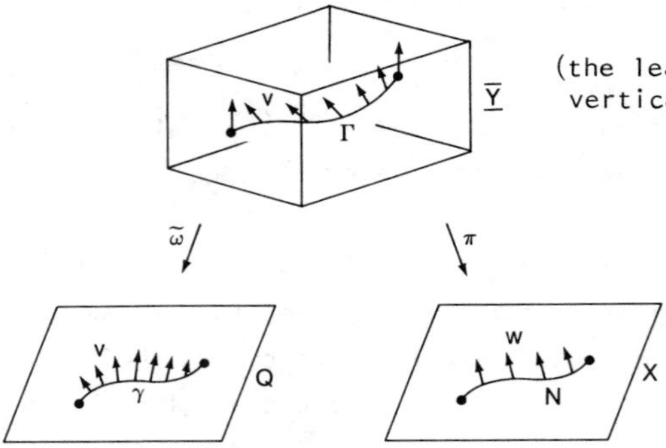

(the leaves of K^* are vertical lines in Y)

To say that the endpoint conditions of N are conjugate does *not* mean $w = 0$ on ∂N. What it does mean is that upstairs on $\partial\Gamma$ the Jacobi field V is tangent to the leaves of the foliation defined by K^* (in the above picture V should be vertical on $\partial\Gamma$). Equivalently, the vector field $v = \tilde{\omega}_* V$ on Q should vanish at the endpoints of γ as in the above figure.

(IV.b.3) Example (geodesics): This is the main classical example, and for our discussion we shall assume some basic elementary facts of Riemannian geometry ([6], [15], [53], and [63]). Let M be a Riemannian manifold and X the unit sphere bundle consisting of pairs

(p,e_1) where $p \in M$ and $e_1 \in T_p(M)$ is a unit vector. Using the projection

$$\pi : X \to M$$

we define the canonical 1-form ω on X by

$$\omega(v) = (e_1, \pi_* v) \qquad v \in T_{(p,e_1)}(X) \quad .$$

(Note: Under the isomorphism $T(M) \cong T^*(M)$ given by the metric, ω is just the canonical 1-form θ on $T^*(M)$.) We also define a Pfaffian system

$$W^* \subset T^*(X)$$

by setting

$$W^*_{(p,e_1)} = \pi^*(e_1^\perp)$$

where $e_1^\perp \subset T_p^*(M)$ is the hyperplane orthogonal to e_1.[2] Then W^* generates a differential ideal I, and (I,ω) gives a Pfaffian system on X whose integral manifolds may be written as

$$t \to \left(p(t), \frac{p'(t)}{\|p'(t)\|} \right) \quad .$$

Here, $p(t)$ describes a curve in M and $\frac{p'(t)}{\|p'(t)\|}$ is its unit tangent vector.

By a *frame field* defined in an open set $U \subset X$ we shall mean that to each point $(p,e_1) \in U$ there is associated a complete orthonormal frame $(p,e_1,\ldots,e_n)$ for $T_p(M)$ where the first vector is the same e_1. In terms of such a frame field there are defined (*up on* U) the dual coframe field $\{\omega^i\}$ and *connection matrix* $\{\omega_j^i\}$, where by definition

$$\begin{cases} \omega^i(v) = (e_i, \pi_* v) & v \in T_{(p,e_1)}(X) \\ De_i = \omega_i^j e_j & . \end{cases} \qquad \text{(IV.b.4)}$$

Here D denotes the *covariant differential* associated to the Riemannian connection. Then it is immediate that

$$\begin{cases} \omega = \omega^1 \\ W^* = \operatorname{span}\{\omega^2,\ldots,\omega^n\} \ . \end{cases} \qquad \text{(IV.b.5)}$$

Using the ranges of indices

$$i \leqq i,j \leqq n, \quad 2 \leqq \rho,\sigma \leqq n \ ,$$

from the 2nd equation in (IV.b.4) it follows that

$$\{\omega^1;\omega^2,\ldots,\omega^n;\omega_1^2,\ldots,\omega_1^n\} = \{\omega;\omega^\rho;\omega_1^\rho\}$$

gives a coframe on $U \subset X$. The 1-forms $\omega^1,\ldots,\omega^n$ are horizontal for $X \to M$ and $\omega_1^2,\ldots,\omega_1^n$ describes how e_1 infinitesimally rotates in $T_p(M)$ when we move in the fibres of $X \to M$. Finally, we recall that for any frame field the connection forms ω_i^j are the unique 1-forms satisfying

$$\begin{cases} \omega_j^i + \omega_i^j = 0 \\ d\omega^i = \omega^j \wedge \omega_j^i \quad .^{(3)} \end{cases} \qquad \text{(IV.b.6)}$$

The *geodesics* on M correspond to the variational problem $(I,\omega;\varphi)$ on X where

$$\varphi = |\omega| \quad .$$

The notation means that

$$\varphi(v) = |\langle\omega,v\rangle| \qquad \text{(IV.b.7)}$$

for v a tangent vector to X. To compute the Euler-Lagrange system (the Euler-Lagrange equations have already been computed in local coordinates in example (I.d.18) (with $U = 0$)), we consider $Z = X \times \mathbb{R}^{n-1}$ where $\mathbb{R}^{n-1}$ has coordinates $\lambda = (\lambda_2,\ldots,\lambda_n)$. Following the algorithm from Chapter I, Section e) we consider the 1-form

$$\psi = \pm\omega + \lambda_\rho\omega^\rho$$

on Z. Here $\pm\omega$ stands for φ as defined by (IV.b.7), which is not really a 1-form in the usual sense, but as we shall now see the formalism carries over (to be precise, one should take $\varphi = L\omega$ where $L = \pm 1$).

Using (IV.b.6) the exterior derivative $\Psi = d\psi$ is given by

$$\Psi = \pm\omega^\rho \wedge \omega_\rho^1 + d\lambda_\rho \wedge \omega^\rho + \lambda_\rho\omega^1 \wedge \omega_1^\rho + \lambda_\rho\omega^\sigma \wedge \omega_\sigma^\rho \quad . \qquad \text{(IV.b.8)}$$

The Cartan system is generated by the Pfaffian equations

$$\begin{cases} \text{(i)} & \partial/\partial\lambda_\rho \lrcorner \Psi = \omega^\rho = 0 \\ \text{(ii)} & \partial/\partial\omega^\rho \lrcorner \Psi = \pm\omega_\rho^1 - d\lambda_\rho + \lambda_\sigma\omega_\rho^\sigma = 0 \\ \text{(iii)} & \partial/\partial\omega_1^\rho \lrcorner \Psi = -\lambda_\rho\omega^1 \mp \omega^\rho = 0 \end{cases} \qquad \text{(IV.b.9)}$$

From (i), (iii) we may infer that $Z_1 = Y = X$ is defined by $\lambda_\rho = 0$. Then using $d\lambda_\rho = 0$ and $\omega_1^\rho + \omega_\rho^1 = 0$, equation (ii) gives

$$\omega_1^\rho = 0 \quad . \tag{IV.b.10}$$

Since by (IV.b.4)

$$De_1 = \omega_1^\rho e_\rho ,$$

(IV.b.10) is the usual condition

$$\frac{D}{dt}\left(\frac{p'(t)}{\| p'(t) \|}\right) = 0 \tag{IV.b.11}$$

that a curve $p(t)$ on M be a geodesic. Geometrically, the condition means that tangent vector to $p(t)$ remains parallel along the curve. For later use we note that by (IV.b.8)

$$\begin{cases} \psi_Y = \pm\omega \\ \Psi_Y = \pm\omega^\rho \wedge \omega_\rho^1 \end{cases}$$

so that

$$\psi_Y \wedge (\Psi_Y)^{n-1} = \pm(n-1)!\omega^1 \wedge \cdots \wedge \omega^n \wedge \omega_1^2 \wedge \cdots \wedge \omega_1^n \neq 0 \ .$$

Thus the variational problem is non-degenerate. Moreover, with the endpoint conditions given by

$$\begin{aligned} K^* &= \mathrm{span}\{\omega^1, \omega^2, \ldots, \omega^n\} \\ &= \pi^*(T^*(M)) \qquad , \end{aligned}$$

the variational problem is well-posed with reduced momentum space $Q = M$. Clearly the endpoint conditions correspond to varying a curve in M keeping its endpoints fixed in the usual sense (as had better be the case!).

To compute the Jacobi equations we shall use the *Cartan structure equation* for the curvature

$$\begin{cases} d\omega_i^j = \omega_i^k \wedge \omega_k^j + \Omega_i^j \\ \Omega_i^j = \frac{1}{2} R_{ik\ell}^j \ \omega^k \wedge \omega^\ell . \end{cases} \tag{IV.b.12}$$

On the momentum space $Y = X$ (= unit sphere bundle of M) the Euler-Lagrange system J is given by (IV.b.9) where $\lambda_\rho = d\lambda_\rho = 0$. We consider an integral manifold $\Gamma \subset X$ with projection $\gamma = \pi(\Gamma) \subset M$

(recalling that in this case $Q = M$), and we parametrize Γ by arclength so that $\omega = ds$. Then (IV.b.9) is

$$\begin{cases} \omega^{\rho} = 0 \\ \omega_1^{\rho} = 0 \end{cases} \qquad \text{(IV.b.13)}$$

Given a vector field V to Y defined along Γ the Jacobi equations are given by (I.b.15) with J replacing I, and they are

$$\begin{cases} \text{(i)} \quad V \lrcorner\, d\omega^{\rho} + d(V \lrcorner\, \omega^{\rho}) \equiv 0 \mod \Gamma \\ \text{(ii)} \quad V \lrcorner\, d\omega_1^{\rho} + d(V \lrcorner\, \omega_1^{\rho}) \equiv 0 \mod \Gamma \end{cases} \qquad \text{(IV.b.14)}$$

By remark ii) following (I.b.15) we may assume that the tangential component to Γ of V vanishes; thus

$$V = V^{\rho}\, \partial/\partial\omega^{\rho} + V_1^{\rho}\, \partial/\partial\omega_1^{\rho} .$$

Using (IV.b.6) and the fact that $\omega^{\rho} \equiv 0 \mod \Gamma$, equations (i) in (IV.b.14) are

$$dV^{\rho} - V_1^{\rho}\omega \equiv 0 \mod \Gamma . \qquad \text{(IV.b.15)}$$

Similarly, using (IV.b.12) equations (ii) in (IV.b.14) are

$$dV_1^{\rho} + \omega_{\sigma}^{\rho} V_1^{\sigma} + R_{1\sigma 1}^{\rho} V^{\sigma} \equiv 0 \mod \Gamma. \qquad \text{(IV.b.16)}$$

To interpret these last two equations we let

$$\begin{aligned} v &= \pi_*(V) \\ &= V^{\rho} e_{\rho} \end{aligned}$$

be the normal vector field along the geodesic $\gamma \subset M$. The tangent to γ is e_1 and we let $R(v,w) \in \mathrm{Hom}(T(M),T(M))$ be the *curvature operator* (cf. [6], [53]) corresponding to the pair v,w of tangent vectors. Then (IV.b.15), (IV.b.16) combine to give the usual *Jacobi equations*

$$\frac{D^2 v}{ds^2} + R(v,e_1)e_1 = 0 \qquad \text{(IV.b.17)}$$

associated to the normal vector field v *along the geodesic* γ.

For example, suppose that M is a surface with adapted frame field (e_1,e_2) along the geodesic γ.

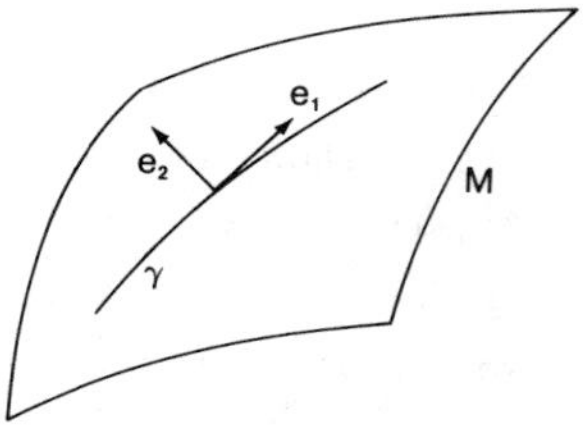

Then $v = \eta e_2$ for some function $\eta(s)$, and we will compute (IV.b.17) using the geodesic property $\omega_1^2 \equiv 0 \mod \gamma$. Thus

$$\frac{D^2 v}{ds^2} = \frac{D}{ds}\left(\frac{D}{ds}(\eta e_2)\right)$$

$$= \frac{D}{ds}(\eta' e_2) \qquad \left(\text{since } \frac{De_2}{ds} = 0\right)$$

$$= \eta'' e_2 \qquad (\quad " \quad)$$

Denoting by R the Gaussian curvature function along γ, (IV.b.17) is

$$\eta'' + R\eta = 0 \quad . \tag{IV.b.18}$$

If we impose the initial conditions $\eta(0) = 0$ and $\eta'(0) > 0$, then standard comparison theorems in O.D.E. give

$$\begin{cases} \eta(s) > 0 & \text{for } s > 0 \text{ in case } R \leqq 0 \\ \eta(s_0) = 0 & \text{for some } s_0 > 0 \text{ in case } R \geqq R_0 > 0 \ . \end{cases}$$

In the 1st case geodesics "spread", in fact exponentially if $R \leqq R_0 < 0$, and there can be no conjugate points. On the other hand, in the 2nd case there must be conjugate points. Of course, this is only an ε-beginning in a highly developed branch of Riemannian geometry (cf. [21], [49], and the references cited there).

(IV.b.19) <u>Example</u>. For a non-classical example we will investigate the Jacobi equations corresponding to the functional

$$\Phi(N) = \frac{1}{2}\int_N \kappa^2\, ds \tag{IV.b.20}$$

defined on curves $N \subset \mathbb{E}^2$ (in accordance with the notational conventions in this section we will reserve γ to denote curves in the reduced momentum space). We have seen in example (I.d.27) that, aside from the degenerate case of straight lines, an arc on a circle is not an extremal of (IV.b.20). However, suppose we add the integral constraint

$$\int_N ds = \ell = \text{constant} \ . \tag{IV.b.21}$$

Variational problems with integral constraints are discussed in Part a of the Appendix, and there it is shown that the Euler-Lagrange equations for the functional (IV.b.20) with constraint (IV.b.21) are the same as the Euler-Lagrange equations for the unconstrained functional (IV.b.20) defined on curves lying in a sphere of radius $\rho(\ell) > 0$ (cf. example (A.a.10)).

Physically we are considering the position assumed by a piece of wire of fixed length ℓ joining two points p,q in the plane with given tangent directions there. The endpoint conditions constitute a 6-dimensional manifold, and the arcs of circles correspond to endpoint conditions lying on a codimension-two submanifold. To see this, fix p and ℓ_p arbitrarily. Then the circles passing through p with tangent direction ℓ_p are parametrized by their radius ρ. Let L_p be the line perpendicular to ℓ_p, and q, ℓ_q another "endpoint"

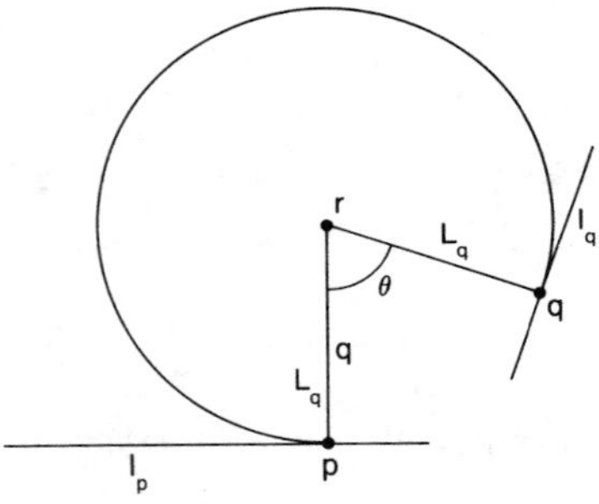

with L_q the corresponding perpendicular. If $r = L_p \cap L_q$ then the conditions that there be a circle giving an extremal of (IV.b.20) with the constraint (IV.b.21) are

$$\begin{cases} \text{(i)} & \overline{pr} = \overline{qr} , \qquad \text{and} \\ \text{(ii)} & \ell = \rho(2\pi - \theta) \qquad . \end{cases}$$

We will investigate the Jacobi equations associated to (IV.b.20), (IV.b.21) at an arc of a circle and will prove the following result:

(IV.b.22) THEOREM. *For each* ρ $(\neq\infty)$ *there exists an* ℓ *such that the endpoint conditions on the arc of length* ℓ *on a circle of radius* ρ *are conjugate.*

In our construction ℓ will have to be large, so that the physical picture may resemble a watch spring.

The proof proceeds in several steps:

Step One: We begin by reviewing the Euler-Lagrange system for the functional (IV.b.20) on a sphere S of radius $1/\rho$. Using the notations in example (I.d.27) we have

$$\begin{cases} X = F(S) \times \mathbb{R} \text{ where } \mathbb{R} \text{ has coordinate } \kappa; \\ Z = X \times \mathbb{R}^2 \text{ where } \mathbb{R}^2 \text{ has coordinates } (\lambda_1, \lambda_2); \\ Y \subset Z \text{ is defined by } \lambda_2 = \kappa; \\ \{\omega, \omega^1, \omega^2, d\lambda_1, d\lambda_2\} \text{ is a coframe on } Y; \end{cases}$$

and by the equation just above (I.d.34) the 2-form Ψ_Y is given by

$$\Psi_Y \equiv \eta_1 \wedge \eta_2 + \eta_3 \wedge \eta_4$$

where

$$\begin{cases} \eta_1 = d\lambda_1 + \left(\lambda_2 \rho - \dfrac{\lambda_2^3}{2}\right)\omega \\ \eta_2 = \theta^1 \\ \eta_3 = d\lambda_2 + \lambda_1 \omega \\ \eta_4 = \theta^2 \end{cases} \qquad \text{(IV.b.23)}$$

The momentum space Y has dimension 5 and the Euler-Lagrange differential system on Y is generated by the Pfaffian equations (cf. (I.d.34) using that $\lambda_2 = \kappa$)

$$\eta_1 = \eta_2 = \eta_3 = \eta_4 = 0 \quad . \qquad \text{(IV.b.24)}$$

Step Two: We now compute the Jacobi equations corresponding to an integral curve Γ of (IV.b.24). For this we use the structure equations for the frame bundle of a surface, reviewed in (I.d.30), to obtain for the exterior derivatives of the forms (IV.b.23) the equations

$$\begin{cases} d\eta_1 \equiv \left(\left(\rho - \frac{3}{2}\lambda_2^2\right) d\lambda_2 - \left(\lambda_2^2 \rho - \dfrac{\lambda_2^4}{2}\right)\theta^1\right) \wedge \omega \\ d\eta_2 \equiv -\theta^2 \wedge \omega \\ d\eta_3 \equiv (d\lambda_1 - \lambda_1 \lambda_2 \theta^1) \wedge \omega \\ d\eta_4 \equiv -(d\lambda_2 + \rho\theta^1 - \lambda_2^2 \theta^1) \wedge \omega \end{cases} \qquad \text{(IV.b.25)}$$

For a vector field

$$V = \alpha\, \partial/\partial\theta^1 + \beta\, \partial/\partial\theta^2 + \varepsilon\, \partial/\partial\lambda_1 + \delta\, \partial/\partial\lambda_2 \tag{IV.b.26}$$

we use (IV.b.25) to compute the variational equations (cf. (I.b.15))

$$d(V \lrcorner\, \eta_i) + V \lrcorner\, d\eta_i \equiv 0 \mod \Gamma$$

to be

$$\begin{cases} d\varepsilon + \left(\delta\left(\rho - \dfrac{3\lambda_2^2}{2}\right) - \alpha\left(\lambda_2^2\rho - \dfrac{\lambda_2^4}{2}\right)\right)\omega \equiv 0 \mod \Gamma \\ d\alpha - \beta\omega \equiv 0 \mod \Gamma \\ d\delta + (\varepsilon - \lambda_1\lambda_2\alpha)\omega \equiv 0 \mod \Gamma \\ d\beta - (\delta + \rho\alpha - \lambda_2^2\alpha)\omega \equiv 0 \mod \Gamma \end{cases} \tag{IV.b.27}$$

These are equivalent to (using $\lambda_2 = \kappa$ and $\lambda_1 = -\kappa'$ on Y)

$$\begin{cases} \alpha' = \beta \\ \beta' = \alpha(\rho - \kappa^2) + \delta \\ \varepsilon' = \alpha\kappa^2\left(\rho - \dfrac{\kappa^2}{2}\right) - \delta\left(\rho - \dfrac{3\kappa^2}{2}\right) \\ \delta' = -\varepsilon - \kappa\kappa'\alpha \end{cases} \tag{IV.b.28}$$

<u>Step Three</u>: We now consider endpoint conditions such that along Γ we have

$$\lambda_2 = \text{constant} > 0 .$$

Then by (IV.b.23) we have

$$\begin{cases} \lambda_1 = 0 \\ \lambda_2 = \sqrt{2\rho} = \kappa \end{cases}$$

The Jacobi equations (IV.b.28) reduce to

$$\begin{cases} \text{(i)} & \alpha' = \beta \\ \text{(ii)} & \beta' = -\rho\alpha + \delta \\ \text{(iii)} & \varepsilon' = 2\rho\delta \\ \text{(iv)} & \delta' = -\varepsilon \end{cases} \tag{IV.b.29}$$

These equations imply the single 4th order equation (in which only terms with an even number of derivatives appear)

$$\alpha^{(iv)} + 3\rho\alpha'' + 2\rho^2\alpha = 0 \quad . \tag{IV.b.30}$$

Conversely, given a solution to (IV.b.30) we may define β,δ,ε by (i), (ii), (iv) in (IV.b.29), and then (iii) holds by virtue of (IV.b.30).

Referring to (IV.b.26), to show the existence of conjugate points we want to find a solution $\alpha(s)$ to (IV.b.30) satisfying

$$\alpha(0) = \alpha(\ell) = \alpha'(0) = \alpha'(\ell) = 0 \ . \tag{IV.b.31}$$

A basis for the vector space of solutions to the linear constant coefficient O.D.E. (IV.b.30) is given by the functions

$$\cos \kappa s,\ \sin \kappa s,\ \cos \frac{\kappa s}{\sqrt{2}},\ \sin \frac{\kappa s}{\sqrt{2}} \quad .$$

Thus we must determine κ,ℓ such that

$$\det \begin{vmatrix} 1 & 0 & \cos \kappa\ell & -\kappa \sin \kappa\ell \\ 0 & \kappa & \sin \kappa\ell & \kappa \cos \kappa\ell \\ 1 & 0 & \cos \frac{\kappa\ell}{\sqrt{2}} & \frac{-\kappa}{\sqrt{2}} \sin \frac{\kappa\ell}{\sqrt{2}} \\ 0 & 1 & \sin \frac{\kappa\ell}{\sqrt{2}} & \frac{\kappa\ell}{\sqrt{2}} \cos \frac{\kappa\ell}{\sqrt{2}} \end{vmatrix} = 0 \tag{IV.b.32}$$

A computation gives that (IV.b.32) is equivalent to

$$F(\kappa\ell) = 2\sqrt{2}\left(1 - \cos \kappa\ell \cos \frac{\kappa\ell}{\sqrt{2}}\right) - 3 \sin \kappa\ell \sin \frac{\kappa\ell}{\sqrt{2}} = 0 \quad . \tag{IV.b.33}$$

To prove the theorem it will suffice to set $\sigma = \kappa\ell$ and determine σ_1,σ_2 such that

$$F(\sigma_1) > 0 \tag{IV.b.34}$$

$$F(\sigma_2) < 0 \ . \tag{IV.b.35}$$

For this we consider the additive subgroup of the circle

$$G = \left\{\frac{2\pi m}{\sqrt{2}}\right\}_{m\in\mathbb{Z}} \subset \mathbb{R}/2\pi\mathbb{Z} \quad .$$

It is well-known that this subgroup is dense (cf. [2]); hence, given $\varepsilon > 0$ we may find $m,n \in \mathbb{Z}$ such that

$$0 \neq \left| \frac{2\pi m}{\sqrt{2}} - 2\pi n \right| < \varepsilon \quad .$$

For $\sigma_1 = 2\pi m$ we have

$$F(\sigma_1) = 2\sqrt{2}\left(1 - \cos\frac{2\pi m}{\sqrt{2}}\right) > 0 \quad ,$$

which gives (IV.b.34).

Similarly, by choosing a point in G close to the equivalence class of $\frac{\pi}{2} - \frac{\pi}{2\sqrt{2}}$, given $\varepsilon > 0$ we may find $m,n \in \mathbb{Z}$ such that

$$0 \neq \left| \left(\frac{2\pi m}{\sqrt{2}} + \frac{\pi}{2\sqrt{2}}\right) - \left(2\pi n + \frac{\pi}{2}\right) \right| < \varepsilon.$$

For $\sigma_2 = 2\pi m + \pi/2$ we have

$$F(\sigma_2) = 2\sqrt{2} - 3\sin\left(\frac{2\pi m}{\sqrt{2}} + \frac{\pi}{2\sqrt{2}}\right) \quad .$$

Since $\sin\left(\frac{2\pi m}{\sqrt{2}} + \frac{\pi}{2\sqrt{2}}\right)$ may be made arbitrarily close to 1 and $2\sqrt{2} < 3$, we will have (IV.b.35) for ε sufficiently small. Q.E.D.

(IV.b.36) <u>Example</u>. Having discussed the Jacobi equations and conjugate points for geodesics and the functional (IV.b.20), we will invert logical order and discuss the Jacobi equations for a classical variational problem with non-degenerate Lagrangian (I.e.18). Following the derivation of the general equations we will specialize to the most important classical example, the minimal surface of revolution.

On $X = J^1(\mathbb{R},\mathbb{R}^m)$ with coordinates $(x;y^\alpha;\dot{y}^\alpha)$ we assume given a Lagrangian $L(x;y^\alpha;\dot{y}^\alpha)$ with $\det\|L_{\dot{y}^\alpha\dot{y}^\beta}\| \neq 0$. In $Z = X \times \mathbb{R}^s$ where $\mathbb{R}^s$ has coordinates $\lambda = (\lambda_\alpha)$ the momentum space Y is defined by

$$\lambda_\alpha = L_{\dot{y}^\alpha} \quad .$$

Of course, in this case $Y \cong \mathbb{R} \times T^*(\mathbb{R}^m)$ and we take $\{dx;\theta^\alpha;d\lambda_\alpha\}$ as a coframe (recall that $\theta^\alpha = dy^\alpha - \dot{y}^\alpha dx$). Then (cf. Chapter I, Section e))

$$\begin{aligned} \psi_Y &= Ldx + \lambda_\alpha\theta^\alpha \\ &= -Hdx + \lambda_\alpha dy^\alpha \end{aligned}$$

where

$$H(x;y^\alpha;\lambda_\alpha) = -L + \dot{y}^\alpha L_{\dot{y}^\alpha}$$

is the Hamiltonian function. The exterior derivative $\Psi_Y = d\psi_Y$ is given by

$$\Psi_Y = -H_{y^\alpha} dy^\alpha \wedge dx - H_{\lambda_\alpha} d\lambda_\alpha \wedge dx + d\lambda_\alpha \wedge dy^\alpha \quad ,$$

and the Cartan system $C(\Psi_Y)$ is generated by the Pfaffian equations

$$\begin{cases} \partial/\partial y^\alpha \lrcorner \Psi_Y = -d\lambda_\alpha - H_{y^\alpha} dx = 0 \\ \partial/\partial\lambda_\alpha \lrcorner \Psi_Y = dy^\alpha - H_{\lambda_\alpha} dx = 0 \quad . \end{cases} \tag{IV.b.37}$$

With the independence condition $\omega = dx \neq 0$ these Pfaffian equations give the Euler-Lagrange system (J,ω) on Y, and the solution curves to (IV.b.37) are obtained by solving the Hamilton equations

$$\begin{cases} \dfrac{d\lambda_\alpha}{dx} = -H_{y^\alpha} \\ \dfrac{dy^\alpha}{dx} = H_{\lambda_\alpha} \quad . \end{cases} \tag{IV.b.38}$$

This is all well-known and has been previously discussed; we only wish to again remark that from the viewpoint of exterior differential systems there is *no choice* but to introduce the Hamiltonian formalism.

Turning to the Jacobi equations, which are the variational equations of (J,ω) on Y, we use the notations

$$\begin{cases} \eta_\alpha = d\lambda_\alpha + H_{y^\alpha} dx \\ \varphi^\alpha = dy^\alpha - H_{\lambda_\alpha} dx \end{cases}$$

so that $J = \{\eta_\alpha;\varphi^\alpha\}$ is the Pfaffian differential system generated algebraically by the η_α's and φ^α's. The exterior derivatives are

$$\begin{cases} d\eta_\alpha = H_{y^\alpha y^\beta} dy^\beta \wedge dx + H_{y^\alpha \lambda_\beta} d\lambda_\beta \wedge dx \\ d\varphi^\alpha = -H_{\lambda_\alpha y^\beta} dy^\beta \wedge dx - H_{\lambda_\alpha \lambda_\beta} d\lambda_\beta \wedge dx \quad . \end{cases} \tag{IV.b.39}$$

We consider a normal vector field

$$V = V^{\alpha}\ \partial/\partial\lambda_{\alpha} + V_{\alpha}\ \partial/\partial y^{\alpha}$$

defined along an integral curve $\Gamma = \{(x;y^{\alpha}(x);\lambda_{\alpha}(x))\}$ of (J,ω). Using (IV.b.39) the variational equations

$$\begin{cases} d(V \lrcorner \eta_{\alpha}) + V \lrcorner d\eta_{\alpha} \equiv 0 \bmod \Gamma \\ d(V \lrcorner \varphi^{\alpha}) + V \lrcorner d\varphi^{\alpha} \equiv 0 \bmod \Gamma \end{cases}$$

are respectively

$$\begin{cases} dV^{\alpha} + \left(H_{y^{\alpha}y^{\beta}}V_{\beta} + H_{y^{\alpha}\lambda_{\beta}}V^{\beta}\right)dx \equiv 0 \bmod \Gamma \\ dV_{\alpha} - \left(H_{\lambda_{\alpha}y^{\beta}}V_{\beta} + H_{\lambda_{\alpha}\lambda_{\beta}}V^{\beta}\right)dx \equiv 0 \bmod \Gamma . \end{cases} \qquad \text{(IV.b.40)}$$

These are the *classical Jacobi equations in Hamiltonian form.* (Note: In Lagrangian form; i.e., on $\mathbb{R} \times T(\mathbb{R}^m)$, they are a little more messy.)

(IV.b.41) Special Case. The classical minimal surface of revolution problem

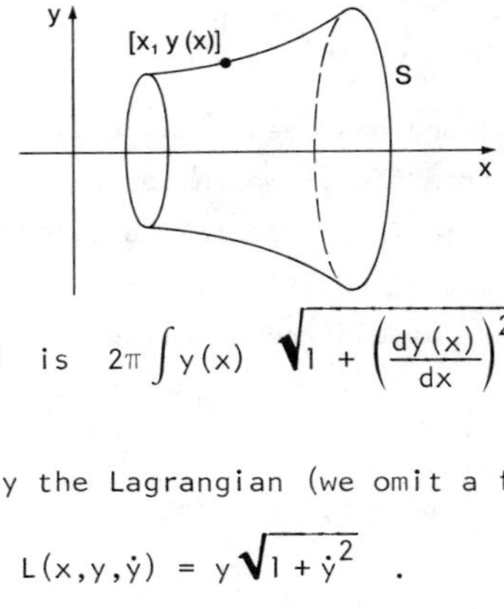

$$\left(\text{Area of } S \text{ is } 2\pi \int y(x) \sqrt{1 + \left(\frac{dy(x)}{dx}\right)^2}\, dx\right)$$

is given on $J^1(\mathbb{R},\mathbb{R})$ by the Lagrangian (we omit a factor of 2π)

$$L(x,y,\dot{y}) = y\sqrt{1+\dot{y}^2} \quad .$$

Solving

$$\lambda = L_{\dot{y}} = \frac{y\dot{y}}{\sqrt{1+\dot{y}^2}}$$

gives

$$\dot{y} = \pm \frac{\lambda}{\sqrt{y^2 - \lambda^2}}$$

and

$$H = -L + \lambda\dot{y}$$

$$= -\sqrt{y^2 - \lambda^2} \quad .$$

Because $H = H(y,\lambda)$ does not depend on x there is a 1st integral

$$H = -a \qquad (a > 0) \ .$$

Since

$$\begin{cases} H_\lambda = \dfrac{\lambda}{H} \\ H_y = -\dfrac{y}{H} \end{cases} , \qquad \text{(IV.b.42)}$$

Hamilton's equations (IV.b.38) are (for the "energy level" a)

$$\begin{cases} \dfrac{d\lambda}{dx} = \dfrac{y}{a} \\ \dfrac{dy}{dx} = \dfrac{\lambda}{a} \end{cases} .$$

These may be integrated to give

$$\begin{aligned} y(x) &= \frac{a}{2}\left(e^{(x-x_0)/a} + e^{-(x-x_0)/a}\right) \\ \lambda(x) &= \frac{a}{2}\left(e^{(x-x_0)/a} - e^{-(x-x_0)/a}\right) \end{aligned} \qquad \text{(IV.b.43)}$$

The two constants a and x_0 are used to pass a solution curve through two given points. These curves are the well-known catenaries in the (x,y) plane

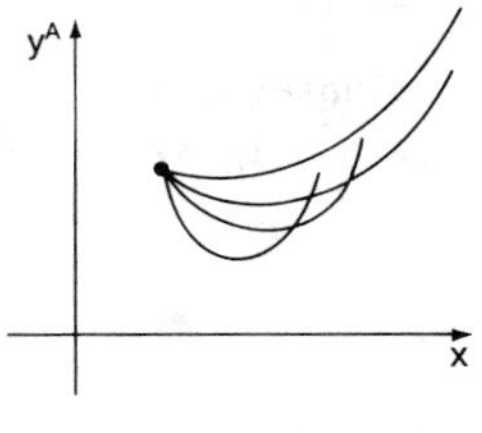

To compute the Jacobi equations we differentiate (IV.b.42) and obtain

$$\begin{cases} H_{\lambda\lambda} = \dfrac{y^2}{H^3} \\ H_{\lambda y} = -\dfrac{\lambda y}{H^3} \\ H_{yy} = \dfrac{\lambda^2}{H^3} \end{cases} .$$

The Jacobi equations (IV.b.40) are then (setting $U = V_1$ and $W = V^1$)

$$\begin{cases} \dfrac{dU}{dx} = -\dfrac{\lambda y}{a^3} U + \dfrac{y^2}{a^3} W \\[2ex] \dfrac{dW}{dx} = -\dfrac{\lambda^2}{a^3} U + \dfrac{\lambda y}{a^3} W \end{cases} . \tag{IV.b.44}$$

Along the solution curve (IV.b.43) we may solve the Jacobi equations (IV.b.44) to obtain

$$\begin{cases} U = a(\lambda(\alpha x + \beta) - y a \alpha) \\ W = a(y(\alpha x + \beta) - \lambda a \alpha) \end{cases} .$$

Since the variational vector field V is $U\, \partial/\partial y + W\, \partial/\partial \lambda$, the endpoint conditions are

$$U(x_1) = U(x_2) = 0 .$$

Using (IV.b.43) it follows that for $\alpha \neq 0$ the function $U(x)$ has exactly two zeroes $x_i(a,\alpha)$ $(i = 1,2)$. In other words, on a general solution curve to the Euler-Lagrange equations for this variational problem there is exactly one pair of conjugate points. This beautiful and instructive example is discussed in detail in, e.g., [13], pages 297-304.

(IV.b.42) <u>Example</u>. We shall derive the general form of the Jacobi equations for a classical 2nd order variational problem with non-degenerate Lagrangian (cf. (I.e.23)).

On $X = J^2(\mathbb{R}, \mathbb{R}^m)$ with coordinates $(x; y^\alpha; \dot{y}^\alpha; \ddot{y}^\alpha)$ we assume given $L(x; y^\alpha; \dot{y}^\alpha; \ddot{y}^\alpha)$ with $\det \| L_{\ddot{y}^\alpha \ddot{y}^\beta} \| \neq 0$. In $Z = X \times \mathbb{R}^{2s}$ where $\mathbb{R}^{2s}$ has coordinates $(\lambda_\alpha; \dot{\lambda}_\alpha)$ the momentum space Y is defined by

$$\dot{\lambda}_\alpha = L_{\ddot{y}_\alpha} .$$

We take $\{dx; \theta^\alpha; \dot{\theta}^\alpha; d\lambda_\alpha; d\dot{\lambda}_\alpha\}$ as a coframe on Y (where we recall that $\theta^\alpha = dy^\alpha - \dot{y}^\alpha dx$, $\dot{\theta}^\alpha = d\dot{y}^\alpha - \ddot{y}^\alpha dx$), so that the basic diagram (IV.b.2) is

$$\begin{array}{ccc} & Y & \\ \tilde{\omega} \swarrow & & \searrow \pi \\ J^1(\mathbb{R}, \mathbb{R}^m) = Q & & X = J^2(\mathbb{R}, \mathbb{R}^m) \end{array} . \tag{IV.b.43}$$

We have (cf. Chapter I, Section e))

$$\psi_Y = Ldx + \lambda_\alpha \theta^\alpha + \dot{\lambda}_\alpha \dot{\theta}^\alpha$$
$$= -Hdx + \lambda_\alpha dy^\alpha + \dot{\lambda}_\alpha d\dot{y}^\alpha$$

where the Hamiltonian is given by

$$H = -L + \lambda_\alpha \dot{y}^\alpha + L_{\ddot{y}^\alpha} \ddot{y}^\alpha \quad .$$

The exterior derivative $\Psi_Y = d\psi_Y$ is

$$\Psi_Y = -H_{y^\alpha} dy^\alpha \wedge dx - H_{\dot{y}^\alpha} d\dot{y}^\alpha \wedge dx - H_{\lambda_\alpha} d\lambda_\alpha \wedge dx - H_{\dot{\lambda}_\alpha} d\dot{\lambda}_\alpha \wedge dx$$
$$+ d\lambda_\alpha \wedge dy^\alpha + d\dot{\lambda}_\alpha \wedge d\dot{y}^\alpha \quad ;$$

consequently the Cartan system is generated by the Pfaffian equations

$$\left\{ \begin{aligned} \eta_\alpha &= -\partial/\partial y^\alpha \lrcorner\, \Psi_Y = d\lambda_\alpha + H_{y^\alpha} dx = 0 \\ \dot{\eta}_\alpha &= -\partial/\partial \dot{y}^\alpha \lrcorner\, \Psi_Y = d\dot{\lambda}_\alpha + H_{\dot{y}^\alpha} dx = 0 \\ \varphi^\alpha &= \partial/\partial \lambda_\alpha \lrcorner\, \Psi_Y = dy^\alpha - H_{\lambda_\alpha} dx = 0 \\ \dot{\varphi}^\alpha &= \partial/\partial \dot{\lambda}_\alpha \lrcorner\, \Psi_Y = d\dot{y}^\alpha - H_{\dot{\lambda}_\alpha} dx = 0 \;. \end{aligned} \right. \qquad \text{(IV.b.44)}$$

Solution curves to the Euler-Lagrange system (J,ω) satisfy Hamilton's equations

$$\left\{ \begin{aligned} \frac{d\lambda_\alpha}{dx} &= -H_{y^\alpha} \\ \frac{d\dot{\lambda}_\alpha}{dx} &= -H_{\dot{y}^\alpha} \\ \frac{dy^\alpha}{dx} &= H_{\lambda_\alpha} \\ \frac{d\dot{y}^\alpha}{dx} &= H_{\dot{\lambda}_\alpha} \quad . \end{aligned} \right. \qquad \text{(IV.b.45)}$$

For a specific example we refer to (II.a.50).

The Jacobi equations for

$$V = V^\alpha \, \partial/\partial\lambda_\alpha + \dot{V}^\alpha \, \partial/\partial\dot{\lambda}_\alpha + V_\alpha \, \partial/\partial y^\alpha + \dot{V}_\alpha \, \partial/\partial \dot{y}^\alpha$$

along a solution curve Γ to (IV.b.45) are

$$\begin{cases} d(V \lrcorner \eta_\alpha) + V \lrcorner d\eta_\alpha \equiv 0 \bmod \Gamma \\ d(V \lrcorner \dot{\eta}_\alpha) + V \lrcorner d\dot{\eta}_\alpha \equiv 0 \bmod \Gamma \\ d(V \lrcorner \varphi^\alpha) + V \lrcorner d\varphi^\alpha \equiv 0 \bmod \Gamma \\ d(V \lrcorner \dot{\varphi}^\alpha) + V \lrcorner d\dot{\varphi}^\alpha \equiv 0 \bmod \Gamma \quad . \end{cases}$$

By (IV.b.44) these are

$$\begin{cases} \dfrac{dV^\alpha}{dx} = -\left(H_{y^\alpha \lambda_\beta} V^\beta + H_{y^\alpha \dot{\lambda}_\beta} \dot{V}^\beta + H_{y^\alpha y^\beta} V_\beta + H_{y^\alpha \dot{y}^\beta} \dot{V}_\beta\right) \\ \dfrac{d\dot{V}^\alpha}{dx} = -\left(H_{\dot{y}^\alpha \lambda_\beta} V^\beta + H_{\dot{y}^\alpha \dot{\lambda}_\beta} \dot{V}^\beta + H_{\dot{y}^\alpha y^\beta} V_\beta + H_{\dot{y}^\alpha \dot{y}^\beta} \dot{V}_\beta\right) \\ \dfrac{dV_\alpha}{dx} = H_{\lambda_\alpha \lambda_\beta} V^\beta + H_{\lambda_\alpha \dot{\lambda}_\beta} \dot{V}^\beta + H_{\lambda_\alpha y^\beta} V_\beta + H_{\lambda_\alpha \dot{y}^\beta} \dot{V}_\beta \\ \dfrac{d\dot{V}_\alpha}{dx} = H_{\dot{\lambda}_\alpha \lambda_\beta} V^\beta + H_{\dot{\lambda}_\alpha \dot{\lambda}_\beta} \dot{V}^\beta + H_{\dot{\lambda}_\alpha y^\beta} V_\beta + H_{\dot{\lambda}_\alpha \dot{y}^\beta} \dot{V}_\beta \quad . \end{cases} \qquad \text{(IV.b.46)}$$

c) <u>Geometry of the Reduced Momentum Space; the 2nd Variation, the Index Form, and Sufficient Conditions for a Local Minimum</u>.

We retain our notations from the preceding two sections and consider a well-posed variational problem $(I,\omega;\varphi)$ on a manifold X. We denote the endpoint conditions (IV.b.1) by A,B, and

$$\mathcal{V}(I,\omega;[A,B]) \subset \mathcal{V}(I,\omega) \qquad \text{(IV.c.1)}$$

will denote the set of integral manifolds $N \subset X$ of (I,ω) satisfying these endpoint conditions. In just a moment this will be made more precise; we refer to the examples in Chapter IV, Section a) where it is seen that we may frequently think of A,B as lying over points $a,b \in X$ and the integral manifolds $N \in \mathcal{V}(I,\omega;[A,B])$ will be integral curves of (I,ω) in X joining a,b and having certain linear conditions imposed on their tangent lines at these endpoints.

We consider the functional

$$\Phi: V(I,\omega;[A,B]) \to \mathbb{R} \tag{IV.c.2}$$

defined by

$$\Phi(N) = \int_N \varphi \quad .$$

As in Chapter IV, Section a) we define the tangent space $T_N(V(I,\omega;[A,B]))$ to consist of the C^∞ normal vector fields to $N \subset X$ that satisfy the variational equations (I.b.15) and endpoint conditions (IV.b.1). We recall that for $v \in C^\infty(N,T(X))$ to satisfy the variational equations means that deformation of N along v preserves, up to order t^2, the condition that N be an integral manifold of (I,ω). We also recall that this condition depends only on the normal projection $[v] \in C^\infty(N,T(X)/T(N))$ of v. However, the endpoint conditions $\langle\omega,v\rangle(a) = \langle\omega,v\rangle(b) = 0$ depend on the actual vector field v, and so the precise definition of $T_N(V(I,\omega;[A,B]))$ is the following:

We let $E = T(X)/T(N)$ *be the normal bundle to* $N \subset X$, *and then* $T_N(V(I,\omega;[A,B])) \subset C^\infty(N,E)$ *is the image of sections* $v \in C^\infty(N,T(X))$ *satisfying (I.b.15) and the endpoint conditions (IV.b.1).*

The differential

$$\delta\Phi(N): T_N(V(I,\omega;[A,B])) \to \mathbb{R} \tag{IV.c.3}$$

of the functional (IV.c.2) has been defined in Chapter IV, Section a), and in the proof of theorem (IV.a.12) (cf. proposition (I.b.5)) we computed it to be given by

$$\delta\Phi(N)(v) = \int_N v \lrcorner d\varphi \quad . \tag{IV.c.4}$$

Also, in theorem (IV.a.12) we have proved that $\delta\Phi(N) = 0$ in case N satisfies the Euler-Lagrange equations (I.d.14). This still leaves open the main questions:

(IV.c.5) Is N a *local minimum* of (IV.c.2), in the sense that $\Phi(N) \leqq \Phi(N')$ for all $N' \in V(I,\omega;[A,B])$ in a neighborhood of N?

(IV.c.6) Is N an *absolute minimum* of (IV.c.2)?

Remarks. We shall not discuss the 2nd question. Regarding the 1st question, it has become customary to distinguish between a *weak local minimum*, meaning that $\Phi(N) \leqq \Phi(N')$ for all $N' \in V(I,\omega;[A,B])$ in a C^2-neighborhood of N (all of our discussions need only C^2 and not C^∞), and a *strong local minimum* meaning that the same inequality holds for all N' in C^0-neighborhood of N. Again we shall not discuss this important distinction (cf. [13], [29], and [31]).

The goal of this section is to formulate and prove a general result guaranteeing that under certain conditions a solution to the Euler-Lagrange equations gives a local minimum for (IV.c.2). Quite naturally this will be an analogue of the usual "2nd derivative test" in elementary calculus, and as such will involve the 2nd variation of the functional Φ. In other words, we must be able to understand something about 2nd derivatives of functionals defined only on integral manifolds of an exterior differential system. A direct approach to this computation seems to be difficult; as we shall see the point turns out to be to study the rather remarkable geometry of the basic diagram (IV.b.2).

To formulate our result we shall make the following:

<u>Assumption</u>. The variational problem $(I,\omega;\varphi)$ is *strongly non-degenerate*, as defined in Chapter I, Section e). Recall that in this case we could introduce a non-degenerate quadratic form $\|A_{\mu\nu}\|$ along integral manifolds N of (I,ω).

(IV.c.7) THEOREM. *Assume that the quadratic form* $\|A_{\mu\nu}\|$ *is positive definite. Then for sufficiently close endpoint conditions, any solution* $N \in V(I,\omega;[A,B])$ *to the Euler-Lagrange equations which remains in a neighborhood of* A *is a local minimum for the functional* Φ.[5]

As indicated above, the proof of this result requires preliminary considerations which in any case are of independent interest, and we shall first turn to them.

i) *Geometry of the momentum space and reduced momentum space.* We consider the basic diagram

$$\begin{array}{ccc} & Y & \\ {\scriptstyle \tilde{\omega}}\swarrow & & \searrow{\scriptstyle \pi} \\ Q & & X \end{array}, \qquad \text{(IV.c.8)}$$

and shall make several observations concerning the relation of the various manifolds. For convenience we first recall the structure equations (I.e.28) and (I.e.29) of $(I,\omega;\varphi)$:

$$\begin{cases} \text{(i)} & d\theta^\rho \equiv 0 \bmod L^* \wedge L^* & \rho = 1,..,s-s_1 \\ \text{(ii)} & d\theta^\mu \equiv -\pi^\mu \wedge \omega \bmod L^* \wedge L^* & \mu = s-s_1+1,..,s \\ \text{(iii)} & d\omega \equiv 0 \bmod L^* \wedge L^* & \\ \text{(iv)} & d\varphi \equiv A_\mu \pi^\mu \wedge \omega \bmod L^* \wedge L^* & \\ \text{(v)} & dA_\mu \equiv A_{\mu\nu}\pi^\nu \bmod L^*, & A_{\mu\nu} = A_{\nu\mu} . \end{cases} \tag{IV.c.9}$$

All of this is on X. Next we recall that

$$Y \subset X \times \mathbb{R}^s$$

is defined by

$$\lambda_\mu = A_\mu \quad ,$$

and that under the assumption $\det\|A_{\mu\nu}\| \neq 0$ we may take

$$\{\omega; \theta^\alpha; d\lambda_\alpha\} \tag{IV.c.10}$$

as a coframe on Y. Moreover, we have

$$\begin{cases} \text{(i)} & \psi_Y = \varphi + \lambda_\alpha \theta^\alpha \\ \text{(ii)} & \Psi_Y \overset{\equiv}{\equiv} (d\lambda_\alpha + \xi_\alpha \omega) \wedge \theta^\alpha \end{cases} \tag{IV.c.11}$$

where "$\overset{\equiv}{\equiv}$" means congruent modulo a linear combination of terms $\theta^\alpha \wedge \theta^\beta$ (cf. (II.b.4)).

Proof of the (ii) in (IV.c.11). Using (IV.c.9) we have

$$\begin{aligned} d\psi_Y &\equiv (A_\mu - \lambda_\mu)\pi^\mu \wedge \omega + d\lambda_\alpha \wedge \theta^\alpha \bmod L^* \wedge L^* \\ &\overset{\equiv}{\equiv} d\lambda_\alpha \wedge \theta^\alpha \bmod L^* \wedge L^* \end{aligned}$$

since $\lambda_\mu = A_\mu$ on Y. Terms in $L^* \wedge L^*$ are a sum of either $\theta^\alpha \wedge \theta^\beta$'s or $\omega \wedge \theta^\alpha$'s, and we may absorb the latter into $\xi_\alpha \omega \wedge \theta^\alpha$'s. Q.E.D.

Actually, in (IV.c.11) we should write $\pi^*\varphi$, $\pi^*\theta^\alpha$ but for simplicity of notation we don't do this. We will however write $(\pi^* I, \omega)$ for the pullback to Y of the exterior differential system (I, ω) on X. We claim that:

(IV.c.12) *The Cauchy characteristic system of* $(\pi^* I, \omega)$ *is the distribution* $F \subset T(Y)$ *given by*

$$F = \operatorname{span}\{\partial/\partial\lambda_\rho\} \quad .$$

This is an immediate consequence of the definition of Cauchy characteristics and the structure equations (IV.c.9).[6] We remark that $F = \ker \pi_*$ is the tangent space to the fibres of $\pi: Y \to X$. It is interesting to observe that the "functions λ_ρ to be determined" appear in this intrinsic way.

Next we consider the endpoint conditions *up on* Y (again omitting the π^*'s)

$$\omega = \theta^1 = \cdots = \theta^s = 0 \quad .$$

These define an integrable sub-bundle, which we shall refer to as the *endpoint sub-bundle*,

$$S \subset T(Y) \qquad \text{(IV.c.13)}$$

of rank s. Since $\dim Y = 2s+1$ and

$$\psi_Y \wedge (\Psi_Y)^s \neq 0 \qquad \text{(IV.c.14)}$$

the closed 2-form Ψ_Y has everywhere the maximal possible rank s, and as a consequence of (IV.c.9) and (IV.c.11) we have

(IV.c.15) *The endpoint sub-bundle* S *is isotropic for the 2-form* Ψ_Y. This means that

$$\langle \Psi_Y, V \wedge W \rangle = 0 \qquad \text{if} \quad V, W \in S \ .$$

Now we turn attention to the fibration

$$\tilde{\omega}: Y \to Q \qquad \text{(IV.c.16)}$$

over the reduced momentum space, where the endpoint sub-bundle $S \subset T(Y)$ is the tangent bundle along the fibres. We claim that:

(IV.c.17) *The* 1st *derived system* $((\pi^* I)_1, \omega)$ *of* $(\pi^* I, \omega)$ *is the pullback to* Y *of a Pfaffian system* (G, ω) *on* Q.

<u>Proof</u>. The 1-forms $\omega, \theta^1, \ldots, \theta^s$ are by definition all horizontal (cf. Chapter 0, Section a) for the fibering (IV.c.16). However, using the structure equations (IV.c.9) we infer that from among these only

$$\omega, \theta^1, \ldots, \theta^{s-s_1}$$

have exterior derivatives that are also horizontal for (IV.c.16). Now on the one hand the 1st derived system of $(\pi^* I, \omega)$ is

$$\theta^1 = \cdots = \theta^{s-s_1} = 0, \ \omega \neq 0 \ ,$$

while on the other hand it is well-known (and easy to prove) that the condition that a differential form η on Y be $\tilde{\omega}^*\xi$ for a form ξ on Q is that both η and $d\eta$ be horizontal for (IV.c.16) (cf. Chapter 0, Section a). This proves (IV.c.17).

Since we shall be working in a neighborhood of a 1-dimensional submanifold $N \subset X$, recalling that N is an interval with endpoints we may draw the following conclusions:

(IV.c.18) *Each integral manifold* $N \subset X$ *of* (I,ω) *is the projection of an integral manifold* $\Gamma \subset Y$ *of* $(\pi^* I,\omega)$. *Moreover,* Γ *is unique up to the* $\mathbb{R}^{s-s_1}$*-action given by the Cauchy characteristics of* $(\pi^* I,\omega)$. *Each integral manifold* $\Gamma \subset Y$ *of* $(\pi^* I,\omega)$ *projects to an integral manifold* $\gamma \subset Q$ *of* (G,ω).

Concerning endpoint conditions, we lift $N \in V(I,\omega)$ to $\Gamma \in V(\pi^* I,\omega)$ and then project via $\tilde{\omega}$ to obtain $\gamma \in V(G,\omega)$. The endpoint conditions A,B for N are independent of the lifting Γ (since $\langle \omega, \partial/\partial\lambda_\rho \rangle = \langle \theta^\alpha, \partial/\partial\lambda_\rho \rangle = 0$) and exactly mean that γ should join two fixed points of the reduced momentum space Q.

All of this general discussion is quite clear in examples (IV.a.17)--(IV.a.20).

ii) *Transferrance of the* 2nd *variation problem to the momentum space.* We now come to one of the main points of the discussion. Because of the generality of the variational problem $(I,\omega;\varphi)$ (even under the assumption of strong non-degeneracy) the situation down on X may be quite complicated. However, due to the two facts:

a) that for $N \in V(I,\omega)$

$$\int_N \varphi = \int_\Gamma \psi_Y \quad , \qquad \pi(\Gamma) = N, \text{ (7)} \qquad \text{(IV.c.19)}$$

b) that ψ_Y has a local Pfaff-Darboux normal form (cf. (0.d.9)), the situation up on Y turns out to be considerably simpler.

In more detail, let $N \in V(I,\omega;[A,B]))$ be a solution of the Euler-Lagrange equations. Then $\delta\Phi(N) = 0$, and we shall denote by

$$\delta^2\Phi(N): \mathrm{Sym}^2(T_N(V(I,\omega;[A,B]))) \to \mathbb{R} \qquad \text{(IV.c.20)}$$

the 2nd *variation* (or *Hessian*) of the functional Φ.[8] Given $v,w \in T_N(V(I,\omega;[A,B]))$ there will be a 2-parameter family

$$N_{s,t} \in \mathcal{V}(I,\omega;[A,B])$$

such that v corresponds to $\partial/\partial s$ and w to $\partial/\partial t$, both evaluated on $N = N_{0,0}$. By definition, the 2nd variation is

$$\delta^2\Phi(N)(v,w) = \frac{\partial^2}{\partial s \partial t}\left(\int_{N_{s,t}} \varphi\right)_{s=t=0} . \qquad \text{(IV.c.21)}$$

We want to obtain a formula for the right hand side, and use this formula to show that under the conditions of theorem (IV.c.7)

$$\delta^2\Phi(N)(v,v) > 0 \qquad \text{for} \quad 0 \neq v \in T_N(\mathcal{V}(I,\omega;[A,B])). \qquad \text{(IV.c.22)}$$

By a standard argument this will be sufficient to establish our result.[9]

We begin by smoothly lifting $\{N_{s,t}\}$ to a family of integral manifolds $\{\Gamma_{s,t}\}$ of $(\pi^* I,\omega)$ on Y and letting V,W be the normal vector fields to $\Gamma = \Gamma_{0,0}$ corresponding respectively to $\partial/\partial s$, $\partial/\partial t$. Then by (IV.c.19) the right hand side of (IV.c.21) is

$$\frac{\partial^2}{\partial s \partial t}\left(\int_{\Gamma_{s,t}} \psi_Y\right)_{s=t=0} . \qquad \text{(IV.c.23)}$$

We will prove the basic formula

$$\delta^2\Phi(N)(v,w) = \int_\Gamma V \lrcorner\, d(W \lrcorner\, \Psi_Y) . \qquad \text{(IV.c.24)}$$

<u>Proof</u>. The left hand side is

$$\frac{\partial}{\partial s}\left(\left(\frac{\partial}{\partial t}\int_{\Gamma_{s,t}} \psi_Y\right)_{t=0}\right)_{s=0}$$

$$= \frac{\partial}{\partial s}\left(\int_{\Gamma_{s,0}} W \lrcorner\, \Psi_Y + d(W \lrcorner\, \psi_Y)\right)_{s=0}$$

by (I.b.5)

$$= \frac{\partial}{\partial s}\left(\int_{\Gamma_{s,0}} W \lrcorner\, \Psi_Y \right)_{s=0}$$

by Stokes' theorem and our endpoint condition $\langle\psi_Y, W\rangle = 0$ on $\partial\Gamma_{s,0}$

$$= \int_{\Gamma} V \lrcorner d(W \lrcorner \Psi_Y) + d(V \lrcorner W \lrcorner \Psi_Y)$$

by (I.b.5) again

$$= \int_{\Gamma} V \lrcorner d(W \lrcorner \Psi_Y) + \langle \Psi_Y, V \wedge W \rangle_{\partial\Gamma}$$

by Stokes' theorem. Now, and this is the main point, *by (IV.c.15) the last term* $\langle \Psi_Y, V \wedge W \rangle_{\partial\Gamma}$ *is zero.* Q.E.D.

Remark. What will emerge is this: Up on Y the essential data consists of:

a) a 1-form ψ_Y of maximal rank;

b) an integrable sub-bundle $S \subset T(Y)$; and

c) an additional sub-bundle S' with $S \subset S' \subset T(Y)$, where S' is the pullback $\tilde{\omega}^*(G^{\perp})$ of the sub-bundle $G^{\perp} \subset T(Q)$ on the reduced momentum space defined by the 1st derived system of $(\pi^* I, \omega)$.

Since $\psi_Y | S = 0$ we may to some extent put a) and b) together in local normal form. More precisely, if S is locally defined by

$$dx = dv^1 = \cdot\cdot = dv^s = 0$$

for suitable functions $x, v^1, \ldots, v^s$, then we must have (cf. (0.d.9))

$$\psi_Y = -H(x,u,v)\,dx + u_\alpha dv^\alpha$$

for appropriate functions $H, u_1, \ldots, u_s$. Since ψ_Y has maximal rank we may assume that $(x; u_\alpha; v^\alpha)$ gives a coordinate system and consider H as a function of these variables.

However, the equations

$$\xi^\rho = A^\rho dx + B^\rho_\alpha dv^\alpha = 0$$

that define S' are generally of a complicated character; what will make everything work is the fact (IV.c.17): *we may assume that* $\xi^\rho = \tilde{\omega}^* \eta^\rho$ *is the pullback of a 1-form from the reduced momentum space* Q.

iii) *The Jacobi equations and the* 2nd *variation.* Classically it is well-known that there is an intimate connection between the Jacobi equations and the 2nd variation (historically this was a fundamental discovery--cf. §315 of [13]). In our general situation it is not immediately clear how the connection should generalize; for example, the 2nd variation pertains to an integral defined on X and the Jacobi equations are defined on Y. However, (IV.c.24) provides the link.

Before explaining this we must digress. We recall that the Euler-Lagrange system (J,ω) on Y is (due to strong non-degeneracy--cf. Theorem (I.e.34)) the Cartan system $(C(\Psi_Y),\omega)$. The Jacobi equations were defined to be the variational equations of (J,ω). Let $\Gamma\subset Y$ be an integral manifold of (J,ω) with normal bundle E. According to the discussion in Chapter I, Section b) the variational equations of (J,ω) are given by a 1st order linear differential operator

$$L: C^{\infty}(\Gamma,E)\to C^{\infty}(\Gamma,E\otimes T^*(\Gamma)) \quad .^{(10)} \qquad \text{(IV.c.25)}$$

On the other hand, since $J=C(\Psi_Y)$ *the tangent lines to* Γ *are just the characteristic directions* $\Psi_Y^{\perp}$ *of* Ψ_Y.[(11)] It follows that Ψ_Y induces a *non-degenerate* pairing

$$\Psi_Y: E\otimes E\to \mathbb{R} \quad . \qquad \text{(IV.c.26)}$$

Definition. For normal vector fields V,W to $\Gamma\subset Y$ we define the *index form* by

$$I(V,W) = \int_{\Gamma} \Psi_Y(W,L\cdot V) \quad , \qquad \text{(IV.c.27)}$$

where the integrand is given by the pairing (IV.c.26).

In suitable local coordinates this is all quite transparent: According to the theorem of Darboux (0.d.8) we may choose local coordinates $(x;u_1..,u_s;v^1,..,v^s)=(x;u_\alpha;v^\alpha)$ on Y so that

$$\Psi_Y = du_\alpha\wedge dv^\alpha \qquad \text{(IV.c.28)}$$

and Γ is given by $u_\alpha=v^\alpha=0$. Then $C(\Psi_Y)$ is the Pfaffian system

$$du_\alpha = dv^\alpha = 0 \ ,$$

and for a normal vector field

$$V = V^\alpha\,\partial/\partial u_\alpha + V_\alpha\,\partial/\partial v^\alpha$$

we have

$$LV = dV^\alpha\otimes\partial/\partial u_\alpha + dV_\alpha\otimes\partial/\partial v^\alpha \quad .$$

Then if

$$W = W^\alpha\,\partial/\partial u_\alpha + W_\alpha\,\partial/\partial v^\alpha$$

it follows that

$$\Psi_Y(W, L\cdot V) = dV_\alpha W^\alpha - dV^\alpha W_\alpha$$

and

$$I(V,W) = \int_\Gamma W^\alpha dV_\alpha - W_\alpha dV^\alpha \quad . \qquad \text{(IV.c.29)}$$

Remark. In these local coordinates the Jacobi equations

$$dV^\alpha = dV_\alpha = 0$$

are absolutely trivial (the solutions are *constant* normal vector fields). If we also put in the endpoint conditions given by the distribution $S \subset T(Y)$ then we still have a normal form. This is a little more subtle and uses (IV.c.15) plus a refinement of the Darboux theorem, according to which if we are given i) a closed 2-form Ψ_Y of maximal rank s on a $(2s+1)$-dimensional manifold, and ii) an integrable distribution $S \subset T(Y)$ of maximal isotropic $(s+1)$-planes for Ψ_Y, then we may choose local coordinates so that (IV.c.28) holds and S is given by $dv^1 = \cdots = dv^s = 0$.

From (IV.c.29) and integration by parts it follows if V,W satisfy the endpoint conditions, so that in particular

$$\langle \Psi_Y, V \wedge W \rangle_{\partial\Gamma} = 0 \ ,$$

then

$$I(V,W) = I(W,V) \ .$$

Thus, the index is a *symmetric bilinear form* on normal vector fields to Γ satisfying the endpoing conditions. In fact we have the following generalization of the main classical result:

(IV.c.30) THEOREM. *The 2nd variation and index are related by*

$$\delta^2\Phi(N)(v,w) = I(V,W) \ .$$

Here, V and W are normal vector fields to any lift Γ of N that project down to v and w, respectively.

Proof of (IV.c.30). In local coordinates for which the normal form (IV.c.28) is valid it is immediate that

$$W \lrcorner\, d(V \lrcorner\, \Psi_Y) = W \lrcorner\, (dV^\alpha \wedge dv^\alpha - dV_\alpha \wedge du_\alpha)$$

$$\equiv (W^\alpha dV_\alpha - W_\alpha dV^\alpha) \bmod C(\Psi_Y) \ .$$

The theorem follows from (IV.c.24) by integrating over Γ using that $C(\Psi_Y) \equiv 0 \bmod \Gamma$. Q.E.D.

Proof of (IV.c.22). By using (IV.c.30) it will suffice to prove that the index

$$I(V,V) > 0 \quad .$$

Here we are working up on Y, and V is any non-zero infinitesimal variation of Γ as an integral manifold of $(\pi^* I,\omega)$ preserving the endpoint conditions. By (IV.c.12) the choice of V lying over a given $v \in T_N(\mathcal{V}(I,\omega;[A,B]))$ is only unique modulo Cauchy characteristics; for convenience we take V to have the form

$$V = V^{\mu}\, \partial/\partial\lambda_{\mu} + V_{\rho}\, \partial/\partial\theta^{\rho} + V_{\mu}\, \partial/\partial\theta^{\mu}. \qquad \text{(IV.c.31)}$$

Using the structure equations (IV.c.9) the conditions that define V (cf. (I.b.15))

$$\begin{cases} d(V \lrcorner\, \theta^{\rho}) + V \lrcorner\, d\theta^{\rho} \equiv 0 \bmod \Gamma \\ d(V \lrcorner\, \theta^{\mu}) + V \lrcorner\, d\theta^{\mu} \equiv 0 \bmod \Gamma \end{cases}$$

are respectively

$$\begin{cases} dV_{\rho} - B^{\alpha}_{\rho} V_{\alpha}\omega \equiv 0 \bmod \Gamma \\ dV_{\mu} - (C_{\mu\nu}V^{\nu} + B^{\alpha}_{\mu}V_{\alpha})\omega \equiv 0 \bmod \Gamma \end{cases} \qquad \text{(IV.c.32)}$$

where we have written

$$\pi^{\mu} \equiv C_{\mu\nu} d\lambda_{\nu} \bmod \operatorname{span}\{\omega,\theta^{\alpha}\} \quad .$$

Recalling that $Y \subset X \times \mathbb{R}^s$ is defined by $\lambda_{\mu} = A_{\mu}$, we have on Y

$$d\lambda_{\mu} \equiv A_{\mu\nu}\pi^{\nu} \bmod \operatorname{span}\{\omega,\theta^{\alpha}\} \quad .$$

Thus $\|C_{\mu\nu}\|$ is the inverse matrix to $\|A_{\mu\nu}\|$, and by our assumption

$$\|A_{\mu\nu}\| > 0 \quad . \qquad \text{(IV.c.33)}$$

Taking x as local coordinate along Γ, (IV.c.32) is the O.D.E. system

$$\begin{cases} \dfrac{dV_{\rho}}{dx} = B^{\alpha}_{\rho} V_{\alpha} \\[2ex] \dfrac{dV_{\mu}}{dx} = C_{\mu\nu}V^{\nu} + B^{\alpha}_{\mu}V_{\alpha} \end{cases} \qquad \text{(IV.c.34)}$$

where $\|B^{\alpha}_{\beta}\|$ is some matrix of functions.(12) If Γ is the interval $a \leqq x \leqq b$, then the endpoint conditions (IV.c.13) are

$$V_{\alpha}(a) = V_{\alpha}(b) = 0 \qquad \alpha = 1,\ldots,s \; . \qquad \text{(IV.c.35)}$$

On the other hand, by (IV.c.11) we have

$$\Psi_Y \equiv (d\lambda_\alpha + \xi_\alpha \omega) \wedge \theta^\alpha \quad .$$

Since (cf. the discussion in Chapter I, Section c))

$$LV = (B^\alpha_\rho V_\alpha)\partial/\partial\theta^\rho + (C_{\mu\nu}V^\nu + B^\alpha_\mu V_\alpha)\partial/\partial\theta^\mu$$

it follows that the integrand in the index form is an expression of the form

$$\Psi_Y(V,LV) = C_{\mu\nu}V^\mu V^\nu + B^\alpha_\mu V_\alpha V^\mu + B^{\alpha\beta}V_\alpha V_\beta \quad .$$

Thus

$$I(V,V) = \int_a^b (C_{\mu\nu}V^\mu V^\nu + B^\alpha_\mu V_\alpha V^\mu + B^{\alpha\beta}V_\alpha V_\beta)\, dx \quad .$$

Recalling that $\|A_{\mu\nu}\|$ is the inverse matrix to $\|C_{\mu\nu}\|$ we have from (IV.c.34) that

$$V^\mu = A_{\mu\nu}\frac{dV_\mu}{dx} + C^\alpha_\mu V_\alpha \ ,$$

and then the above integral is of the form

$$I(V,V) = \int_a^b \left(A_{\mu\nu}\frac{dV_\mu}{dx}\frac{dV_\nu}{dx} + D^{\alpha\mu}\frac{dV_\mu}{dx}V_\alpha + E^{\alpha\beta}V_\alpha V_\beta\right)dx \qquad \text{(IV.c.36)}$$

By (IV.c.33) we will have, for any vector $\xi = (\xi_\mu)$,

$$A_{\mu\nu}\xi_\mu\xi_\nu \geqq \delta\|\xi\|^2 \qquad \text{(IV.c.37)}$$

for some $\delta > 0$. If we use the Cauchy-Schwarz inequality in the "rob Peter to pay Paul" form

$$2|(\xi,\zeta)| \leqq \varepsilon\|\xi\|^2 + \frac{1}{\varepsilon}\|\zeta\|^2$$

for vectors ξ,ζ, then (IV.c.36) and (IV.c.37) give, for suitable constants $c_1 > 0$ and c_2

$$I(V,V) \geqq c_1 \sum_\mu \int_a^b \left(\frac{dV_\mu}{dx}\right)^2 dx - c_2 \sum_\alpha \int_a^b V_\alpha^2\, dx \ . \qquad \text{(IV.c.38)}$$

We now use the well-known

(IV.c.39) LEMMA. *Let* $f(x) \in C^1[a,b]$ *satisfy* $f(a) = 0$. *Then*

$$\int_a^b f(x)^2\, dx \leqq \frac{(b-a)^2}{2}\left(\int_a^b f'(x)^2\, dx\right) \quad .$$

Proof of Lemma. We have

$$(f(x))^2 = \left(\int_a^x f'(t)\,dt\right)^2$$
$$\leqq (x-a)\left(\int_a^x f'(t)^2\,dt\right)$$

by the Cauchy-Schwarz inequality

$$\leqq (x-a)\left(\int_a^b f'(t)^2\,dt\right) .$$

Now integrate both sides for $a \leqq x \leqq b$. Q.E.D.

We now may complete the proof of the theorem. Choosing our end-point conditions sufficiently close and using the lemma together with (IV.c.34) gives

$$\begin{cases} \int_a^b V_\rho^2\,dx \leqq \varepsilon \sum_\alpha \int_a^b V_\alpha^2\,dx \\ \int_a^b V_\mu^2 \leqq \varepsilon\left(\sum_\nu \int_a^b \left(\frac{dV_\nu}{dx}\right)^2 dx + \sum_\alpha \int_a^b V_\alpha^2\,dx\right) . \end{cases} \quad \text{(IV.c.40)}$$

It follows that the 2nd term on the right hand side of (IV.c.38) may be absorbed in the first; i.e., we will have

$$I(V,V) \geqq c_3 \sum_\mu \int_a^b \left(\frac{dV_\mu}{dx}\right)^2 dx$$

for some constant $c_3 > 0$. If all terms $\frac{dV_\mu}{dx} = 0$ it again follows from (IV.c.40) that $V \equiv 0$. Thus the index $I(V,V) > 0$. Q.E.D.

Examples. This theorem covers the classical cases in the calculus of variations, as for instance in example (I.e.18) when

$$\left\| L_{\dot{y}^\alpha \dot{y}^\beta} \right\| > 0$$

and in example (I.e.23) when

$$\left\| L_{\ddot{y}^\alpha \ddot{y}^\beta} \right\| > 0 .$$

More interestingly it applies to the functional

$$\Phi(N) = \frac{1}{2} \int_N (\kappa_1(s)^2 + \cdots + \kappa_\ell(s)^2)\,ds \quad \text{(IV.c.41)}$$

defined on curves $N \subset \mathbb{E}^n$ and where the endpoint conditions are given by fixing the ℓth osculating spaces at ∂N (cf. example (IV.a.20)).

For "$\ell = 0$" this is just the case of geodesics (actually we should take kinetic energy--but cf. (II.a.23) and example (IV.b.3)).

For $\ell = n-1$ the conditions of theorem (IV.c.7) are met in the setting of Chapter II, Section b).

For $1 \leqq \ell \leqq n-2$, if we formulate the variational problem up on $F(\mathbb{E}^n) \times \mathbb{R}^{n-1}$ (as was done in Chapter II, Section b) in the case $n = 3$, $\ell = 1$) then the quadratic form $\|A_{\mu\nu}\|$ is only positive semi-definite (for the functional corresponding to $L(\kappa_1,\ldots,\kappa_{n-1})$ we have $A_{\mu\nu} = \frac{\partial^2 L}{\partial\kappa_\mu \partial\kappa_\nu}$). However, the variational problem descends to one on the manifold $X_\ell = F(\ell,\mathbb{E}^n) \times \mathbb{R}^\ell$ where $F(\ell,\mathbb{E}^n)$ is the *Stiefel manifold* of all partial frames $(x;e_1,\ldots,e_\ell)$, and on X_ℓ the quadratic form associated to the variational problem is positive definite. In other words, although it is more convenient to *compute* the Euler-Lagrange system of (IV.c.41) up on $F(\mathbb{E}^n) \times \mathbb{R}^{n-1}$ (this is because $F(\mathbb{E}^n)$ is a Lie group), we must descend the variational problem by allowing ourselves to "spin arbitrarily the irrelevant part of the Frenet frame" in order to apply theorem IV.c.7.

d) Fields and the Hamilton-Jacobi Equation; Further Sufficient Conditions for a Local Minimum.

Two of the most important concepts in the classical calculus of variations are a field (sometimes called a geodesic field) and the Hamilton-Jacobi equation. We shall briefly discuss each of these in our general setting and shall then give a few examples and one application, which we state now (the proof appears at the end of this section).[13]

(IV.d.1) THEOREM. *Let* $(I,\omega;\varphi)$ *be a strongly non-degenerate variational problem on a manifold* X, *and let* $N \subset X$ *be an integral manifold of* (I,ω) *satisfying the conditions:*

i) N *is a solution of the Euler-Lagrange equations;*

ii) *the quadratic form* $\|A_{\mu\nu}\|$ *is positive definite along* N; *and*

iii) *no two points of* N *are conjugate.*

Then N *gives a local minimum of the functional (IV.c.3).*

Remarks. This result, which is fundamental for classical 1st order variational problems, includes theorem (IV.c.7) as a special case.

This may be seen by noting that no solution to the Jacobi differential system on Y can satisfy the endpoint conditions A, B if A and B are sufficiently close (this is a consequence of positivity and transversality in the definition of well-posed endpoint conditions).

The proof of the theorem will follow our discussion of fields and the Hamilton-Jacobi equation; again the point is to understand the geometry of the basic diagram

$$\begin{array}{ccc} & Y & \\ {\scriptstyle\tilde{\omega}}\swarrow & & \searrow{\scriptstyle\pi} \\ Q & & X \end{array} \quad . \qquad \text{(IV.d.2)}$$

We first discuss fields. Let $(I,\omega;\varphi)$ be a well-posed variational problem on X with Euler-Lagrange differential system (J,ω) on Y. We recall that J is the Cartan system $C(\Psi_Y)$ of the 2-form

$$\Psi_Y = d\psi_Y \quad . \qquad \text{(IV.d.3)}$$

We also recall that our generators $\theta^1,..,\theta^s = \underbrace{\theta^1,..,\theta^m}; \underbrace{\theta^{m+1},..,\theta^s}$ for I have been chosen in such a way that

i) $\lambda_{m+1} = \cdot\cdot = \lambda_s = 0$ on Y;

ii) $K^* = \operatorname{span}\{\omega,\theta^1,..,\theta^m\}$ is a completely integrable sub-bundle of $T^*(Y)$ whose corresponding foliation is given by the fibres of $\tilde{\omega}$; and

iii) $\psi_Y = \varphi + \lambda_\rho\theta^\rho \qquad 1 \leqq \rho \leqq m$.

We do not yet assume that $(I,\omega;\varphi)$ is strongly non-degenerate, so that for example the functional (IV.c.41) is included in our discussion.

So far we have made only occasional use of the relation (IV.d.3); now it will come into full force.

Definitions. i) By a *field* for $(I,\omega;\varphi)$ we shall mean a submanifold

$$S \subset Y$$

such that the restriction Ψ_S of Ψ_Y to S satisfies

$$\begin{cases} \Psi_S = 0 \\ \omega\wedge\theta^1\wedge\cdot\cdot\wedge\theta^m|S \neq 0 \end{cases} . \qquad \text{(IV.d.4)}$$

ii) The restriction ψ_S of ψ_Y to S is called *Hilbert's invariant integral*.

We shall give four observations on this definition.

Remarks. i) Let $I(\Psi_Y)$ be the differential ideal generated by the closed 2-form Ψ_Y and set

$$\Omega = \omega \wedge \theta^1 \wedge \cdots \wedge \theta^m \quad .$$

Then a field is simply an integral manifold of the exterior differential system $(I(\Psi_Y),\Omega)$.

ii) In the old literature differential form and "integral" meant the same thing (a differential form is "something you integrate"). An "invariant integral" simply meant a closed form, since by Stokes' theorem the path of integration could be deformed keeping the boundary fixed without changing the value of the integral. In the case at hand it follows that

$$d\psi_S = 0$$

for a field S, which explains the origin of the name Hilbert's invariant integral. We shall be primarily concerned with the situation when S is simply-connected, in which case we may define a function G on S by

$$G(s) = \int_{s_0}^{s} \psi_S \qquad \text{(IV.d.5)}$$

where $s_0 \in S$ is a base point.

iii) We recall our notation

$$\Psi_Y^{\perp} = \{(y,v) \in T(Y) : v \lrcorner (\Psi_Y(y)) = 0\}$$

for the line sub-bundle of $T(Y)$ given by the field of characteristic directions of Ψ_Y, and make the following observation:

(IV.d.6) *Let* $S \subset Y$ *be a field. Then the line bundle* $\Psi_Y^{\perp}$ *is tangent to* S. *In particular,* S *is fibered by integral curves of the Euler-Lagrange system* (J,ω).

Proof. The proof is elementary exterior algebra. For $y \in S$ we have $T_y(Y) \cong \mathbb{R}^{2m+1}$ and $\Psi_Y(y)^m \neq 0$. Suppose that $L \subset T_y(Y)$ is a linear subspace of *co*dimension m such that

$$(\Psi_Y(y))_L = 0 \quad . \qquad \text{(IV.d.7)}$$

If we assume that L is defined by

$$\alpha_1 = \cdots = \alpha_m = 0$$

where the $\alpha_i \in T_y^*(Y)$ are linearly independent, then (IV.d.7) implies that there exist $\beta^i \in T_y^*(Y)$ such that

$$\Psi_Y(y) = \alpha_1 \wedge \beta^1 + \cdots + \alpha_k \wedge \beta^k \quad .$$

Since $\Psi_Y(y)$ has rank m it follows that the α_i, β^i are linearly independent. Thus $\Psi_Y(y)$ is in the usual symplectic normal form. In particular

$$\{v \lrcorner \Psi_Y(y) : v \in T_y(Y)\} = \text{span}\{\alpha_i, \beta^i\} \subset T_y^*(Y) \quad ,$$

and consequently the characteristic direction

$$\Psi_Y^{\perp}(y) = (\text{span}\{\alpha_i, \beta^i\})^{\perp} \subset T_y(Y)$$

is contained in L. Q.E.D.

<u>Note</u>. A plane $L \subset T_y(Y)$ satisfying (IV.d.7) is said to be *isotropic* for the alternating bilinear form $\Psi_Y(y)$. Maximal isotropic planes (i.e., $\dim L = m+1$) are sometimes called *Lagrangian subspaces*.

(iv) For our last remark following the definition of a field and Hilbert's invariant integral, we shall show that the restriction (J_S, ω) of the Euler-Lagrange differential system (J, ω) to a field S is given by

$$\theta^1 = \cdots = \theta^m = 0, \qquad \omega \neq 0 \tag{IV.d.8}$$

(it is understood that all differential forms are restricted to S). We may write this as

$$J_S = (\pi^* I)_S \quad . \tag{IV.d.9}$$

For the proof, as previously noted by (IV.d.6) the field S is foliated by integral curves of the Pfaffian differential system J_S. Since $\{\theta^1, \ldots, \theta^m, \omega\}$ gives a coframe on S, the field is also foliated by the integral curves of $(\pi^* I)_S$. Finally, since it is always the case that

$$(\pi^* I)_S \subset J_S$$

(this is because $\pi^* I \subset J$), we conclude the equality (IV.d.9).

<u>Definition</u>. A function g defined on an open subset $R \subset Q$ of the reduced momentum space is a solution to the *Hamilton-Jacobi equation* associated to $(I, \omega; \varphi)$ in case there is a cross-section

$$s : R \to Y \qquad\qquad \tilde{\omega} \circ s = \text{identity}$$

such that

$$dg = s^*(\psi_Y) \quad . \qquad \text{(IV.d.10)}$$

We shall give three observations on this definition (the 2nd of these is lengthy).

Remarks. i) From

$$0 = d^2g = s^*(\Psi_Y)$$

it follows that the image

$$s(R) = S \subset Y$$

is a field and that there is a function G on S such that

$$G(s) = \int_{s_0}^{s} \psi_S + \text{(constant)}$$

where $\psi_S = (\psi_Y)_S$. In fact, G is defined to be the unique function on S with $s^*G = g$, and it follows from (IV.d.10) that

$$dG - \psi_S = 0 \quad .$$

Thus a solution to the Hamilton-Jacobi equation defines a field $S \subset Y$ such that ψ_S is exact (i.e., in fancy terms $\psi_S = 0$ in the deRham cohomology group $H^1_{DR}(S,\mathbb{R})$).

Conversely, given a field $S \subset Y$ such that $S \to Q$ is injective and ψ_S is exact we obtain a solution to the Hamilton-Jacobi equation.

ii) A variant of this construction is an analogue in our setting of the classical action function, which we now pause to discuss.

(IV.d.11) Interlude. *Discussion of the action function for certain non-degenerate variational problems.*

Suppose that $(I,\omega;\varphi)$ is a strongly non-degenerate variational problem on a manifold X and let $q_0 \in Q$. From each point y in the fibre F_{q_0} of $\tilde{\omega}: Y \to Q$ there issues a unique solution curve Γ_y to the Euler-Lagrange differential system (J,ω) on Y, and we set $\gamma_y = \tilde{\omega}(\Gamma_y)$. The map

$$F_{q_0} \to \mathbb{P}T_{q_0}(Q)$$

given by

$$y \to \{\text{tangent direction to } \gamma_y \text{ at } q_0\}$$

is well-defined, and for the purposes of this discussion we will assume it is a diffeomorphism (think of the case of geodesics).

This is the picture in the case $n=2$.

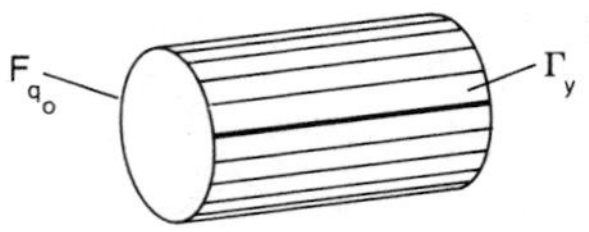

$F_{q_0} = S^1$ of tangent directions at q_0

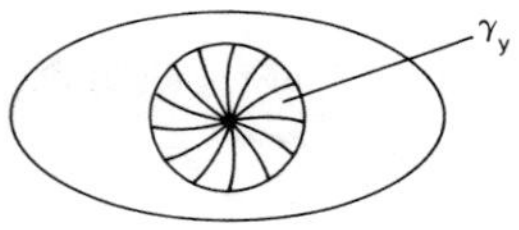

Under the map $\tilde{\omega}$, F_{q_0} is contracted to the point q_0, and the lines Γ_y on the cylinder map down to the curves γ_y issuing from q_0.

We let S be a neighborhood of F_{q_0} in $\cup_{y \in F_{q_0}} \Gamma_y$, chosen sufficiently small so that the map

$$\tilde{\omega}: S \backslash F_{q_0} \to Q \backslash \{q_0\}$$

is a diffeomorphism onto its image. We let U be the image $\tilde{\omega}(S)$, so that U is a neighborhood of $q_0 \in Q$ and we have a cross-section

$$s: U \backslash \{q_0\} \to S \backslash F_{q_0} \qquad (s = \tilde{\omega}^{-1}) \quad .$$

<u>Note</u>. In case $(I,\omega;\varphi)$ arises from geodesics on a Riemannian metric on a manifold M (cf. Chapter IV, Section b)), the reduced momentum space $Q = M$ and U may be thought of as a neighborhood of q_0 such that every $q \in U$ is joined to q_0 by a unique geodesic. Assuming that the metric is complete, we may take U to be a geodesic ball whose radius is the distance to the nearest conjugate point.

Setting $\Psi_S = (\Psi_Y)_S$, we claim that

$$\Psi_S = 0 \quad . \tag{IV.d.12}$$

In particular this implies that $S \backslash F_{q_0}$ is a field lying over $U \backslash \{q_0\}$.

<u>Proof of (IV.d.12)</u>. Let V be a vector field spanning the field of characteristic directions $\Psi_Y^{\perp}$ of Ψ_Y. Then:

$$\begin{cases} \text{(i)} & V \lrcorner \Psi_Y = 0 & \text{(by definition)} \\ \text{(ii)} & V \text{ is tangent to } S & \text{(by construction of } S) \\ \text{(iii)} & L_V(\Psi_S) = 0 & \text{(by (i) and (0.a.1))} . \end{cases}$$

Using (iii) it follows by integration that

$$\exp(tV)^* \Psi_S = \Psi_S ,$$

so that it will suffice to verify that

$$<\Psi_Y;W,W'> = 0$$

for a pair of tangent vectors $W, W' \in T_y(S)$ at a point $y \in F_{q_0}$. But then

$$T_y(S) \cong T_y(F_{q_0}) \oplus \Psi_S(y)^{\perp} ,$$

and (IV.d.12) follows from (IV.c.15) and (i). Q.E.D.

Definition. We define the *action function* $A(q)$ $(q \in U)$ by

$$A(q) = \int_{q_0}^{q} s^*(\varphi_S) ,$$

where the integral is taken along the unique curve γ_y joining q_0 to q.

Remark. The notation means that γ_y is the projection of an integral curve Γ_y of the Euler-Lagrange system on Y.

(IV.d.13) PROPOSITION. *The action function is a solution to the Hamilton-Jacobi equation.*

Proof. We have essentially given the argument during the course of the above discussion. Since Γ_y is an integral curve of $(\pi^* I, \omega)$ the action function is

$$\begin{aligned} A(q) &= \int_{\Gamma_y} \varphi_S \\ &= \int_{\Gamma_y} \psi_S \qquad (\psi_S = \psi_Y | S). \end{aligned}$$

On the other hand, since by (IV.d.9) and (IV.d.12)

$$\begin{cases} (\psi_S)_{F_{q_0}} = 0 \\ d\psi_S = \Psi_S = 0 \end{cases}$$

we have

$$A(q) = s^*\left(\int_{y_0}^{y} \psi_S\right)$$

where $y_0 \in F_{q_0}$ is fixed and the integral is taken along *any* path from y_0 to y. Thus

$$dA = s^*(G)$$

where the function G on S is defined by

$$G(y) = \int_{y_0}^{y} \psi_S .$$

This completes the proof of proposition (IV.d.13).

(iii) By proposition (IV.d.13) (or, more precisely, by a generalization of this proposition when q_0 is replaced by an arbitrary "non-characteristic" initial hypersurface), we may determine a general solution to the classical Hamilton-Jacobi equation (cf. example (IV.d.14) below) if we have solved the Euler-Lagrange differential system, which in the classical case amounts to solving Hamilton's canonical equations. As explained in [2], it is Jacobi's theorem that conversely knowing a general solution to the Hamilton-Jacobi equations allows one to integrate Hamilton's equation by quadratures, and this has provided the most effective method for doing this in problems of classical mechanics (loc. cit.).

For a general strongly non-degenerate variational problem it is reasonable to also expect that solving the Hamilton-Jacobi equation would allow one to integrate the Euler-Lagrange system by quadratures, and that this would provide an effective method for treating specific examples going substantially beyond the techniques of this text.

(IV.d.14) Example. We consider a non-degenerate classical variational problem (cf. (I.e.18))

$$\Phi = \int L\left(x, y(x), \frac{dy(x)}{dx}\right) dx$$

where $y(x) = (y^1(x), \ldots, y^m(x))$ and

$$\det \left\| L_{\dot{y}^\alpha \dot{y}^\beta} \right\| \neq 0 .$$

Then X has coordinates $(x; y^\alpha; \dot{y}^\alpha)$ with (I, ω) being given by

$$\begin{cases} \theta^\alpha = dy^\alpha - \dot{y}^\alpha dx = 0 \\ \omega = dx \neq 0 \end{cases},$$

and $Z = X \times \mathbb{R}^m$ has coordinates $(x;y^\alpha;\dot{y}^\alpha;\lambda_\alpha)$ with $Y \subset Z$ being defined by

$$L_{\dot{y}^\alpha} = \lambda_\alpha \quad .$$

Moreover, $(x;y^\alpha;\lambda_\alpha)$ are local coordinates on Y, and we may use either $\{\omega;\theta^\alpha;d\lambda_\alpha\}$ or $\{dx;dy^\alpha;d\lambda_\alpha\}$ as a coframe on Y. We also recall that

$$\begin{aligned} \psi_Y &= L\omega + \lambda_\alpha \theta^\alpha \\ &= -Hdx + \lambda_\alpha dy^\alpha \end{aligned}$$

where

$$H(x;y^\alpha;\lambda_\alpha) = -L + \dot{y}^\alpha \lambda_\alpha$$

is the Hamiltonian.

A submanifold $S \subset Y$ of dimension $m+1$ on which

$$\omega \wedge \theta^1 \wedge \cdots \wedge \theta^m = dx \wedge dy^1 \wedge \cdots \wedge dy^m \neq 0$$

is locally given by

$$(x,y) \to (x,y,\lambda(x,y)) \qquad (y = (y^\alpha) \text{ and } \lambda = (\lambda_\alpha)). \tag{IV.d.15}$$

Since

$$\Psi_Y = -dH \wedge dx + d\lambda_\alpha \wedge dy^\alpha$$

the conditions that S be a field are

$$\begin{cases} \text{(i)} & \dfrac{\partial \lambda_\alpha}{\partial y^\beta} = \dfrac{\partial \lambda_\beta}{\partial y^\alpha} \\ \text{(ii)} & \dfrac{\partial H}{\partial y^\alpha} + \dfrac{\partial \lambda_\alpha}{\partial x} + \dfrac{\partial H}{\partial \lambda^\beta} \dfrac{\partial \lambda_\beta}{\partial y^\alpha} = 0 \quad . \end{cases}$$

In (ii) the derivatives of H are evaluated at points $(x,y,\lambda(x,y))$.

In a non-degenerate classical variational problem the reduced momentum space is just $J^0(\mathbb{R},\mathbb{R}^m) = \mathbb{R}^{m+1}$ with coordinates $(x;y^\alpha)$. A cross-section of

$$\tilde{\omega}: Y \to Q$$

is given by (IV.d.15) where $\lambda(x,y)$ is defined over an open set $R \subset \mathbb{R}^{m+1}$. The conditions that $S = s(R)$ be a field are given by (i) and (ii) above. In case R is simply-connected by (i) we may determine a function $g(x,y)$, which is uniquely defined up to a function of x alone, such that

$$\lambda_\alpha(x,y) = \frac{\partial g(x,y)}{\partial y^\alpha} .$$

Then (ii) is

$$\frac{\partial}{\partial y^\alpha}\left(\frac{\partial g(x,y)}{\partial x} + H\left(x,y,\frac{\partial g(x,y)}{\partial y}\right)\right) = 0$$

where

$$\frac{\partial g(x,y)}{\partial y} = \left(\frac{\partial g(x,y)}{\partial y^1},\ldots,\frac{\partial g(x,y)}{\partial y^m}\right).$$

To have a solution $g(x,y)$ to the Hamilton-Jacobi equation means that we are given over some open set $R\subset\mathbb{R}^{m+1}$ a cross-section (IV.d.15) such that

$$dg(x,y) = s^*(\psi_Y)$$

where

$$s(x,y) = (x,y,\lambda(x,y)) .$$

The condition $dg = s^*(\psi_Y)$ is

$$\frac{\partial g(x,y)}{\partial x}dx + \frac{\partial g(x,y)}{\partial y^\alpha}dy^\alpha = -H(x,y,\lambda(x,y))dx + \lambda_\alpha(x,y)dy^\alpha .$$

It follows that

$$\lambda_\alpha(x,y) = \frac{\partial g(x,y)}{\partial y^\alpha} ,$$

and

$$\frac{\partial g(x,y)}{\partial x} + H\left(x,y,\frac{\partial g(x,y)}{\partial y}\right) = 0 .$$

This is the usual form of the Hamilton-Jacobi equation for a classical variational problem.

We refer to [2], [5], and [29] for examples of how the classical Hamilton-Jacobi equation may be used to integrate the Euler-Lagrange equations arising from problems of classical mechanics.

Remark. Referring to the interlude (IV.d.11), and especially to proposition (IV.d.13), we consider the action function

$$A(x,y) = \int_{(x_0,y_0)}^{(x,y)} s^*(\varphi)$$

for a classical variational problem where

$$\begin{cases} \varphi = L(y,\dot{y})\,dx \\ \left\| L_{\dot{y}^\alpha \dot{y}^\beta} \right\| > 0 \end{cases} .$$

If $L = T-U$ corresponds to a mechanical system, $A(x,y)$ represents the least action required when the system evolves from (x_0,y_0) to (x,y).

Definition. The levels sets $V_{\underline{c}}$ defined by

$$A(\underline{x},y) = \text{constant } \underline{c} \qquad (x = \underline{x} \text{ fixed})$$

are hypersurfaces in $\mathbb{R}^m$ called *wave fronts*.

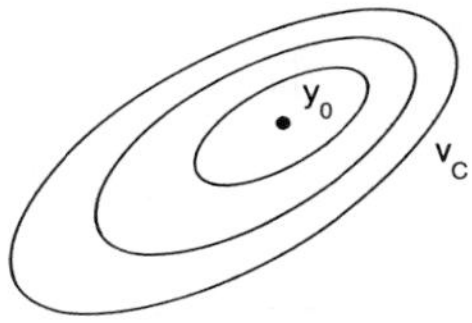

The interior of V_c represents all points $y \in \mathbb{R}^m$ reachable from y_0 by action less or equal to $\underline{c}$ in time less or equal to $\underline{x}$.

We consider the Euler-Lagrange differential system on $Z = J^1(\mathbb{R},\mathbb{R}^m) \times \mathbb{R}^m$. Recalling that Z has coordinates $(x;y^\alpha;\dot{y}^\alpha;\lambda_\alpha)$ we consider a solution curve

$$x \to (x;y^\alpha(x);\dot{y}^\alpha(x);\lambda_\alpha(x))$$

with $y(0) = y_0$. Then $x \to y(x)$ is a curve in $\mathbb{R}^m$ with tangent vector

$$\dot{y}(x) = \dot{y}^\alpha(x)\ \partial/\partial y^\alpha = \frac{dy(x)}{dx} ,$$

and using the action function we may interpret the quantities $\lambda_\alpha(x)$ as follows: First note that

$$\lambda(x) = \lambda_\alpha(x)\,dy^\alpha \in T^*_{y(x)}(\mathbb{R}^m)$$

gives a hyperplane (still denoted by $\lambda(x)$) in each tangent space $T_{y(x)}(\mathbb{R}^m)$. Recalling that

$$\lambda_\alpha(x) = L_{\dot{y}^\alpha}(y(x),\dot{y}(x)) ,$$

it follows that if we define the *indicatrix* to be the hypersurface

$$I(y,c) \subset T_y(\mathbb{R}^m)$$

given by the level sets

$$I(y,c) = \{(y,\dot{y}) : L(y,\dot{y}) = c\}$$

of the Lagrangian, then on the one hand $\lambda(x)$ *is the tangent hyperplane to the indicatrix passing through* $dy(x)/dx$

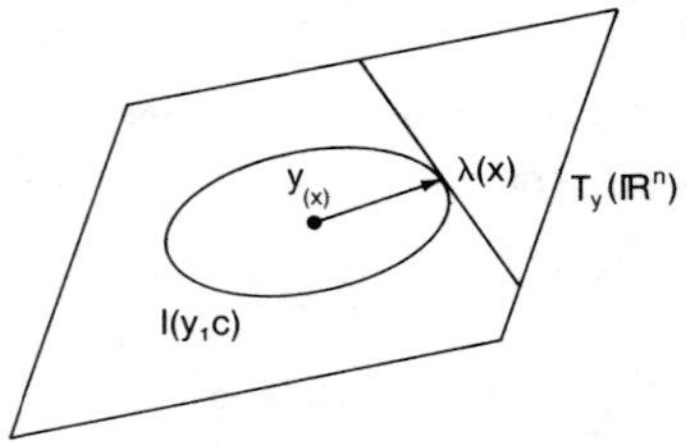

$$\begin{cases} c = L(y(x),\dot{y}(x)) \\ \dot{y}(x) = dy(x)/dx \quad . \end{cases}$$

On the other hand, referring to the Hamilton-Jacobi equation we see that

$$\lambda_\alpha(x) = \frac{\partial A(x,y)}{\partial \lambda_\alpha} \quad .$$

It follows that $\lambda(x)$ *is also the tangent hyperplane to the wave front passing through* $y(x)$.

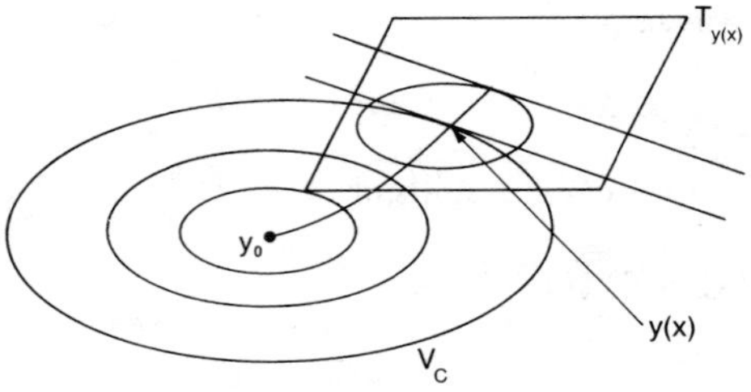

<u>Conclusion</u>. *The wave front sets and solution curves to the Euler-Lagrange equations are related by the following geometric property: The tangent hyperplane to* V_c *at* $y(x)$ *is equal to the tangent hyperplane to the indicatrix* $I(y(x),c)$ $(c = L(y(x),\dot{y}(x)))$ *at the point* $dy(x)/dx$.

From this geometric construction it is more or less clear that finding a general solution to the Hamilton-Jacobi equation and integrating the Euler-Lagrange equations comes to essentially the same thing (Jacobi's theorem, cf. [2]).

(IV.d.17) Example. We consider a non-degenerate classical 2nd order variational problem (cf. (I.e.23))

$$\Phi = \int L\left(x, y(x), \frac{dy(x)}{dx}, \frac{d^2y(x)}{dx^2}\right) dx$$

where $y = (y^1, \ldots, y^m)$ and

$$\det \| L_{\ddot{y}^\alpha \ddot{y}^\beta} \| \neq 0 \quad .$$

Then X has coordinates $(x; y^\alpha; \dot{y}^\alpha; \ddot{y}^\alpha)$ with (I, ω) being given by

$$\begin{cases} \theta^\alpha = dy^\alpha - \dot{y}^\alpha dx = 0 \\ \dot{\theta}^\alpha = d\dot{y}^\alpha - \ddot{y}^\alpha dx = 0 \\ \omega = dx \neq 0 \end{cases} ,$$

and $Z = X \times \mathbb{R}^{2m}$ has coordinates $(x; y^\alpha; \dot{y}^\alpha; \ddot{y}^\alpha; \lambda_\alpha; \dot{\lambda}_\alpha)$ with $Y \subset Z$ being defined by

$$L_{\ddot{y}^\alpha} = \dot{\lambda}_\alpha \quad .$$

Thus $(x; y^\alpha; \dot{y}^\alpha; \lambda_\alpha; \dot{\lambda}_\alpha)$ are local coordinates on Y and we may use either $\{\omega; \theta^\alpha; \dot{\theta}^\alpha; d\lambda_\alpha; d\dot{\lambda}_\alpha\}$ or $\{dx; dy^\alpha; d\dot{y}^\alpha; d\lambda_\alpha; d\dot{\lambda}_\alpha\}$ as a coframe. The basic 1-form is

$$\begin{aligned} \psi_Y &= L\omega + \lambda_\alpha \theta^\alpha + \dot{\lambda}_\alpha \dot{\theta}^\alpha \\ &= -H dx + \lambda_\alpha dy^\alpha + \dot{\lambda}_\alpha d\dot{y}^\alpha \end{aligned}$$

where

$$H = -L + \lambda_\alpha \dot{y}^\alpha + \dot{\lambda}_\alpha \ddot{y}^\alpha$$

is the Hamiltonian. A submanifold $S \subset Y$ of dimension $2m+1$ on which

$$\omega \wedge \theta^1 \wedge \cdots \wedge \theta^m \wedge \dot{\theta}^1 \wedge \cdots \wedge \dot{\theta}^m = dx \wedge dy^1 \wedge \cdots \wedge dy^m \wedge d\dot{y}^1 \wedge \cdots \wedge d\dot{y}^m \neq 0$$

is locally given by

$$(x, y, \dot{y}) \to (x, y, \dot{y}, \lambda(x, y, \dot{y}), \dot{\lambda}(x, y, \dot{y})) \quad . \qquad \text{(IV.d.18)}$$

The conditions that S be a field are

$$\begin{cases} \text{(i)} & \dfrac{\partial\lambda_\alpha}{\partial y^\beta} = \dfrac{\partial\lambda_\beta}{\partial y^\alpha} \\[2ex] \text{(ii)} & \dfrac{\partial\dot\lambda_\alpha}{\partial \dot y^\beta} = \dfrac{\partial\dot\lambda_\beta}{\partial \dot y^\alpha} \\[2ex] \text{(iii)} & \dfrac{\partial\lambda_\alpha}{\partial \dot y^\beta} = \dfrac{\partial\dot\lambda_\beta}{\partial y^\alpha} \\[2ex] \text{(iv)} & \dfrac{\partial\lambda_\alpha}{\partial x} + \dfrac{\partial H}{\partial\lambda_\beta}\dfrac{\partial\lambda_\beta}{\partial y^\alpha} + \dfrac{\partial H}{\partial\dot\lambda_\beta}\dfrac{\partial\dot\lambda_\beta}{\partial y^\alpha} + \dfrac{\partial H}{\partial y^\alpha} = 0 \\[2ex] \text{(v)} & \dfrac{\partial\dot\lambda_\alpha}{\partial x} + \dfrac{\partial H}{\partial\lambda_\beta}\dfrac{\partial\lambda_\beta}{\partial \dot y^\alpha} + \dfrac{\partial H}{\partial\dot\lambda_\beta}\dfrac{\partial\dot\lambda_\beta}{\partial \dot y^\alpha} + \dfrac{\partial H}{\partial \dot y^\alpha} = 0 \end{cases}$$

where the derivatives of H are evaluated on S. Equations (i)-(iii) give

$$(\lambda,\dot\lambda) = \left(\frac{\partial g(x,y,\dot y)}{\partial y}, \frac{\partial g(x,y,\dot y)}{\partial \dot y}\right)$$

for some function $g = g(x,y,\dot y)$, and then equations (iv), (v) are the derivatives

$$\begin{cases} \dfrac{\partial}{\partial y^\alpha}\left(\dfrac{\partial g}{\partial x} + H\left(x,y,\dot y, \dfrac{\partial g}{\partial y}, \dfrac{\partial g}{\partial \dot y}\right)\right) = 0 \\[2ex] \dfrac{\partial}{\partial \dot y^\alpha}\left(\dfrac{\partial g}{\partial x} + H\left(x,y,\dot y, \dfrac{\partial g}{\partial y}, \dfrac{\partial g}{\partial \dot y}\right)\right) = 0 \end{cases}$$

of the equation

$$\frac{\partial g(x,y,\dot y)}{\partial x} + H\left(x,y,\dot y, \frac{\partial g(x,y,\dot y)}{\partial y}, \frac{\partial g(x,y,\dot y)}{\partial \dot y}\right) = 0 \quad . \quad \text{(IV.d.19)}$$

The reduced momentum space is $Q = J^1(\mathbb{R},\mathbb{R}^m)$ with coordinates $(x;y^\alpha;\dot y^\alpha)$, and a cross-section of $Y \to Q$ over an open set $R \subset Q$ is given by (IV.d.18). A function $g(x,y,\dot y)$ defined in R is a solution to the Hamilton-Jacobi equation in case there is such a cross-section $s: R \to Y$ with

$$dg = s^*(\psi_Y) \quad .$$

This equation is

$$\frac{\partial g(x,y,\dot y)}{\partial x}\,dx + \frac{\partial g(x,y,\dot y)}{\partial y^\alpha}\,dy^\alpha + \frac{\partial g(x,y,\dot y)}{\partial \dot y^\alpha}\,d\dot y^\alpha$$

$$= -H(x,y,\dot y,\lambda,\dot\lambda)\,dx + \lambda_\alpha dy + \dot\lambda_\alpha dy^\alpha$$

where $\lambda = \lambda(x,y,\dot y)$, $\dot\lambda = \dot\lambda(x,y,\dot y)$. It follows that (IV.d.19) is the Hamilton-Jacobi equation for 2nd order classical variational problems. We note that it is formally the same equation as for a classical 1st order variational problem.

Special Case. Suppose that in $\mathbb{R}^5$ with coordinates $(x,y,\dot y,\lambda,\dot\lambda)$ we have a Hamiltonian function $H = H(y,\dot y,\lambda,\dot\lambda)$ and we want to show that the corresponding Hamiltonian equations are completely integrable. By a theorem of Jacobi (page 260 of [2]) this will be the case if we can find a solution $g = g(y,\dot y,\eta,\dot\eta)$ to the equation

$$H\left(y,\dot y,\frac{\partial g}{\partial y},\frac{\partial g}{\partial \dot y}\right) = K(\eta,\dot\eta) \qquad \text{(IV.d.20)}$$

satisfying

$$\det\begin{Vmatrix} g_{y\eta} & g_{y\dot\eta} \\ g_{\dot y\eta} & g_{\dot y\dot\eta}\end{Vmatrix} \neq 0 \quad .$$

(such a g is called a *complete integral*).

In our discussion of the functional (II.a.43) we encountered the Hamiltonian

$$H = \frac{1}{2y}\left(\dot\lambda^2 G^5 - \frac{1}{G}\right) + \lambda\dot y \quad .$$

In addition to H we also found the 1st integral

$$V = -xH + \lambda y \quad ,$$

which unfortunately depends on x (if V_x were zero, then by corollary 1 on page 272 of [2] the corresponding Hamiltonian equations would be completely integrable). It is therefore natural to look for solutions to (IV.d.20). We have been unable to find a complete integral, but will now show how to find a solution $g(y,y,\eta)$ expressed as an abelian integral depending on three functions of one parameter η.

The traditional method of solving the Hamilton-Jacobi equation is by separation of variables (cf. pages 261-264 of [2]). We follow a modified approach. Set

$$\begin{cases} g(y,\dot{y},\eta) = k(y,\eta) + \ell(\dot{y},\eta) \\ k' = \dfrac{\partial k}{\partial y} \\ \ell' = \dfrac{\partial \ell}{\partial \dot{y}} \end{cases}$$

and consider the pair of O.D.E.'s depending on the parameter η and arbitrary function $L(\eta)$

$$\begin{cases} -2yk' = L \\ (\ell')^2 \dfrac{G^5}{\dot{y}} - \dfrac{1}{\dot{y}G} = L \quad . \end{cases} \tag{IV.d.21}$$

A solution to this system gives a solution to (IV.d.20) with $K=0$, since

$$\begin{aligned} H\left(y,\dot{y},\frac{\partial g}{\partial y},\frac{\partial g}{\partial \dot{y}}\right) &= H(y,\dot{y},k',\ell') \\ &= \frac{L\dot{y}}{2y} - \frac{L\dot{y}}{2y} \\ &= 0 \quad . \end{aligned}$$

The 1st equation in (IV.d.21) is

$$\frac{dk}{dy} = -\frac{L}{2y} \quad ,$$

which has the integral

$$k(y,\eta) = -\frac{L(\eta)}{2}\log y + M(\eta) \quad .$$

The 2nd equation in (IV.d.21) is

$$\frac{d\ell}{d\dot{y}} = \frac{1}{(1+\dot{y}^2)^3}\sqrt{L\dot{y}\sqrt{1+\dot{y}^2}+1} \quad .$$

To "integrate" this we consider the rational 1-form

$$\omega = \frac{\rho d\sigma}{(1+\sigma^2)^3}$$

on the algebraic curve

$$(\rho^2-1)^2 = L^2\sigma^2(1+\sigma^2) \quad .$$

This is a curve of genus three (for general choice of $L(\eta)$), and the abelian integral

$$\ell = \int \omega + N(\eta)$$

gives the general solution to the 2nd equation in (IV.d.21).

(IV.d.22) Example. We shall find and outline the determination of a complete integral for the Hamilton-Jacobi equation associated to the functional

$$\Phi = \frac{1}{2} \int_N \kappa^2 \, ds \tag{IV.d.23}$$

defined on curves $N \subset \mathbb{E}^2$. For this we follow the notation of examples (I.d.27) and (i) in Chapter II, Section b). Thus Y has a coframe $\{\omega, \theta^1, \theta^2, d\lambda_1, d\lambda_2\}$ where

$$\begin{cases} \omega = \omega^1 = (dx, e_1) \\ \theta^1 = \omega^2 = (dx, e_2) \\ \theta^2 = \omega_1^2 - \kappa\omega & \qquad \omega_1^2 = (de_1, e_2) \\ \lambda_2 = \kappa & . \end{cases}$$

Moreover,

$$\begin{aligned} \psi_Y &= \frac{\kappa^2}{2}\omega + \lambda_1\theta^1 + \lambda_2\theta^2 \\ &= -\frac{(\lambda_2)^2}{2}\omega^1 + \lambda_1\omega^2 + \lambda_2\omega_1^2 \quad . \end{aligned} \tag{IV.d.24}$$

Relative to a fixed reference frame we identify $\mathbb{E}^2$ with $\mathbb{R}^2$ and set

$$\begin{cases} x = (u, v) \\ e_1 = (\cos\varphi, \sin\varphi) \\ e_2 = (-\sin\varphi, \cos\varphi) \quad . \end{cases}$$

Then

$$\begin{cases} \omega^1 = \cos\varphi \, du + \sin\varphi \, dv \\ \omega^2 = -\sin\varphi \, du + \cos\varphi \, dv \\ \omega_1^2 = d\varphi \quad . \end{cases}$$

The fibration $Y \to Q$ is given by the foliation

$$\omega = \theta^1 = \theta^2 = 0 \quad .$$

This is

$$du = dv = d\varphi = 0 ,$$

so that we may take (u,v,φ) as coordinates on the reduced momentum space.

We shall not write out the equations that determine a field $S \subset Y$, but rather shall go directly to the Hamilton-Jacobi equations (solving these will give a field). Over a simply-connected region R of (u,v,φ)-space (recall that $\varphi \in \mathbb{R}/2\pi\mathbb{Z}$ is an angular coordinate) a cross-section of $Y \to Q$ is given by

$$\begin{cases} \lambda_1 = \lambda_1(u,v,\varphi) \\ \lambda_2 = \lambda_2(u,v,\varphi) . \end{cases} \tag{IV.d.25}$$

We want to determine this cross-section so that for some function $g(u,v,\varphi)$

$$dg = s^*(\psi_Y) \tag{IV.d.26}$$

where ψ_Y is given by (IV.d.24). Plugging in the formulas for $\omega^1, \omega^2, \omega_1^2$ gives

$$s^*(\psi_Y) = \left(\frac{-(\lambda_2)^2}{2} \cos\varphi - \lambda_1 \sin\varphi\right)du + \left(\frac{-(\lambda_2)^2}{2} \sin\varphi + \lambda_1 \cos\varphi\right)dv + \lambda_2 d\varphi .$$

Thus (IV.d.26) is

$$\begin{cases} \text{(i)} & \lambda_2 = g_\varphi \\ \text{(ii)} & \frac{-g_\varphi^2}{2} \cos\varphi - \lambda_1 \sin\varphi = g_u \\ \text{(iii)} & \frac{-g_\varphi^2}{2} \sin\varphi + \lambda_1 \cos\varphi = g_v \end{cases} .$$

The linear combination

$$-\sin\varphi \cdot \text{(ii)} + \cos\varphi \cdot \text{(iii)}$$

gives

$$\text{(iv)} \qquad \lambda_1 = -\sin\varphi\, g_u + \cos\varphi\, g_v .$$

The linear combination

$$\cos\varphi \cdot \text{(ii)} + \sin\varphi \cdot \text{(iii)}$$

gives the P.D.E.

$$\frac{g_\varphi^2}{2} = -\cos\varphi\, g_u - \sin\varphi\, g_v \tag{IV.d.27}$$

whose solution determines by (i), (iv) a solution of the Hamilton-Jacobi equation associated to the functional (IV.d.23).

It is instructive to integrate (IV.d.27) using the classical *method of characteristics* (cf. [13], [45]). In the space of variables (u,v,φ,g,p,q,r) we consider the submanifold U defined by

$$r^2 = -2(\cos\varphi\, p + \sin\varphi\, q) \quad . \tag{IV.d.28}$$

Then $\dim U = 6$, and on U we consider the 1-form

$$\theta = -dg + p\,du + q\,dv + r\,d\varphi$$

(thus we think of $p = g_u$, $q = g_v$, $r = g_\varphi$), and look for integral manifolds of the Pfaffian differential system

$$\theta = 0, \qquad du \wedge dv \wedge d\varphi \neq 0 \quad . \tag{IV.d.29}$$

Such an integral manifold is locally of the form

$$(u,v,\varphi) \to (u,v,\varphi,g(u,v,\varphi),g_u,g_v,g_\varphi)$$

where $g(u,v,\varphi)$ satisfies the Hamilton-Jacobi equation (IV.d.27).

We set

$$h = \sqrt{2(-p\cos\varphi - q\sin\varphi)} \qquad (=r\,|\,U)$$

and compute the exterior derivative $\Theta = d\theta$ to be given by

$$\Theta = dp \wedge (du + h_p\,d\varphi) + dq \wedge (dv + h_q\,d\varphi) \quad .$$

The *characteristic vector field* V [15] is uniquely determined up to non-zero multiples by the conditions

$$\begin{cases} V \lrcorner\, \theta = 0 \\ V \lrcorner\, \Theta = 0 \end{cases} .$$

Setting

$$V = \alpha\, \partial/\partial u + \beta\, \partial/\partial v + \gamma\, \partial/\partial\varphi + \delta\, \partial/\partial g + \lambda\, \partial/\partial p + \mu\, \partial/\partial q$$

these equations are

$$\begin{cases} \delta = \alpha p + \beta q + \gamma h \\ \lambda(du+h_p\,d\varphi) + \mu(dv + h_q\,d\varphi) - dp(\alpha + h_p\gamma) - dq(\beta + h_q\gamma) = 0 \end{cases} .$$

Normalizing by taking $\gamma = -1$ we find that

$$V = h_p\ \partial/\partial u + h_q\ \partial/\partial v - \partial/\partial\varphi + (ph_p + qh_q - h)\,\partial/\partial g \quad .$$

Since h is homogeneous of degree 1/2 in (p,q) we have

$$V = -\frac{2\cos\varphi}{h}\ \partial/\partial u - \frac{2\sin\varphi}{h}\ \partial/\partial v - \partial/\partial\varphi - \frac{h}{2}\ \partial/\partial g \quad .$$

It is convenient to take $\varphi = -t$ as parameter on the integral curves of V. The O.D.E. system that gives the characteristic curves of the Hamilton-Jacobi equation(16) is thus

$$\begin{cases} \dot{u} = \dfrac{\sqrt{2}\cos t}{\sqrt{-p\cos t - q\sin t}} \\[2ex] \dot{v} = \dfrac{2\sin t}{\sqrt{-p\cos t - q\sin t}} \\[2ex] \dot{g} = \dfrac{\sqrt{p\cos t - q\sin t}}{\sqrt{2}} \end{cases} \qquad \text{(IV.d.30)}$$

together with $\dot{p} = \dot{q} = 0$. Thus we want to solve these equations when p,q are *constants*.

Note that

$$2\dot{g} + p\dot{u} + q\dot{v} = 0 \quad ,$$

so that (IV.d.30) has the 1st integral

$$2g + pu + qv = \text{constant.}$$

Using this it will suffice to solve the 1st two equations. Writing

$$\begin{cases} \cos t = \frac{1}{2}\,(e^{it} + e^{-it}) \\[1ex] \sin t = \frac{1}{2i}\,(e^{it} - e^{-it}) \end{cases}$$

it will clearly suffice to solve a general equation (for a complex-valued function $w(z)$ of a complex variable z)

$$\frac{dw(z)}{dz} = \frac{\alpha e^z}{\sqrt{\beta e^z + \gamma e^{-z}}}$$

where α,β,γ are constants. Setting $\zeta = e^z$ this is

$$dw(\zeta) = \frac{\alpha d\zeta}{\sqrt{\beta\zeta + \gamma\zeta^{-1}}} .$$

We indicate briefly how this equation may be integrated (of course, one way is to go to almost any book of tables). The algebraic curve

$$\eta^2 = \beta\zeta + \gamma\zeta^{-1}$$

is a 2-sheeted covering of the ζ-sphere with branch points at $\zeta = \pm i\sqrt{\frac{\gamma}{\beta}}$. By the Riemann-Hurwitz formula ([36]) it is therefore biholomorphic to the Riemann sphere with coordinate ξ (it is easy to make this explicit). We may therefore consider $\omega = \frac{\alpha d\zeta}{\eta}$ as a meromorphic 1-form on the ξ-sphere, and then the indefinite integral $\int\omega$ may be evaluated by a partial fraction expansion:

$$\int\omega = \sum_i R_i(\xi)\log(\xi - \xi_i) + R(\xi) ,$$

where $R(\xi)$, $R_i(\xi)$ are rational functions of ξ. Again this is easy to make explicit in the simple case at hand.

Unwinding all of this leads to an explicit integration of the system (IV.d.30). The formulas obtained involve only elementary functions.

To solve the Hamilton-Jacobi equation we prescribe the function $g(u,v,0)$ arbitrarily on the initial surface $\varphi = 0$ subject only to the condition $g_u < 0$. This gives the surface (cf. (IV.d.28))

$$(u,v) \to (u,v,0,g(u,v,0),g_u,g_v,\sqrt{-2g_u}) \qquad \text{(IV.d.31)}$$

in U. From the point (IV.d.31) we flow out time φ along the integral curve of V obtained by our integration of (IV.d.30). The 4th coordinate $g(u,v,\varphi)$ gives our desired solution to the equation (IV.d.27).

Remark. It is curious that the integration of the Euler-Lagrange equations associated to (IV.d.23) involves elliptic functions while the integration of the Hamilton-Jacobi equation involves only elementary functions. In each case "solving" means relative to a coordinate system, and it is not clear just how the coordinate systems are related.

Proof of Theorem (IV.d.1.). We consider the basic diagram (IV.d.2), and let $\Gamma \subset Y$ be the solution to the Euler-Lagrange system

(J,ω) corresponding to the solution N of the Euler-Lagrange equations via theorem (I.e.9). Denote by $\gamma = \tilde{\omega}(\Gamma)$ the projection of Γ into the reduced momentum space. We claim that:

(IV.d.32) *The assumption that* Γ *has no conjugate points allows us to find a field* $S \subset Y$ *containing* Γ *and lying diffeomorphically over a simply connected neighborhood* R *of* γ .

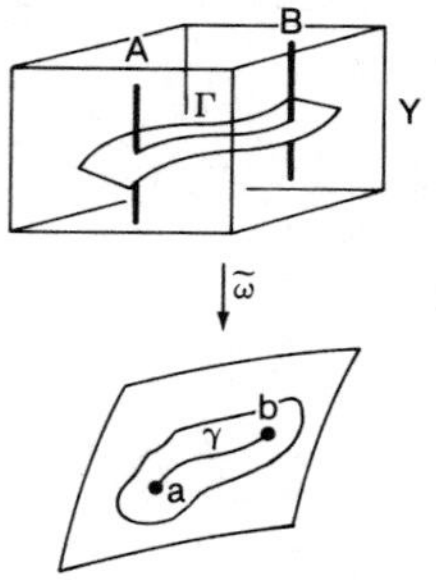

In this figure the endpoint conditions A,B are represented on Y by the fibres of $\tilde{\omega}$ lying over the endpoints a,b of γ. Although it is not relevant to the *present* proof, we recall (cf. (IV.c.17)) that γ is an integral manifold of the (unique) differential system (G,ω) that pulls back under $\tilde{\omega}^*$ to the 1st derived system of $(\pi^* I,\omega)$ on Y.(17)

Proof of (IV.d.32). We use the method of characteristics, as in the example just discussed. Recall that Γ is an integral curve of (J,ω). Locally, Ψ_Y has the Darboux normal form (IV.c.28), and so given $y \in \Gamma$ we may find a little piece of m-dimensional manifold T transverse to Γ and such that the restriction

$$(\Psi_Y)_T = 0 \quad . \tag{IV.d.33}$$

We may also assume that T projects diffeomorphically onto its image in Q.

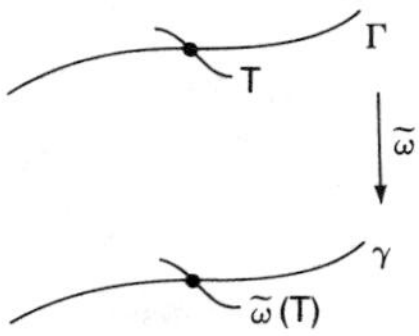

(Here we should picture T as having higher dimension.) Let V be a vector field spanning the characteristic line bundle $\Psi_Y^{\perp}$, and define S to be the V-flow of T. More precisely, we may assume that Γ is the V-flow of the point $y \in \Gamma \cap T$ for $-\delta \leqq t \leqq \delta$, and for small $\varepsilon > 0$ we let S be the V-flow of T for $-\delta-\varepsilon \leqq t \leqq \delta+\varepsilon$.

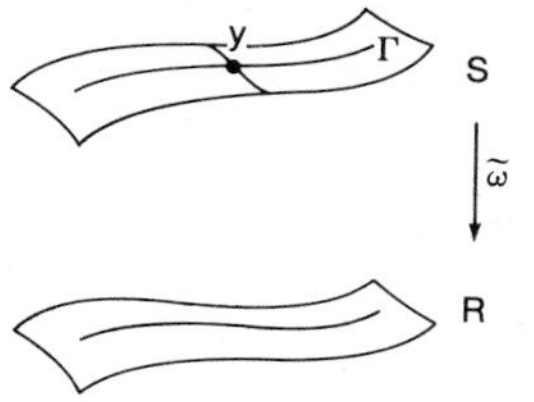

We must show that:

i) $\Psi_Y|S \equiv 0$

ii) $S \to R$ is a diffeomorphism (for ε sufficiently small).

The first is easy. Using the same argument as in the proof of (IV.d.12)[18] it will suffice to show that for any point $\tilde{y} \in T$ and tangent vectors W, W' to T

$$\begin{cases} \langle \Psi_Y(\tilde{y}), (W,W') \rangle = 0 \\ \langle \Psi_Y(\tilde{y}), (V,W) \rangle = 0 \end{cases} . \tag{19}$$

The first is by our construction (IV.d.33) of T, while the second is a consequence of $V \lrcorner \Psi_Y \equiv 0$.

For ii), we may assume that the V-flow of any point $\tilde{y} \in T$ for $-\delta-\varepsilon \leqq t \leqq \delta+\varepsilon$, which we recall is an integral curve of (J,ω), contains no pair of conjugate points. If we call this curve $\Gamma_{\tilde{y}}$ and let $\gamma_{\tilde{y}} = \tilde{\omega}(\Gamma_{\tilde{y}})$, then the $\gamma_{\tilde{y}}$ give a *non-intersecting* family of curves that foliate R .

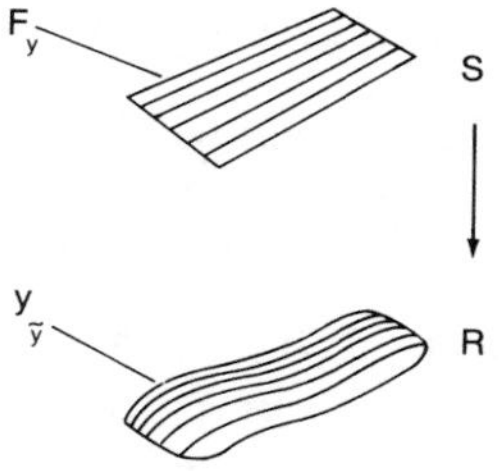

From this ii) is clear.

Remark. The standard example of conjugate points is the antipodal points on S^2 for the variational problem corresponding to geodesics in the usual metric. In this case (cf. example (IV.b.3))

$$\begin{aligned} &Y \cong \{\text{bundle of unit tangent vectors}\} \\ &\downarrow \tilde{\omega} \\ &Q \cong S^2 \end{aligned}$$

is the usual projection. Given Γ corresponding to a geodesic connecting the North and South poles we may embed Γ in an integral manifold S of $(\Psi_Y, \omega \wedge \theta^1)$ as before, but the map $\tilde{\omega}: S \to R \subset S^2$ looks like the following picture in which the ends of S are pinched to points:

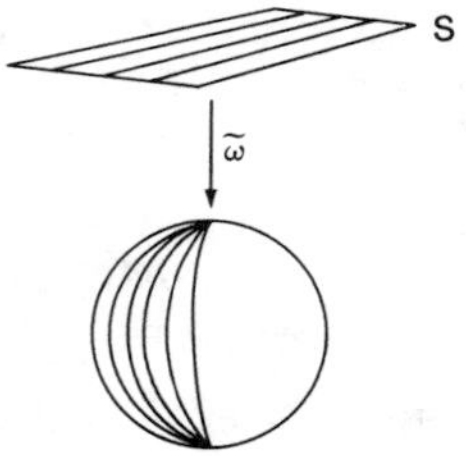

Thus $\gamma = \tilde{\omega}(\Gamma)$ cannot be embedded in a field.

Using (IV.d.32) we are now ready to reduce the proof of (IV.d.1) to a computation along N. Let $N_1 \subset X$ be a nearby integral manifold of (I,ω) with the same endpoint conditions as N, and let $\Gamma_1 \subset Y$ be an integral manifold of $(\pi^* I,\omega)$ lying over N_1 (cf. (IV.c.18)). Denote by $\gamma_1 = \tilde{\omega}(\Gamma_1)$ the projection of Γ_1 to Q; thus, γ and γ_1 have the same endpoints in the usual sense. Finally, denote by

$$\tilde{\Gamma}_1 = s(\gamma_1)$$

the unique curve in the field S lying over γ_1. Although it is *not* the case that $\tilde{\Gamma}_1$ is an integral manifold of the differential system $(\pi^* I,\omega)$, it follows from (IV.c.17) and (IV.c.18) that

(IV.d.34) $\tilde{\Gamma}_1 \subset Y$ *is an integral manifold of* $(\pi^* I_1,\omega)$ *where* I_1 *is the* 1st *derived system of* I.[20]

In terms of our bases this means that

$$\theta^\rho \equiv 0 \bmod \tilde{\Gamma}_1 \qquad \rho = 1,..,s-s_1 . \qquad \text{(IV.d.35)}$$

Of course, both Γ and Γ_1 are integral manifolds of $(\pi^* I,\omega)$; in particular

$$\theta^\alpha \equiv 0 \bmod \Gamma \qquad \alpha = 1,..,s . \qquad \text{(IV.d.36)}$$

The picture is something like

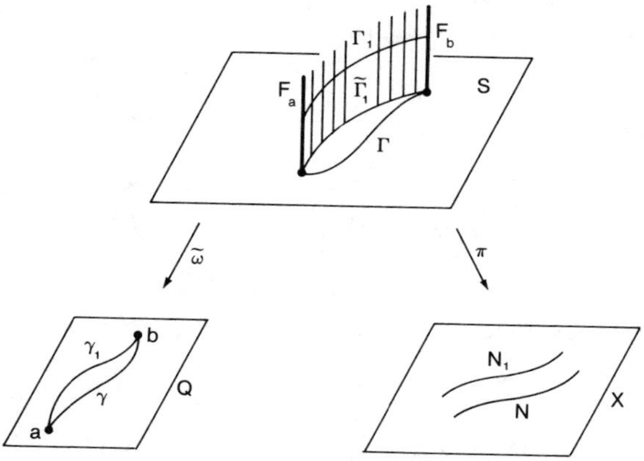

Here the fibres of $\tilde{\omega}$ are the vertical lines F_q. The figure is supposed to suggest that the endpoint conditions are given by F_a, F_b, and Γ_1, $\tilde{\Gamma}_1$ are each cross-sections of the fibre space $F_{\gamma_1} = \cup_{q \in \gamma_1} F_q$.

Now, *and this is the crux of the argument*, we consider

$$\Delta = \int_{N_1} \varphi - \int_N \varphi \quad . \tag{IV.d.37}$$

We want to show that

$$\Delta \geqq 0 \tag{IV.d.38}$$

provided that N_1 is sufficiently close to N. The following computation leading to (IV.d.38) is the crucial step:

$$\Delta = \int_{\Gamma_1} \varphi - \int_{\Gamma} \varphi$$

$$= \int_{\Gamma_1} \varphi - \int_{\Gamma} \psi_Y$$

since

$$\psi_Y = \varphi + \lambda_\alpha \theta^\alpha$$

and (IV.d.36) holds

$$= \int_{\Gamma_1} \varphi - \int_{\tilde{\Gamma}_1} \psi_Y$$

since S is a field and therefore $d\psi_S = 0$ (this is where we use Hilbert's invariant integral). We thus have

$$\Delta = \int_{\gamma_1} \eta$$

where

$$\eta = s_1^*(\varphi) - s^*(\psi_Y)$$

with $s_1 : \gamma_1 \to \Gamma_1$ being the section corresponding to Γ_1. To prove (IV.d.38) we must therefore show that

$$\eta \geqq 0 \qquad \text{on} \quad \gamma_1 \tag{IV.d.39}$$

provided that N_1 is sufficiently close to N. What we shall do is prove (IV.d.39) in two cases that show clearly what is going on, and then explain why it is true in general.

<u>Case (i)</u>. We consider a classical variational problem on $J^1(\mathbb{R}, \mathbb{R}^m)$ with coordinates $(x; y^\alpha; \dot{y}^\alpha)$ and Lagrangian $L(x, y, \dot{y})$ with

$$\| L_{\dot{y}^\alpha \dot{y}^\beta} \| > 0 \quad . \tag{IV.d.40}$$

In coordinates the basic diagram (IV.d.2) is

$$\begin{array}{ccc} & Y \subset \{(x; y^\alpha; \dot{y}^\alpha; \lambda_\alpha)\} & \\ & \tilde{\omega} \swarrow \quad \searrow \pi & \\ J^0(\mathbb{R}, \mathbb{R}^m) = \{(x; y^\alpha)\} & & \{(x; y^\alpha; \dot{y}^\alpha)\} = J^1(\mathbb{R}, \mathbb{R}^m) \end{array} .$$

We recall that $Y \subset J^1(\mathbb{R}, \mathbb{R}^m) \times \mathbb{R}^m$ is defined by

$$\lambda_\alpha = L_{\dot{y}^\alpha}(x, y, \dot{y}) \quad ,$$

and that

$$\psi_Y = L dx + \lambda_\alpha (dy^\alpha - \dot{y}^\alpha dx) \quad .$$

Now N_1 is a 1-jet

$$\left(x; y^\alpha(x); \frac{dy^\alpha(x)}{dx} \right) \tag{IV.d.41}$$

of a function $y(x)$ (we write $y(x)$ instead of $y_1(x)$). Suppose that in $(x; y^\alpha; \dot{y}^\alpha)$ coordinates[21] the section s is

$$(x, y^\alpha) \to (x; y^\alpha; s^\alpha(x, y)) \quad ,$$

and for notational simplicity set

$$s^{\alpha}(x) = s^{\alpha}(x, y(x)) \quad .$$

Then

$$s_1^*(\varphi) = L\left(x, y(x), \frac{dy(x)}{dx}\right) dx \qquad \text{(IV.d.42)}$$

and

$$s^*(\psi_Y) = L(x, y(x), s(x))dx + L_{\dot{y}}(x, y(x), s(x)) \left(\frac{dy^{\alpha}(x)}{dx} - s^{\alpha}(x)\right) dx. \qquad \text{(IV.d.43)}$$

It follows from (IV.d.42), (IV.d.43) that

$$\eta = E\left(x, y(x), \frac{dy(x)}{dx}, s(x)\right) dx$$

where

$$E(x,y,u,v) = L(x,y,u) - L(x,y,v) - L_{\dot{y}^{\alpha}}(x,y,v)(u^{\alpha} - v^{\alpha}) \qquad \text{(IV.d.44)}$$

is the so-called *Weierstrass* E-*function* (cf. [13], [29]). The point is that by Taylor's theorem

$$E(x,y,u,v) = L_{\dot{y}^{\alpha}\dot{y}^{\beta}}(x,y,v)(u^{\alpha} - v^{\alpha})(u^{\beta} - v^{\beta}) + 0(\|u - v\|^2) \quad .$$

In particular, E vanishes to 2nd order on the locus $u = v$, while by (IV.d.40)

$$E(x,y,u,v) \geqq 0 \quad , \quad \|u - v\| < \varepsilon \quad .$$

It is then clear that (IV.d.39) holds when N_1 is sufficiently close to N.[22]

Case (ii) We consider a classical 2nd order variational problem on $J^2(\mathbb{R}, \mathbb{R}^m)$ with coordinates $(x; y^{\alpha}; \dot{y}^{\alpha}; \ddot{y}^{\alpha})$ and Lagrangian $L(x, y, \dot{y}, \ddot{y})$ with

$$\|L_{\ddot{y}^{\alpha}\ddot{y}^{\beta}}\| > 0 \quad . \qquad \text{(IV.d.45)}$$

In coordinates the basic diagram (IV.d.2) is

$$\begin{array}{ccc} & Y \subset \{(x; y^{\alpha}; \dot{y}^{\alpha}; \ddot{y}^{\alpha}; \lambda_{\alpha}; \dot{\lambda}_{\alpha})\} & \\ \swarrow^{\tilde{\omega}} & & \searrow^{\pi} \\ J^1(\mathbb{R}, \mathbb{R}^m) = \{(x; y^{\alpha}; \dot{y}^{\alpha})\} & & \{(x; y^{\alpha}; \dot{y}^{\alpha}; \ddot{y}^{\alpha})\} = J^2(\mathbb{R}, \mathbb{R}^m) \end{array} \quad ,$$

where $Y \subset J^2(\mathbb{R}, \mathbb{R}^m) \times \mathbb{R}^{2m}$ is defined by

$$\dot{\lambda}_{\alpha} = L_{\ddot{y}^{\alpha}}(x, y, \dot{y}, \ddot{y}) \quad .$$

We also have

$$\psi_Y = Ldx + \lambda_\alpha(dy^\alpha - \dot{y}^\alpha dx) + \dot{\lambda}_\alpha(d\dot{y}^\alpha - \ddot{y}^\alpha dx) \quad .$$

The system (G,ω) on the reduced momentum space $J^1(\mathbb{R},\mathbb{R}^m)$ that pulls back to the 1st derived system of $(\pi^* I,\omega)$ is the canonical system

$$\begin{cases} \theta^\alpha = dy^\alpha - \dot{y}^\alpha dx = 0 \\ \omega = dx \neq 0 \end{cases} .$$

Thus, and this is the crucial point beyond the classical case:

$$\psi_Y \equiv Ldx + L_{\ddot{y}^\alpha}(d\dot{y}^\alpha - \ddot{y}^\alpha)dx \quad \text{mod} \quad \tilde{\omega}^* G \tag{IV.d.46}$$

Now N_1 is a 2-jet

$$\left(x; y^\alpha(x); \frac{dy^\alpha(x)}{dx}; \frac{d^2y^\alpha(x)}{dx^2}\right)$$

of a function $y(x)$. Suppose that in the local coordinates $(x;y^\alpha;\dot{y}^\alpha;\ddot{y}^\alpha;\lambda_\alpha)$ on Y we write the section s giving the field (IV.d.32) as

$$(x;y^\alpha;\dot{y}^\alpha) \to (x;y^\alpha;\dot{y}^\alpha;s^\alpha(x,y,\dot{y});r_\alpha(x,y,\dot{y})) \quad .$$

For notational simplicity we set

$$s^\alpha(x) = s^\alpha\left(x, y(x), \frac{dy(x)}{dx}\right) .$$

Then since $\varphi = Ldx$ we have

$$s_1^*(\varphi) = L\left(x, y(x), \frac{dy(x)}{dx}, \frac{d^2y(x)}{dx^2}\right) dx . \tag{IV.d.47}$$

On the other hand, since $\gamma_1 \subset J^1(\mathbb{R},\mathbb{R}^m)$ is the 1-jet of $y(x)$ *and is hence an integral manifold of* G, by (IV.d.46)

$$s^*(\psi_Y) = L\left(x, y(x), \frac{dy(x)}{dx}, s(x)\right) dx$$

$$+ L_{\ddot{y}^\alpha}\left(x, y(x), \frac{dy(x)}{dx}, s(x)\right)\left(\frac{d^2y^\alpha(x)}{dx^2} - s^\alpha(x)\right) dx . \tag{IV.d.48}$$

It follows from (IV.d.47), (IV.d.48) that

$$\eta = E\left(x, y(x), \frac{dy(x)}{dx}, \frac{d^2y(x)}{dx^2}, s(x)\right)$$

where

$$E(x,y,\dot{y},u,v) = L(x,y,\dot{y},u) - L(x,y,\dot{y},v) - L_{\ddot{y}^\alpha}(x,y,\dot{y},v)(u^\alpha - v^\alpha)$$

is a *Weierstrass* E-*function for the* 2nd *order variational problem.* Again, just as in case (i) the assumption (IV.d.45) gives $\eta \geqq 0$ and completes the proof of the theorem.

Remarks on the General Case. The argument is really the same as the 2nd case just given. The essential point is that, using (IV.d.35), (IV.d.46) becomes

$$\psi_Y \equiv \varphi + A_\mu \theta^\mu \mod G \quad .$$

Then a computation similar to that in Chapter IV, Section c) shows that η vanishes to 2nd *order* when $N_1 = N$ and therefore the dominant term for N_1 close to N is a 2nd derivative, which then turns out to be positive due to $\|A_{\mu\nu}\| > 0$.

e) Mixed Endpoint Conditions and the Classical Problem of Lagrange.

Classical variational problems with constraints are traditionally solved by the method of Lagrange multipliers. The theory developed in this text offers an alternative approach, one that turns out to involve working with fewer variables and that may therefore sometimes have computational advantages in examples. In this section we shall explain how this goes, and shall also compare our approach with the traditional Lagrange multiplier method: In a word, it turns out that the quantities λ_α we have used throughout the text are *not* the same as Lagrange multipliers, and in fact play quite a different role in the theory (cf. (IV.e.41)).

i) Well Posed Mixed Variational Problems.

In Chapter IV, Section a) we have introduced and discussed the class of well-posed variational problems. This class contains most of our examples and forms a very natural setting for the deeper aspects of the theory (2nd variation, Hamilton-Jacobi equation, etc.). However, the classical Lagrange and Mayer problems (cf. [5], [13]) require different types of endpoint conditions; ones that are *not* symmetric in the sense that the role of the two endpoints cannot be interchanged. We shall introduce these mixed endpoint conditions in a general setting, and then later shall apply them to the classical Lagrange problem.

Let $(I,\omega;\varphi)$ be a variational problem on a manifold X, and make the following

(IV.e.1) Assumption. (I,ω) *is a Pfaffian system in good form that is locally embeddable, and*

$$\varphi = L\omega$$

where L *is a function on* X.

Remark. By (I.c.10) this assumption is the same as saying that $(I,\omega;\varphi)$ is *locally* equivalent to a classical variational problem with constraints. However, the admissable changes of coframe constitute a larger group than in classical variational problems with constraints (even if we allow contact transformations).

We choose generators $\{\theta^\alpha\}$ for the differential ideal I such that the 1st derived system is generated by the subset $\{\theta^\rho\}$. Thus the differential system is generated by the Pfaffian equations

$$\begin{cases} \theta^\rho = 0 & \rho = 1,..,s-s_1 \\ \theta^\mu = 0 & \mu = s-s_1+1,..,s \\ \omega \neq 0 & \end{cases} \qquad \text{(IV.e.2)}$$

where s_1 is the Cartan integer. Denoting as usual by $\{\theta^\alpha\} \subset A^*(X)$ the *algebraic* ideal generated by the 1-forms θ^α the structure equations of (IV.e.2) are

$$\begin{cases} d\theta^\rho \equiv 0 \bmod\{\theta^\alpha\} \\ d\theta^\mu \equiv -\pi^\mu \wedge \omega \bmod\{\theta^\alpha\} , \end{cases} \qquad \text{(IV.e.3)}$$

where the 1-forms π^μ are linearly independent modulo span$\{\theta^\alpha,\omega\}$.

Let $N \in V(I,\omega)$ be an integral manifold of (IV.e.2), diffeomorphic to an interval $[a,b] = \{a \leqq x \leqq b\}$. We shall define a subspace (cf. Chapter IV, Section a) for an explanation of similar notations)

$$T_N(V(I,\omega;[A,B'])\subset T_N(V(I,\omega))$$

given by infinitesimal variations of N as an integral manifold of (I,ω) that satisfy a set of mixed endpoint conditions.

Definition. A vector field $v \in C^\infty(N,T(X))$ satisfying (I.b.15) is said to give an *infinitesimal variation of* N *satisfying mixed endpoint conditions* in case

$$\begin{cases} \text{(i)} & (v \lrcorner \varphi)_{\partial N} = 0 \\ \text{(ii)} & (v \lrcorner \theta^\mu)_{\partial N} = 0 \\ \text{(iii)} & (v \lrcorner \theta^\rho)_a = 0 \end{cases} \quad . \tag{IV.e.4}$$

Explanation. Each of

$$v \lrcorner \varphi, \quad v \lrcorner \theta^\rho, \quad v \lrcorner \theta^\mu$$

are functions on N. The boundary ∂N consists of the endpoints a,b of the interval [a,b]. Equations (i), (ii) mean that the corresponding functions should vanish at both endpoints, while (iii) means that the function $v \lrcorner \theta^\rho$ should only vanish at the first endpoint a.

Let $\{N_t\} \subset V(I,\omega)$ be a 1-parameter family of integral manifolds of (I,ω) with $N_0 = N$ and whose infinitesimal variation v satisfies the mixed endpoint conditions. In a moment we shall explain

i) what it means that the variational problem $(I,\omega;\varphi)$ should be *non-degenerate* (this will reduce to the previous concept introduced in Chapter I, Section e)); and

ii) what it means that N should satisfy the Euler-Lagrange differential system with mixed endpoint conditions. Using these concepts we shall prove the

(IV.e.5) PROPOSITION. *Suppose that* $(I,\omega;\varphi)$ *is non-degenerate and that* N *satisfies the Euler-Lagrange differential system with mixed endpoint conditions. Then*

$$\frac{d}{dt}\left(\int_{N_t} \varphi\right)_{t=0} = 0$$

for all curves $\{N_t\}$ *in* $V(I,\omega)$ *whose infinitesimal variations satisfy the mixed endpoint conditions (IV.e.4).*

Symbolically we may write this as:

$$v \in T_N(V(I,\omega;[A,B'])) \Rightarrow (\delta\Phi)(v) = 0 \quad .$$

Proof. Using the notational convention (II.b.4) the structure equations (IV.e.3) may be written out as

$$\begin{cases} d\theta^\rho \equiv -A^\rho_{\alpha\mu}\theta^\alpha \wedge \pi^\mu + B^\rho_\alpha \theta^\alpha \wedge \omega \\ d\theta^\mu \equiv -\pi^\mu \wedge \omega + A^\mu_{\nu\alpha}\pi^\nu \wedge \theta^\alpha \end{cases} \quad . \tag{IV.e.6}$$

In the second equation we may have to replace π^μ in (IV.e.3) by $\pi^\mu + C^\mu_\alpha \theta^\alpha$, but this is an allowable substitution. We write

$$\begin{cases} dL \equiv L_\mu \pi^\mu \mod\{\theta^\alpha, \omega\} \\ dL_\mu \equiv L_{\mu\nu}\pi^\nu \mod\{\theta^\alpha, \omega\} \end{cases}$$

and say that $(I, \omega; \varphi)$ is *non-degenerate* in case

$$\det \| L_{\mu\nu} \| \neq 0 \quad . \tag{IV.e.7}$$

This agrees with our previous definition when the additional integrability conditions (I.e.28) are satisfied.

Following the algorithm in Chapter I, Section e) for the computation of the Euler-Lagrange differential system, we consider on the manifold $X \times \mathbb{R}^s$ the 1-form

$$\psi = \varphi + \lambda_\alpha \theta^\alpha$$

Since

$$\begin{aligned} d\varphi &= dL \wedge \omega + L d\omega \\ &\equiv L_\mu \pi^\mu \wedge \omega + A_\alpha \theta^\alpha \wedge \omega + B_{\mu\alpha}\pi^\mu \wedge \theta^\alpha \ , \end{aligned}$$

where the second step uses our assumption that (I, ω) is locally embeddable, the exterior derivative $\Psi = d\psi$ is given by

$$\begin{aligned} \Psi \equiv\ & (L_\mu - \lambda_\mu)\pi^\mu \wedge \omega + \left(d\lambda_\mu - (A_\mu + \lambda_\rho B^\rho_\mu)\omega + (B_{\nu\mu} + \lambda_\rho A^\rho_{\mu\nu})\pi^\nu\right) \wedge \theta^\mu \\ &+ \left(d\lambda_\rho - (A_\rho + \lambda_\sigma B^\sigma_\rho)\omega + (B_{\rho\mu} + \lambda_\sigma A^\sigma_{\mu\rho})\pi^\mu\right) \wedge \theta^\rho + \lambda_\mu A^\mu_{\nu\alpha}\pi^\nu \wedge \theta^\alpha \end{aligned} \tag{IV.e.8}$$

We set

$$\begin{cases} C_\mu = A_\mu + \lambda_\rho B^\rho_\mu \\ C_\rho = A_\rho + \lambda_\sigma B^\sigma_\rho \\ D_{\nu\mu} = B_{\nu\mu} + \lambda_\rho A^\rho_{\mu\nu} + \lambda_\gamma A^\gamma_{\nu\mu} \\ D_{\mu\rho} = B_{\rho\mu} + \lambda_\sigma A^\sigma_{\mu\rho} + \lambda_\nu A^\nu_{\mu\rho} \end{cases}$$

and write (IV.e.8) as

$$\begin{aligned} \Psi \equiv\ & (L_\mu - \lambda_\mu)\pi^\mu \wedge \omega + (d\lambda_\mu - C_\mu \omega + D_{\nu\mu}\pi^\nu) \wedge \theta^\mu \\ &+ (d\lambda_\rho - C_\rho \omega + D_{\mu\rho}\pi^\mu) \wedge \theta^\rho \ . \end{aligned} \tag{IV.e.9}$$

The Cartan system $C(\Psi)$ is generated by the Pfaffian equations (cf. (I.d.16))

$$\begin{cases} \text{(i)} & \partial/\partial\lambda_\alpha \lrcorner \Psi = \theta^\alpha = 0 \\ \text{(ii)} & \partial/\partial\pi^\mu \lrcorner \Psi \equiv (L_\mu - \lambda_\mu)\omega = 0 \\ \text{(iii)} & -\partial/\partial\theta^\mu \lrcorner \Psi \equiv d\lambda_\mu + C_\mu\omega + D_{\nu\mu}\pi^\nu = 0 \\ \text{(iv)} & -\partial/\partial\theta^\rho \lrcorner \Psi \equiv d\lambda_\rho + C_\rho\omega + D_{\mu\rho}\pi^\mu = 0 \end{cases} \tag{IV.e.10}$$

where, as usual, $\equiv$ denotes congruence modulo $\{\theta^\alpha\}$.

To say that N is a solution of the Euler-Lagrange differential system means that we can determine functions $\lambda_\alpha(x)$ such that (IV.e.10) are satisfied with the independence condition $\omega \neq 0$. By (ii) we must have

$$L_\mu = \lambda_\mu \quad ,$$

and so the endpoint conditions on the functions $\lambda_\mu(x)$ are already determined by the endpoints of N. By non-degeneracy we may take $\{\omega;\theta^\alpha;d\lambda_\mu\}$ as a coframe on X. We may view (i), (iii), (iv) as an O.D.E. system for the curve $N \subset X$ together with the functions $\lambda_\rho(x)$ that is *linear* in the $\lambda_\rho(x)$. To say that N *is a solution of the Euler-Lagrange differential system with mixed endpoint conditions* shall *by definition* mean that there is a solution to (IV.e.10) satisfying

$$\lambda_\rho(b) = 0 \; . \tag{IV.e.11}$$

We now complete the proof of the proposition. By the basic computation (I.b.5)

$$\begin{aligned} \frac{d}{dt}\left(\int_{N_t} \varphi\right)_{t=0} &= \int_N v \lrcorner d\varphi + d(v \lrcorner \varphi) \\ &= \int_N v \lrcorner d\varphi \end{aligned}$$

by Stokes' theorem and $(v \lrcorner \varphi)_{\partial N} = 0$. By the Euler-Lagrange equations

$$\begin{aligned} v \lrcorner d\varphi &\equiv -v \lrcorner d(\lambda_\alpha\theta^\alpha) \\ &\equiv -v \lrcorner (d\lambda_\alpha \wedge \theta^\alpha + \lambda_\alpha d\theta^\alpha) \\ &\equiv d\lambda_\alpha(v \lrcorner \theta^\alpha) + \lambda_\alpha d(v \lrcorner \theta^\alpha) \ \text{mod } N \end{aligned}$$

where the last step follows from (I.b.15). Thus

$$v \lrcorner d\varphi \equiv d(\lambda_\alpha (v \lrcorner \theta^\alpha))$$

and consequently again by Stokes' theorem

$$\frac{d}{dt}\left(\int_{N_t} \varphi\right)_{t=0} = (\lambda_\alpha (v \lrcorner \theta^\alpha))_{\partial N} .$$

Our boundary conditions are (cf. (IV.e.4) and (IV.e.11)

$$\begin{cases} (v \lrcorner \theta^\mu)(a) = (v \lrcorner \theta^\mu)(b) = 0 \\ (v \lrcorner \theta^\rho)(a) = 0, \quad \lambda_\rho(b) = 0 . \end{cases}$$

It follows that

$$(\lambda_\alpha (v \lrcorner \theta^\alpha))_{\partial N} = 0 . \qquad \text{Q.E.D.}$$

Definitions. i) Suppose that $(I,\omega;\varphi)$ is a non-degenerate variational problem satisfying the assumption (IV.e.1). With the endpoint conditions (IV.e.4) we shall say that $(I,\omega;\varphi)$ *gives a well-posed mixed variational problem.*

ii) The *Euler-Lagrange equations* for this problem are the usual Euler-Lagrange equations with the additional endpoint conditions (IV.e.11).(23)

ii) The Lagrange Problem (cf.[4], [5], [13], [40], and [57]).

This problem is traditionally stated as follows:

In the space $J^1(\mathbb{R},\mathbb{R}^m)$ *with coordinates* $(x;y^1,\ldots,y^m;\dot{y}^1,\ldots,\dot{y}^m) = (x,y,\dot{y})$, *determine the extremals of the functional*

$$\Phi = \int L\left(x,y(x), \frac{dy(x)}{dx}\right) dx \tag{IV.e.12}$$

defined on parametrized curves $x \to y(x) \in \mathbb{R}^m$ *subject to a set of constraints*

$$g^\rho\left(x,y(x). \frac{dy(x)}{dx}\right) = 0 . \tag{IV.e.13}$$

The constraints are said to be finite (as opposed to infinitesimal) or *holonomic* in case the functions $g^\rho = g^\rho(x,y)$ do not depend on $\dot{y}$ (cf. the note on the next page). If additionally

$$g^\rho = g^\rho(y)$$

depends only on y, then we may consider the Lagrange problem as a classical variational problem defined on the submanifold

$$M = \{y \in \mathbb{R}^m : g^\rho(y) = 0\}$$

of $\mathbb{R}^m$. The Euler-Lagrange equations for this problem have been discussed in Chapter I, Section d).

We use the additional ranges of indices

$$\begin{cases} 1 \leqq \mu, \nu \leqq \ell \\ \ell+1 \leqq \rho, \sigma \leqq m \end{cases} .$$

If the constraints are holonomic then locally we may assume they are of the form

$$y^\rho = h^\rho(x;y^\mu) . \qquad \text{(IV.e.14)}$$

The natural endpoint conditions are to give the values

$$y^\mu(a), \ y^\mu(b)$$

where $a \leqq x \leqq b$, and then use the equations (IV.e.14) to determine the remaining endpoint conditions. For a classical variational problem posed on a submanifold $M \subset \mathbb{R}^m$ these are just the natural endpoint conditions of Chapter IV, Section a).

In general the constraints (IV.e.13) are said to be *non-holonomic*. In the opposite extreme to the holonomic case the matrix

$$\left\| \frac{\partial g^\rho(x,y,\dot{y})}{\partial \dot{y}^\alpha} \right\|$$

has maximal rank $m-\ell$. Locally we may assume the constraints are given by

$$\dot{y}^\rho = h^\rho(x;y^\alpha;\dot{y}^\mu) . \qquad \text{(IV.e.15)}$$

Thus the constrained curves are given by solutions to an O.D.E. system. From this point of view the natural endpoint conditions are to give the values

$$y^\mu(a), \quad y^\rho(a), \quad y^\mu(b) \qquad \text{(IV.e.16)}$$

and use the equations (IV.e.15) to determine the remaining endpoint values $y^\rho(b)$.

Note. The word non-holonomic means non-integrable. To explain this suppose that the constraints are linear of the form:

$$g^{\rho}(x,y,\dot{y}) = A^{\rho}_{\alpha}(y)\dot{y}^{\alpha} .$$

Then the Pfaffian equations

$$\theta^{\rho} = A^{\rho}_{\alpha}(y)\,dy^{\alpha} = 0$$

define a distribution in $\mathbb{R}^m$, and solutions to the O.D.E. system (IV.e.15) are parametrized curves tangent to this distribution (we encountered an example of this in (II.a.40)). In general this distribution will be non-integrable, and this is the origin of the word non-holonomic.

(IV.e.17) Example. The most classical problem in the calculus of variations is the brachistochrone:
Determine the curve γ joining two fixed points A,B such that a particle P travelling from A to B under the influence of gravity will take the shortest time.

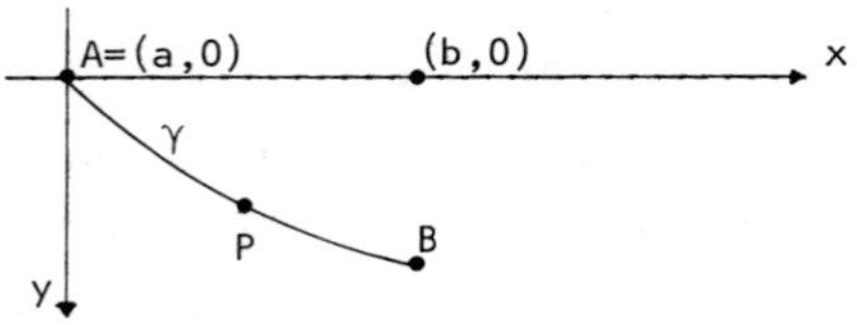

In every elementary course on the calculus of variations it is proved that the solution curve to this problem is a cycloid (cf. also Appendix b). Briefly, if we give γ by $x \to (x,y(x))$, then at time t the particle P will have coordinates

$$(x(t),y(x(t)))$$

and velocity

$$v(t) = \sqrt{1 + \left(\frac{dy}{dx}\right)^2}\;\frac{dx}{dt} .$$

By the law of conservation of total energy

$$\frac{mv^2}{2} = gy$$

where m is the mass of P and g is the gravitational constant. (Note that y is measured *downward*.) This gives

$$dt = c \frac{\sqrt{1 + \left(\frac{dy}{dx}\right)^2}}{\sqrt{y}} dx \qquad c > 0 ,$$

so that what is required is to minimize the functional

$$\Phi = \int_a^b L\left(x, y(x), \frac{dy(x)}{dx}\right) dx$$

where

$$L(x,y,\dot{y}) = \frac{\sqrt{1+\dot{y}^2}}{\sqrt{y}}$$

with the endpoint values $y(a)$, $y(b)$ being given.

Note. From the present point of view it would be more natural to work in the space $J^1(\mathbb{R},\mathbb{R}^2)$ having variables $(t,x,y,\dot{x},\dot{y})$ with the functional

$$\Phi = \int dt$$

and constraint

$$\frac{m}{2}(\dot{x}^2 + \dot{y}^2) = gy .$$

However, the endpoint conditions (IV.e.16) are not the desired ones.

Alternatively, we could work in the space with variables $(x,y,v,\dot{y},\dot{v})$ (where v stands for velocity) and seek to minimize the functional

$$\Phi = \int \frac{\sqrt{1+\dot{y}^2}}{v} dx$$

subject to the holonomic constraint

$$\frac{mv^2}{2} = gy .$$

In this case the endpoint conditions below (IV.e.14) are the desired ones for the problem. We shall now see an extension of this approach.

An interesting variant of the brachistochrone is to assume that P travels in a retarding medium where the force of retardation is a function of the velocity. Then energy is lost according to a rule

$$\frac{d}{dt}\left(\frac{mv^2}{2} + gy\right) = -R(v) \qquad \text{(IV.e.18)}$$

where $R(v) > 0$. *The brachistichrone problem in a retarding medium is therefore naturally posed as a Lagrange problem with holonomic constraints.*

In more detail, in the space $J^1(\mathbb{R},\mathbb{R}^2)$ with coordinates $(x,y,v,\dot{y},\dot{v})$ we consider the constrained classical variational problem given by the following data:

$$\text{(i)} \quad \Phi = \int L\left(x, y(x), v(x), \frac{dy(x)}{dx}, \frac{dv(x)}{dx}\right) dx \qquad \text{(functional)}$$

$$\text{(ii)} \quad L(x,y,v,\ddot{y},\dot{v}) = \frac{\sqrt{1+\dot{y}^2}}{v} \qquad \text{(Lagrangian)}$$

$$\text{(iii)} \quad mv\dot{v} + g\dot{y} + \frac{\sqrt{1+\dot{y}^2}\; R(v)}{v} = 0 \qquad \text{(constraint)}$$

(IV.e.19)

with the endpoint values

$$y(a), \quad y(b), \quad v(a)$$

being given. Here "." stands for "d/dx" and v stands for velocity, so that

$$dt = \frac{\sqrt{1+\dot{y}^2}}{v}\, dx$$

and

$$\frac{d}{dt}\left(\frac{mv^2}{2} + gy\right) = (mv\dot{v} + g\dot{y})\,\frac{dx}{dt}$$

$$= \frac{(mv\dot{v} + g\dot{y})\,v}{\sqrt{1+\dot{y}^2}} .$$

We shall continue this example below.

(IV.e.20) Example. We consider a classical 2nd order variational problem given by a functional

$$\Phi = \int L\left(x, y(x), \frac{dy(x)}{dx}, \frac{d^2y(x)}{dx^2}\right) dx \qquad \text{(IV.e.21)}$$

As mentioned several times previously this may be considered as a 1st order problem with a non-holonomic constraint. Explicitly, the space $J^2(\mathbb{R},\mathbb{R}^m)$ with coordinates $(x,y,\dot{y},\ddot{y})$ is naturally embedded in $J^1(\mathbb{R},\mathbb{R}^{2m})$ with coordinates $(x,y,u,\dot{y},\dot{u})$ by the inclusion mapping

$$(x,y,\dot{y},\ddot{y}) \mapsto (x,y,\dot{y},\dot{y},\ddot{y}) .$$

Thus $J^2(\mathbb{R},\mathbb{R}^m) \subset J^1(\mathbb{R},\mathbb{R}^{2m})$ is defined by the equation

$$\dot{y} - u = 0 , \qquad \text{(IV.e.22)}$$

and the extremals of (IV.e.21) are given by the extremals of the constrained functional

$$\begin{cases} \tilde{\Phi} = \int L\left(x, y(x), u(x), \dfrac{dy(x)}{dx}, \dfrac{du(x)}{dx}\right) dx \\ \dfrac{dy(x)}{dx} = u(x) \end{cases} \qquad \text{(IV.e.23)}$$

where

$$L(x,y,u,\dot{y},\dot{u}) = L(x,y,u,\dot{u}) \quad . \qquad \text{(IV.e.24)}$$

In Chapter IV, Section a) we pointed out that, assuming that $\det\|L_{\ddot{y}^\alpha \ddot{y}^\beta}\| \neq 0$, (IV.e.21) gave a well-posed variational problem with the endpoint conditions

$$y(a), \quad \frac{dy}{dx}(a), \quad y(b), \quad \frac{dy}{dx}(b) \quad \text{fixed.} \qquad \text{(IV.e.25)}$$

It is curious to note that the natural endpoint conditions (IV.e.16) for the constrained 1st order variational problem (IV.e.23) do *not* give (IV.e.25), since the former endpoint conditions refer to only fixing the $3m$ constants

$$y(a), \quad \frac{dy}{dx}(a), \quad \frac{dy}{dx}(b)$$

and then determining $y(b)$ by

$$y(b) = \int_a^b \frac{dy(x)}{dx}\, dx + y(a) \quad .$$

This will be explained below (cf. IV.e.48)).

Returning to the general discussion, Lagrange problems with constraints are traditionally solved by the method of Lagrange multipliers. We shall comment on this below. Here we wish to point out that the formalism of exterior differential systems provides an alternative approach (*not* an equivalent one) that sometimes has the practical advantage of working in a space with fewer variables. We shall now explain this.

We consider a Lagrange problem (IV.e.12), (IV.e.13) where

$$\operatorname{rank}\left\|\frac{\partial g^\rho(x,y,\dot{y})}{\partial \dot{y}^\sigma}\right\| = m - \ell \qquad \text{(IV.e.26)}$$

is maximal. Under the assumption that

$$\det\left\|\frac{\partial^2 L}{\partial\dot{y}^\mu\partial\dot{y}^\nu}\right\| \neq 0 \qquad \text{(VI.e.27)}$$

we shall set up the Lagrange problem as a well-posed mixed variational problem $(I,\omega;\varphi)$. Of course, it is understood that the natural endpoint conditions (IV.e.16) for the Lagrange problem will correspond to the mixed endpoint conditions (IV.e.4) for $(I,\omega;\varphi)$.

For X we take the submanifold of $J^1(\mathbb{R},\mathbb{R}^m)$ given by the constraint equations

$$g^\rho(x,y,\dot{y}) = 0 \quad ,$$

and for (I,ω) we take the restriction to X of the canonical system on $J^1(\mathbb{R},\mathbb{R}^m)$ (cf. (0.e.2)). Since all our formulations are intrinsic we are free to choose convenient generators for the Pfaffian system I, and for reasons to be explained momentarily we take these to be

$$\begin{cases} \theta^\mu = dy^\mu - \dot{y}^\mu dx = 0 \\ \theta^\rho = g^\rho_{\dot{y}^\sigma}(dy^\sigma - \dot{y}^\sigma)dx + g^\rho_{\dot{y}^\mu}(dy^\mu - \dot{y}^\mu dx) = 0 \end{cases} \quad . \qquad \text{(IV.e.28)}$$

By our rank assumption (IV.e.26) these are independent. Moreover, since on X

$$0 = dg^\rho \equiv g^\rho_{\dot{y}^\sigma} d\dot{y}^\sigma + g^\rho_{\dot{y}^\mu} d\dot{y}^\mu \mod\{dy^\alpha, dx\} \quad ,$$

if we set

$$\omega = dx, \quad \pi^\mu = d\dot{y}^\mu$$

then the structure equations of (IV.e.28) are

$$\begin{cases} d\theta^\mu \equiv -\pi^\mu \wedge \omega \mod\{\theta^\alpha\} \\ d\theta^\rho \equiv 0 \mod\{\theta^\alpha\} \end{cases} \quad .$$

In other words, our choice (IV.e.28) of generators for I was made simply to have the 1st derived system generated by

$$\text{span}\{\theta^\rho\} \; .$$

At this point it is clear that the assumption (IV.e.1) and structure equations (IV.e.3) are fulfilled. Moreover, it is also evident from (IV.e.28) that the mixed endpoint conditions (IV.e.4) are exactly the natural endpoint conditions (IV.e.16) in the Lagrange problem. Finally, the non-degeneracy condition (IV.e.7) is easily seen to be equivalent to (IV.e.27).[24] By proposition (IV.e.5) we have established the first part of the following result:

(IV.e.29) THEOREM. *The Lagrange variational problem (IV.e.12), (IV.e.13) satisfying (IV.e.26) and (IV.e.27) gives a well-posed mixed variational problem* $(I,\omega;\varphi)$. *The solutions to the Euler-Lagrange equations associated to* $(I,\omega;\varphi)$ *give extremals for the Lagrange problem. Finally, the Euler-Lagrange differential system* (J,ω) *is globally in Hamiltonian form.*

Proof. We only need to prove the last statement, and essentially this only requires that we recall the construction of (J,ω). On $Z = X \times \mathbb{R}^m$ where $\mathbb{R}^m$ has coordinates $\lambda = (\lambda_1,\ldots,\lambda_m)$ we consider the 1-form

$$\psi = \varphi + \lambda_\alpha \theta^\alpha$$

with exterior derivative

$$\Psi = d\psi.$$

Then (J,ω) is the Pfaffian system on $Y \subset Z$ canonically constructed from the Cartan system $C(\Psi)$ (cf. Chapter I, Section e)--basically (J,ω) is the involutive prolongation of $(C(\Psi),\omega)$, as explained in Chapter I, Section c)).

Now, and this is the essential point, a new set of generators for I leads to the *same* exterior differential system (J,ω). As new generators we simply take

$$\tilde{\theta}^\alpha = dy^\alpha - \dot{y}^\alpha\, dx \quad .$$

Denoting by $(\tilde{\lambda}_\alpha)$ the corresponding new variables we have

$$\begin{aligned}\psi &= \varphi + \tilde{\lambda}_\alpha \tilde{\theta}^\alpha \\ &= -H dx + \tilde{\lambda}_\alpha dy^\alpha\end{aligned}$$

where

$$H = -L(x,y,\dot{y}) + \tilde{\lambda}_\alpha \dot{y}^\alpha \; . \qquad \text{(IV.e.30)}$$

From the proof of proposition (IV.e.5) we infer that:

(i) The momentum space $Y \subset Z$ is defined by

$$\lambda_\mu = L_{\dot{y}^\mu} \quad ,$$

and therefore

$$\dim Y = 2m + 1 \;;$$

ii) $\{\omega; dy^{\alpha}; d\tilde{\lambda}_{\alpha}\}$ gives a coframe on Y.

Consequently, quite visibly the 1-form

$$\psi_Y = -\tilde{H}dx + \tilde{\lambda}_{\alpha}dy^{\alpha}$$

has maximal rank (i.e., $\psi_Y \wedge (d\psi_Y)^m \neq 0$) and the Euler-Lagrange system is just the Cartan system of $\Psi_Y = d\psi_Y$. Q.E.D.

One advantage of the construction where dim $\mathcal{X}$ is smaller than in the traditional approach (see below) is that fewer 1st integrals are required to integrate the Euler-Lagrange system. For instance, it follows from the Hamiltonian property of the Euler-Lagrange system that:

If $L = L(y,\dot{y})$ does not depend on x, $m = 2$, and we have one 1st integral V independent of H, then the Euler-Lagrange system is a completely integrable Hamiltonian system.

(IV.e.31) Example. We continue our discussion of the brachistochrone in a retarding medium (cf. example (IV.e.17)). From (IV.e.9) it is clear that the vector fields

$$\partial/\partial x, \quad \partial/\partial y$$

each leave invariant the Lagrangian (ii) and constraint (iii). Hence they induce infinitesimal symmetries of the variational problem $(I,\omega;\varphi)$. By Noether's theorem (II.a.10) there are two 1st integrals H, V. These are easily seem to in general be independent. If we assume that the retarding force is a rational function $R(v)$ of the velocity, then we have the following result first found by Euler:

(IV.e.32) *The Euler-Lagrange equations of the brachistochrone in a retarding medium are algebraically integrable by quadratures.*

The explicit integration of these equations, as set up by determining the Pfaffian system (IV.e.28) and following our usual algorithm, is somewhat messy. For the purpose of obtaining formulas it appears to be advantageous to formulate the problem in parametric form, and this will be done following a discussion of the classical Lagrange multiplier method.

iii) The Classical Approach to the Lagrange Problem.

We consider a classical variational problem with constraints as given by (IV.e.12) and (IV.e.13). To solve the problem of determining the extremals for this problem (for the endpoint conditions we have described above) Euler and Lagrange devised the following elegant:

(IV.e.33) Lagrange Multiplier Rule: *Consider the unconstrained classical variational problem with Lagrangian*

$$\tilde{L} = L + \Lambda_\rho g^\rho \tag{IV.e.34}$$

where the Λ_ρ *are functions of* x *to be determined. Then the extremals of the functional (IV.e.12) with constraints (IV.e.13) are given by solution curves to the Euler-Lagrange equations*

$$\frac{d}{dx}(\tilde{L}_{\dot{y}^\alpha}) = \tilde{L}_{y^\alpha} \tag{IV.e.35}$$

together with the constraining equations (IV.e.13).

First we will indicate how this method is supposed to work in practice in the maximally non-holonomic case (IV.e.15). For suitable functions F_μ, G_ρ, equations (IV.e.13) and (IV.e.34) may be written as

$$\begin{cases} \text{(i)} & \dfrac{dy^\rho(x)}{dx} = h^\rho\left(x; y^\alpha(x); \dfrac{dy^\mu(x)}{dx}\right) \\ \text{(ii)} & (L_{\dot{y}^\mu\dot{y}^\nu} + L_{\dot{y}^\mu\dot{y}^\rho} h^\rho_{\dot{y}^\nu}) \dfrac{d^2y^\nu(x)}{dx^2} = F_\mu\left(x; y^\alpha(x); \dfrac{dy^\mu(x)}{dx}; \Lambda_\rho(x); \dfrac{d\Lambda_\rho(x)}{dx}\right) \\ \text{(iii)} & L_{\dot{y}^\rho\dot{y}^\mu} \dfrac{d^2y^\mu(x)}{dx^2} + L_{\dot{y}^\rho\dot{y}^\sigma} \dfrac{d}{dx}\left(h^\sigma\left(x; y^\alpha(x); \dfrac{dy^\mu(x)}{dx}\right)\right) + \dfrac{d\Lambda_\rho(x)}{dx} \\ & = G_\rho\left(x; y^\alpha(x) \quad \dfrac{dy^\mu(x)}{dx}; \lambda_\rho(x)\right) \quad . \end{cases}$$

Under suitable non-degeneracy assumptions equations (ii) may be solved for $d^2y^\mu(x)/dx^2$. Plugging this into (iii) gives a system of the form

$$\begin{cases} \text{(i)} & \dfrac{dy^\rho(x)}{dx} = h^\rho\left(x;y^\alpha(x);\ \dfrac{dy^\mu(x)}{dx}\right) \\ \text{(ii)} & \dfrac{d^2y^\mu(x)}{dx^2} = H^\mu\left(x;y^\alpha(x);\ \dfrac{dy^\mu(x)}{dx};\ \Lambda_\rho(x);\ \dfrac{d\Lambda_\rho(x)}{dx}\right) \\ \text{(iii)} & I_\rho\left(x;y^\alpha(x);\ \dfrac{dy^\mu(x)}{dx}\ ;\ \Lambda_\sigma(x);\ \dfrac{d\Lambda_\sigma(x)}{dx}\right) = 0 \end{cases}$$

We then assume that (iii) may be solved for the $d\Lambda_\sigma(x)/dx$ and plug the resulting solution back into (ii). The end result is a system of the form

$$\begin{cases} \text{(i)} & \dfrac{dy^\rho(x)}{dx} = h^\rho\left(x;y^\alpha(x);\ \dfrac{dy^\mu(x)}{dx}\right) \\ \text{(ii)} & \dfrac{d^2y^\mu(x)}{dx^2} = J^\mu\left(x;y^\alpha(x);\ \dfrac{dy^\mu(x)}{dx}\ ;\ \Lambda_\rho(x)\right) \\ \text{(iii)} & \dfrac{d\Lambda_\rho(x)}{dx} = K_\rho\left(x;y^\alpha(x);\ \dfrac{dy^\mu(x)}{dx}\ ;\ \Lambda_\sigma(x)\right) \end{cases} \tag{IV.e.36}$$

These are the Euler-Lagrange equations for the functional (IV.e.12) with constraints (IV.e.13).

The endpoint conditions (IV.e.16) are the natural ones for the functions $y^\rho(x)$, $y^\mu(x)$ appearing in the O.D.E. system (IV.e.36). The endpoint conditions for the $\Lambda_\rho(x)$ are given by requiring that

$$\Lambda_\sigma(b)g^\sigma_{\dot{y}^\rho}\left(x,y(b),\ \frac{dy}{dx}\ (b)\right) + L_{\dot{y}^\rho}\left(b,y(b),\ \frac{dy}{dx}\ (b)\right) = 0\ . \tag{IV.e.37}$$

We will explain the reason for this choice during the proof of the following result:

(IV.e.38) *A solution curve to (IV.e.35) and (IV.e.15) satisfying the endpoint conditions (IV.e.16) and (IV.e.37) is an extremal for the functional (IV.e.12) with the constraints (IV.e.13).*

<u>Proof</u>. Let $y(x,t)$ $(a \leqq x \leqq b,\ 0 \leqq t \leqq \varepsilon)$ be a variation of $y(x) = y(x,0)$ satisfying

$$y^\mu(a,t) = A^\mu,\quad y^\mu(b,t) = B^\mu,\quad y^\rho(a,t) = C^\rho$$

and the constraint equations

$$g^\rho\left(x,y(x,t),\ \frac{dy(x,t)}{dx}\right) = 0$$

(we are assuming these can be locally solved in the form (IV.e.15)). Assume also that there are functions $\Lambda_\rho(x)$ such that (IV.e.35) and (IV.e.37) are satisfied. We want to show that

$$\frac{d}{dt}\left(\int_a^b L\left(x,y(x,t),\ \frac{dy(x,t)}{dx}\right) dx\right)_{t=0} = 0\ . \tag{IV.e.39}$$

The proof is essentially the same as that for the classical

Euler-Lagrange equations. Setting $y_t(x) = (\partial y(x,t)/\partial t)_{t=0}$ the left hand side of (IV.e.39) is

$$\int_a^b \left(y_t^\alpha L_{y^\alpha} + \frac{dy_t^\alpha}{dx} L_{\dot{y}^\alpha} \right) dx \quad .$$

Integration by parts using

$$y_t^\mu(a) = y_t^\mu(b) = y_t^\rho(a) = 0$$

gives

$$\int_a^b y_t^\alpha \left(L_{y^\alpha} - \frac{d}{dx}\left(L_{\dot{y}^\alpha}\right)\right) dx + \left(y_t^\rho L_{\dot{y}^\rho} \right)\Big|_{x=b} \quad .$$

By (IV.e.35) this is

$$- \int_a^b y_t^\alpha \left(\Lambda_\rho g^\rho_{y^\alpha} - \frac{d}{dx}\left(\Lambda_\rho g^\rho_{\dot{y}^\alpha}\right)\right) dx + \left(y_t^\rho L_{\dot{y}^\rho} \right)\Big|_{x=b} \quad .$$

Another integration by parts using (IV.e.37) (this is where these end-point conditions come in) gives for the left hand side of (IV.e.39)

$$- \int_a^b \Lambda_\rho \left(y_t^\alpha g^\rho_{y^\alpha} + \frac{dy_t^\alpha}{dx} g^\rho_{\dot{y}^\alpha} \right) dx \quad . \qquad \text{(IV.e.40)}$$

On the other hand, from the constraint equations we have

$$0 = \frac{d}{dt}\left(g^\rho\left(x, y(x,t), \frac{dy}{dx}(x,t)\right)\right)_{t=0}$$

$$= y_t^\alpha g^\rho_{y^\alpha} + \frac{dy_t^\alpha}{dx} g^\rho_{\dot{y}^\alpha} \quad ,$$

so that (IV.e.40) is zero. This establishes (IV.e.39) and completes the proof of (IV.e.38). Q.E.D.

(IV.e.41) Remark. A natural question is whether or not (IV.e.38) may be deduced directly from theorem (IV.e.29); the answer to this question is affirmative (and not too difficult to establish). However, it is interesting to note that *the parameters* λ_α *we have used throughout this text are not the Lagrange multipliers* Λ_ρ *in the classical case*: the λ_α are the *linear* fibre coordinates in $W^* \subset T^*(X)$ while the Λ_ρ correspond to a choice of generators for the ideal of functions vanishing on $X \subset T^*(M)$.

(IV.e.42) Example. The simplest Lagrange problem is to determine the equations of geodesics parametrized proportionally to arc-length. For convenience we consider only the Euclidean case. In the space $\mathbb{R}^{2m+1}$ with coordinates $(s;y^1,\ldots,y^m;\dot{y}^1,\ldots,\dot{y}^m) = (s,y,\dot{y})$ we consider the functional

$$\Phi = \int ds \tag{IV.e.43}$$

with the constraint

$$\|\dot{y}\| = \sqrt{(\dot{y}^1)^2 + \cdots + (\dot{y}^m)^2} = c = \text{constant} > 0 .$$

Following the rule (IV.e.33) we set

$$\tilde{L} = 1 + \Lambda g$$

where

$$g = \|\dot{y}\| - c .$$

The equations (IV.e.35) are

$$\frac{d}{ds}\left(\frac{\Lambda \dot{y}^\alpha}{\|\dot{y}\|}\right) = 0,$$

and these give

$$\frac{\Lambda \dot{y}^\alpha}{\|\dot{y}\|} = c^\alpha = \text{constant} .$$

Taking into account the constraint $g = 0$ we have respectively

$$\begin{cases} \Lambda \dot{y}^\alpha = b^\alpha = \text{constant} \\ \Lambda = \sqrt{(b^1)^2 + \cdots + (b^m)^2}/c = \text{constant} \\ y^\alpha(s) = a^\alpha s + A^\alpha \end{cases} .$$

The constant c is determined by the a^α, which are then fixed by the endpoint conditions.

A similar method applies to geodesics on any Riemannian manifold.

(IV.e.44) Example. We continue our discussion of the brachistochrone in a retarding medium (cf. (IV.e.17) and (IV.e.31)). For the purposes of explicitly integrating the equations it is convenient to set the problem up as a classical variational problem with constraints where x,y,v are functions of a parameter τ, derive the Euler-Lagrange equations, and then take $\tau = cs$ where s is arclength and c is a suitable constant.

For a simpler example of this method we reconsider the functional (IV.e.43) as given on a space with variables $(\tau, y, \dot{y})$ by

$$\Phi = \int \|\dot{y}\| \, d\tau$$

and where there are no constraints. The Euler-Lagrange equations are

$$\frac{d}{d\tau}\left(L_{\dot{y}^\alpha}\right) = 0 \qquad ;$$

i.e.,

$$\frac{d}{d\tau}\left(\frac{\dot{y}^\alpha}{\|\dot{y}\|}\right) = 0 \qquad .$$

We may now first integrate and then set $\tau = cs$

$$\Rightarrow \|\dot{y}\| = \text{constant}$$
$$\Rightarrow y^\alpha = a^\alpha = \text{constant},$$

which may then be integrated in the obvious way.

For the brachistochrone in a retarding medium we work in a space with variables $(\tau; x, y, v; \dot{x}, \dot{y}, \dot{v})$ and consider the functional

$$\Phi = \int \frac{\sqrt{\dot{x}^2 + \dot{y}^2}}{v} \, d\tau$$

with the constraint

$$g(v, \dot{x}, \dot{y}, \dot{v}) = v\dot{v} + \delta\dot{y}\sqrt{\dot{x}^2 + \dot{y}^2}\,\tilde{R}(v) \; . \tag{IV.e.45}$$

Here we have multiplied (iii) in (IV.e.19) by $1/m$, set $\delta = g/m$, and replaced $R(v)$ by $\tilde{R}(v) = \frac{R(v)}{mv}$. Following the prescription (IV.e.33) we consider the Lagrangian

$$\tilde{L} = \frac{\sqrt{\dot{x}^2 + \dot{y}^2}}{v} + \Lambda g \qquad .$$

Setting

$$K = \frac{1}{v\sqrt{\dot{x}^2 + \dot{y}^2}} + \frac{\Lambda \tilde{R}(v)}{\sqrt{\dot{x}^2 + \dot{y}^2}}$$

the Euler-Lagrange equations (IV.e.35) are

$$\begin{cases} \text{(i)} & \frac{d}{d\tau}(K\dot{x}) = 0 \\ \text{(ii)} & \frac{d}{d\tau}(K\dot{y} + \Lambda\delta) = 0 \\ \text{(iii)} & v\frac{d\Lambda}{d\tau} = (\dot{x}^2 + \dot{y}^2)K_v \end{cases} \tag{IV.e.46}$$

(the third equation follows from $\tilde{L}_{\dot{v}} = \Lambda v$ and $\tilde{L}_v = \Lambda\dot{v} + (\dot{x}^2 + \dot{y}^2)K)$. We now set $\tau = cs$ so that (i), (ii) give

$$\begin{cases} K\frac{dx}{ds} = a \\ K\frac{dy}{ds} = b - \Lambda\delta \end{cases} \tag{IV.e.47}$$

where

$$K = c\frac{1}{v} + \Lambda\tilde{R}(v) \quad .$$

Using the preceding two equations we infer that

$$K^2 = c^2(a^2 + (b - \Lambda\delta)^2)$$

using which we may solve for v as a function of Λ; i.e.,

$$v = V(\Lambda, a, b, c)$$

(this formula means that v is one branch of an algebraic function). Now (iii) in (IV.e.46) and (IV.e.47) give

$$\begin{cases} \text{(i)} & dx = \frac{a\,ds}{K} \\ \text{(ii)} & dy = \frac{(b-\Lambda\delta)ds}{K} \\ \text{(iii)} & ds = \frac{v\,d\Lambda}{K_v} \quad . \end{cases}$$

In (iii) we may integrate to determine $s = s(\Lambda)$ as a function of Λ. Then (i) and (ii) give

$$\begin{cases} dx = f(\Lambda)d\Lambda \\ dy = h(\Lambda)d\Lambda \end{cases}$$

for suitable functions $f(\Lambda)$, $h(\Lambda)$. One more integration gives x, y as functions of Λ, and then the inverse function $\Lambda = \Lambda(s)$ of $s = s(\Lambda)$ gives an integration of the Euler-Lagrange equations by quadratures.

We note that even in the simple case when $R(v) = \lambda v$ is proportional to the velocity the explicit integration will be complicated. Most likely a more profitable approach is to use (IV.e.32) together with the following remarks:

(i) The generic level set

$$H = c$$

of the Hamiltonian function consists of the real points of a 2-dimensional complex abelian variety $A_{\mathbb{C}}$ (this follows from theorem (IV.e.29) and the general results in Chapter 10 of [2]);

(ii) the Hamiltonian vector field is an element in the Lie algebra $\mathfrak{a}_{\mathbb{C}}$ of translation invariant vector fields on $A_{\mathbb{C}}$; and

(iii) the explicit integration of the Euler-Lagrange equations may be carried out by noting that $A_{\mathbb{C}}$ is birationally equivalent to the Jacobian variety of a hyperelliptic algebraic curve (cf. [36])

$$\xi^2 = \prod_{i=1}^{6} (\eta - a_i)$$

of genus two. On this curve C we consider the abelian integrals (here $p, q \in C$)

$$u_1(p) = \int_q^p \frac{d\eta}{\xi}$$

$$u_2(p) = \int_q^p \frac{\eta d\eta}{\xi} .$$

By the Jacobi inversion theorem (loc. cit.), the equations

$$\begin{cases} u_1(0) + \alpha_1 t = u_1(p_1(t)) + u_2(p_2(t)) \\ u_2(0) + \alpha_2 t = u_2(p_1(t)) + u_2(p_2(t)) \end{cases}$$

may be uniquely and holomorphically solved for a divisor

$$D(t) = p_1(t) + p_2(t)$$

where

$$p_i(t) = (\eta_i(t), \xi_i(t)) \qquad (i = 1,2) .$$

This curve in $A_{\mathbb{C}}$ represents the solution to the Euler-Lagrange equations.

(IV.e.48) Example. We will again discuss the functional (IV.e.21) and explain why the natural Lagrange endpoint conditions (IV.e.16) are not the ones desired for this particular problem.

Namely, we consider the Lagrange multiplier rule (IV.e.33) applied to the constrained functional (IV.e.23), (IV.e.24). For simplicity of notation we consider the case $m = 2$ so that the new Lagrangian (IV.e.34) is given by

$$\tilde{L} = L(x,y,u,\dot{u}) + \Lambda(\dot{y} - u) \quad .$$

The Euler-Lagrange equations (IV.e.35) are

$$\begin{cases} \frac{d}{dx}(L_{\dot{u}}) = L_u - \Lambda \\ \frac{d}{dx}(\Lambda) = L_y \quad . \end{cases}$$

Recalling the embedding $J^2(\mathbb{R},\mathbb{R}^m) \subset J^1(\mathbb{R},\mathbb{R}^{2m})$, these may be written in the coordinates $(x,y,\dot{y},\ddot{y})$ as

$$\begin{cases} (i) \quad \frac{d}{dx}(L_{\ddot{y}}) = L_{\dot{y}} - \Lambda \\ (ii) \quad \frac{d}{dx}(\Lambda) = L_y \quad . \end{cases} \tag{IV.e.49}$$

If we now *differentiate* (i) and substitute in (ii) we obtain the usual 2nd order Euler-Lagrange equation (cf. (I.d.23))

$$\frac{d^2}{dx^2}(L_{\ddot{y}}) - \frac{d}{dx}(L_{\dot{y}}) + L_y = 0 \tag{IV.e.50}$$

for which the natural endpoint conditions are (IV.e.25). However, (IV.e.50) only implies (IV.e.49) up to adding a constant to (i), which explains why the endpoint conditions (IV.e.16) contain three and not four constants.

(iv) Some Related Examples.

We shall conclude by discussing some recreational examples, related to the brachistochrone in a retarding medium, that may be conveniently integrated using the formalism of exterior differential systems.

(IV.e.51) Example. It is well-known that airplanes travel essentially along great circles. It is also a common experience that, especially on long flights, the east-west trip takes approximately 6/5 as long as the west-east trip along the same route. The reason for this is, of course, that, especially at high altitudes, the prevailing wind currents run from west to east. These same wind currents also imply that the optimal path for airplane flight is not exactly a great circle. We shall show that the Euler-Lagrange differential system for determining the optimal path is algebraically integrable by quadratures (at least if we make simplifying assumptions). This will be done by discussing a series of simpler examples leading up to the one of interest.

1st Simpler Example. We imagine that we are in a boat in a lake or river without current and wish to get from point A to point B in the shortest time. Assuming that the boat travels with constant velocity it is clear that the optimal path is a straight line. We shall now prove this by a method that will generalize to more interesting situations.

For this consider parametrized paths

$$\tau \to (x(\tau), y(\tau)), \qquad 0 \leqq \tau \leqq 1 ,$$

from A to B .

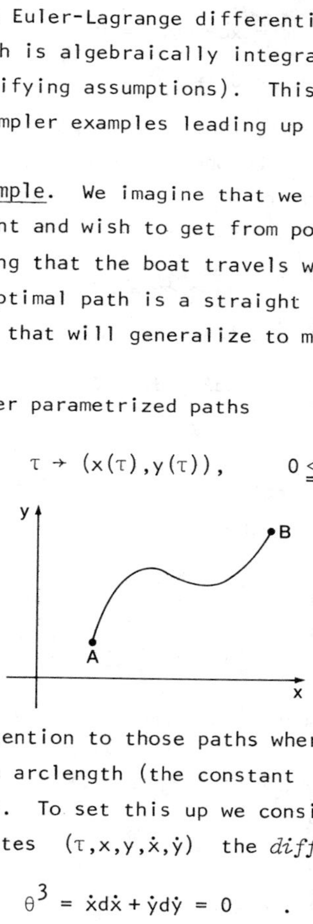

We want to restrict attention to those paths where $d\tau$ is a constant multiple $\mu^{-1}ds$ of the arclength (the constant μ will be the length of the particular path). To set this up we consider in the space $J^1(\mathbb{R},\mathbb{R}^2)$ with coordinates $(\tau,x,y,\dot{x},\dot{y})$ the *differential constraint*

$$\theta^3 = \dot{x}d\dot{x} + \dot{y}d\dot{y} = 0 .$$

Curves $\tau \to (\tau, x(\tau), y(\tau), \frac{dx(\tau)}{d\tau}, \frac{dy(\tau)}{d\tau})$ along which $\theta^3 = 0$ satisfy

$$\frac{d}{d\tau}\left(\left\| \left(\frac{dx(\tau)}{d\tau}, \frac{dy(\tau)}{d\tau} \right) \right\|\right) = 0 ,$$

and therefore

$$ds/d\tau = \mu = \text{constant}$$

along such curves. This suggests that we consider the exterior differential system (I,ω) given by

$$\begin{cases} \tilde{\theta}^1 = dx - \dot{x}d\tau = 0 \\ \tilde{\theta}^2 = dy - \dot{y}d\tau = 0 \\ \tilde{\theta}^3 = \dot{x}d\dot{x} + \dot{y}d\dot{y} = 0 \\ \omega = d\tau \neq 0 \end{cases} , \tag{IV.e.52}$$

whose integral curves are 1-jets of curves parametrized by a constant multiple of arclength. In the open set where $\dot{y} \neq 0$ it is more convenient to take as generators of (I,ω) the Pfaffian equations (cf. (IV.e.28) where a similar choice was made)

$$\begin{cases} \theta^1 = dx - \dot{x}d\tau = 0 \\ \theta^2 = \dot{x}(dx - \dot{x}d\tau) + \dot{y}(dy - \dot{y}d\tau) \\ \theta^3 = \dot{x}d\dot{x} + \dot{y}d\dot{y} = 0 \\ \omega = d\tau \neq 0 \end{cases} . \tag{IV.e.53}$$

Setting $\pi = d\dot{x}$ we have a coframe $\{\omega,\theta^1,\theta^2,\theta^3,\pi\}$ on the open subset $\{\dot{y} \neq 0\}$ of $J^2(\mathbb{R},\mathbb{R}^2)$. Note that

$$\begin{cases} d\dot{y} = -\dfrac{\dot{x}}{\dot{y}}\pi \\ dy - \dot{y}d\tau = \dfrac{1}{\dot{y}}\theta^2 - \dfrac{\dot{x}}{\dot{y}}\theta^1 \end{cases} .$$

Using these we easily compute the structure equations of (IV.e.53) to be

$$\begin{cases} d\theta^1 = -\pi \wedge \omega \\ d\theta^2 = \dfrac{\mu^2}{\dot{y}^2}\pi \wedge \theta^1 - \dfrac{\dot{x}}{\dot{y}^2}\pi \wedge \theta^2 + \omega \wedge \theta^3 \\ d\theta^3 = 0 \end{cases} \tag{IV.e.54}$$

where $\mu = \sqrt{\dot{x}^2 + \dot{y}^2}$. Thus θ^2, θ^3 generate the 1st derived system of (I,ω).

Returning to our variational problem our assumption is that the velocity v of the boat satisfies

$$v = \frac{ds}{dt} = c \quad ,$$

where t is time and c is a constant. Hence if we set

$$\varphi = dt = \frac{ds}{c} = \frac{\mu}{c}\, d\tau \quad ,$$

then we want to minimize the functional

$$\Phi(N) = \int_N \varphi$$

on integral manifolds N of (I,ω) satisfying the endpoint conditions described above. Following our usual prescription, we consider on $J^1(\mathbb{R},\mathbb{R}^2)\times\mathbb{R}^3$ the 1-form

$$\psi = \varphi + \lambda_1\theta^1 + \lambda_2\theta^2 + \lambda_3\theta^3 \; .$$

Using (IV.e.54) the exterior derivative $\Psi = d\psi$ is given by the 2-form

$$\begin{aligned}\Psi \;=\; & \frac{1}{\mu c}\;\theta^3\wedge\omega + d\lambda_\alpha\wedge\theta^\alpha - \lambda_1\pi\wedge\omega \\ & + \;\lambda_2\left(\frac{\mu^2}{\dot y^2}\,\pi\wedge\theta^1 - \frac{\dot x}{\dot y^2}\,\pi\wedge\theta^2 + \omega\wedge\theta^3\right) \qquad .\end{aligned}$$

The Cartan system is generated by the Pfaffian equations (cf. (I.d.16))

$$\left\{\begin{aligned} &\partial/\partial\lambda_\alpha \,\lrcorner\, \Psi = \theta^\alpha = 0 \\ &\partial/\partial\pi \,\lrcorner\, \Psi \equiv \lambda_1\omega = 0 \\ &\partial/\partial\theta^1 \,\lrcorner\, \Psi \equiv -d\lambda_1 - \frac{\lambda_2\mu^2}{\dot y^2}\,\pi = 0 \\ &\partial/\partial\theta^2 \,\lrcorner\, \Psi \equiv -d\lambda_2 + \frac{\lambda_2\dot x}{\dot y^2}\,\pi = 0 \\ &\partial/\partial\theta^3 \,\lrcorner\, \Psi \equiv -d\lambda_3 + \left(\frac{1}{\mu c} - \lambda_2\right)\omega = 0 \; . \end{aligned}\right. \qquad \text{(IV.e.55)}$$

It follows that along solution curves to the Euler-Lagrange system

$$\left\{\begin{aligned} &\mu = \text{constant}, \quad \lambda_1 = 0, \quad \pi = 0, \\ &\lambda_2 = \text{constant}, \quad d\lambda_3 + \left(\lambda_2 - \frac{1}{\mu c}\right)\omega = 0 \quad . \end{aligned}\right.$$

We infer that

$$\begin{cases} t = \frac{s}{c} \\ x = \alpha s + \beta \\ y = \gamma s + \delta \end{cases}$$

along solution curves to (IV.e.55), which establishes our claim about the optimal path from A to B.

Note. The method of this example gives a general technique for restricting the parametrization of competing curves in variational problems.

2nd Simpler Example. We next imagine a river having a current moving with constant velocity c_1. Given a boat that travels with constant velocity c_2 in still water we want again to determine the path from A to B that requires the least time of travel.

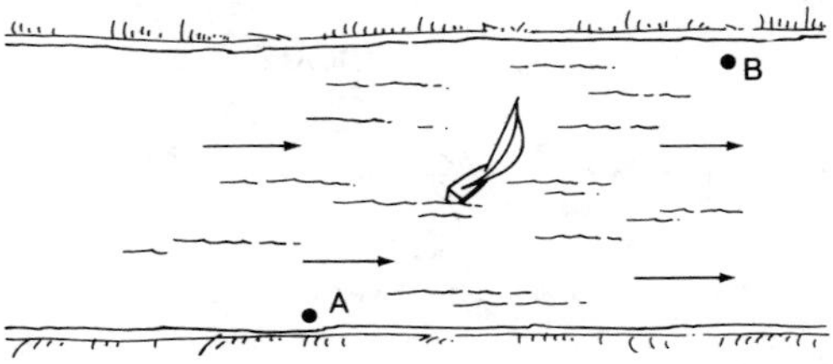

Again it is well-known experimentally that the optimal path is a straight line, a fact that we shall now prove. For this we imagine the river as part of $\mathbb{R}^2$ with coordinates (x,y) and where the current has velocity vector $V_1 = c_1\, \partial/\partial x$. If V is the velocity vector of the boat then we have a picture

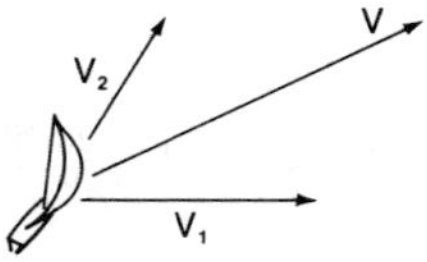

where V_2 is the direction in which the boat is being steered.

To set the problem up mathematically we again use the space $J^1(\mathbb{R},\mathbb{R}^2)$ with coordinates $(\tau,x,y,\dot{x},\dot{y})$, and then

$$V = \left(\frac{dx}{d\tau}, \frac{dy}{d\tau}\right) \frac{d\tau}{dt}$$

$$V - V_1 = \left(\frac{dx}{d\tau}\frac{d\tau}{dt} - c_1, \frac{dy}{d\tau}\frac{d\tau}{dt}\right) = V_2 .$$

Our assumption that the boat travels with constant velocity c_2 *relative to the surrounding water* is then expressed by

$$\|V_2\|^2 = \left(\frac{dx}{d\tau}\frac{d\tau}{dt} - c_1\right)^2 + \left(\frac{dy}{d\tau}\frac{d\tau}{dt}\right)^2 = c_2^2 .$$

We view this as a quadratic equation for $d\tau/dt$ and consider a solution

$$d\tau/dt = f(\dot{x},\dot{y},c_1,c_2) .$$

Note that f is homogeneous of degree -1 in $(\dot{x},\dot{y})$. Setting $L = 1/f$ gives

$$dt = L(\dot{x},\dot{y},c_1,c_2)d\tau$$

where L is homogeneous of degree $+1$ in $(\dot{x},\dot{y})$.

Proceeding as in the previous example we want to determine the extremals of the variational problem $(I,\omega;\varphi)$ where (I,ω) is given by

$$\begin{cases} \tilde{\theta}^1 = dx - \dot{x}d\tau = 0 \\ \tilde{\theta}^2 = dy - \dot{y}d\tau = 0 \\ \tilde{\theta}^3 = dL = 0 \\ \omega = d\tau \neq 0 \end{cases} \tag{IV.e.56}$$

(we note that integral manifolds of (I,ω) are given by 1-jets of curves along which L is constant), and where

$$\varphi = dt = Ld\tau \quad . \tag{IV.e.57}$$

Remark. In first approximation we could consider the variational problem $(\tilde{I},\omega;\varphi)$ where $(\tilde{I},\omega)$ is the canonical Pfaffian system on $J^1(\mathbb{R},\mathbb{R}^2)$ given by $\tilde{\theta}^1 = \tilde{\theta}^2 = 0$, $\omega \neq 0$ and φ is given by (IV.e.57). But since

$$L(\lambda\dot{x},\lambda\dot{y},c_1,c_2) = \lambda L(\dot{x},\dot{y},c_1,c_2)$$

it follows that

$$\det\begin{Vmatrix} L_{\dot{x}\dot{x}} & L_{\dot{x}\dot{y}} \\ L_{\dot{y}\dot{x}} & L_{\dot{y}\dot{y}} \end{Vmatrix} \equiv 0$$

and consequently the standard variational problem is degenerate. This degeneracy may be removed by fixing the parametrization as we have done.

As before it is convenient to work in the open set where $L_{\dot{y}} \neq 0$ and replace (IV.e.56) by

$$\begin{cases} \theta^1 = dx - \dot{x}d\tau = 0 \\ \theta^2 = L_{\dot{x}}(dx - \dot{x}d\tau) + L_{\dot{y}}(dy - \dot{y}d\tau) = 0 \\ \theta^3 = dL = 0 \\ \omega = d\tau \neq 0 \end{cases} \qquad \text{(IV.e.58)}$$

Noting that

$$dL_{\dot{x}},\ dL_{\dot{y}} \equiv 0 \ \mathrm{mod}\{\pi, \theta^3\} \qquad \text{(IV.e.59)}$$

we set

$$\begin{cases} \pi = d\dot{x} \\ dL_{\dot{x}} - \dfrac{L_{\dot{x}}}{L_{\dot{y}}} dL_{\dot{y}} \equiv \alpha\pi \ \mathrm{mod}\{\theta^\alpha\} \\ \dfrac{dL_{\dot{y}}}{L_{\dot{y}}} \equiv \beta\pi \ \mathrm{mod}\{\theta^\alpha\} \end{cases}$$

where α, β are functions of $\dot{x}, \dot{y}$. The structure equations of $(I, \omega; \varphi)$ are

$$\begin{cases} d\theta^1 = -\pi \wedge \omega \\ d\theta^2 \equiv \alpha\pi \wedge \theta^1 + \beta\pi \wedge \theta^2 + \omega \wedge \theta^3 \\ d\theta^3 = 0 \\ d\varphi = \theta^3 \wedge \omega \end{cases} \qquad \text{(IV.e.60)}$$

As before, using (IV.e.60) the Euler-Lagrange differential system associated to $(I, \omega; \varphi)$ turns out to be generated by the Pfaffian equations

$$\begin{cases} \theta^\alpha = 0 \\ \lambda_1 \omega = 0 \\ -d\lambda_1 - \lambda_2 \alpha\pi = 0 \\ -d\lambda_2 - \lambda_2 \beta\pi = 0 \\ -d\lambda_3 + (1 - \lambda_2)\omega = 0 . \end{cases} \qquad \text{(IV.e.61)}$$

It follows that along a general solution curve

$$\begin{cases} L = \text{constant}, & \lambda_1 = 0, \quad \pi = 0, \\ \lambda_2 = \text{constant}, & d\lambda_3 + (\lambda_2 - 1)\omega = 0 . \end{cases}$$

In particular

$$\pi = d\dot{x} = -\frac{L_{\dot{y}}}{L_{\dot{x}}}\, d\dot{y} = 0$$

implies that the extremal is a straight line.

3rd Simpler Example. We keep the same situation as in the 2nd simpler example but where the current is given by a vector field

$$V_1 = f(y)\ \partial/\partial x$$

depending on y. We ask again for the path of shortest time of travel between points A, B, and will show that

(IV.e.62) *The Euler-Lagrange differential system associated to this problem is integrable by quadratures.*

Proof. We retain the preceding notations. Our assumption is that

$$\|V_2\|^2 = \left(\frac{dx}{d\tau}\frac{d\tau}{dt} - f(y(\tau))\right)^2 + \left(\frac{dy}{d\tau}\frac{d\tau}{dt}\right)^2 = c_2^2 \qquad \text{(IV.e.63)}$$

is constant. Let $1/L$ be solution of (IV.e.63) considered as a quadratic equation for $d\tau/dt$. Then

$$\begin{cases} L = L(\dot{x},\dot{y},y,c_2) \\ L(\lambda\dot{x},\lambda\dot{y},y,c_2) = \lambda L(\dot{x},\dot{y},y,c_2) . \end{cases}$$

It is possible to explicitly integrate the Euler-Lagrange differential system following the method of the preceding example. However, for the purposes of proving (IV.e.62) it is easier to proceed as follows: Consider the classical variational problem with Lagrangian

$$\tilde{L}(\dot{x},\dot{y},y) = L(\dot{x},\dot{y},y,c_2)^{\nu} \qquad (\nu \neq 1) .$$

This variational problem is easily seen to be non-degenerate for general ν, and the Hamiltonian

$$H = (\nu - 1)\tilde{L}$$

is a 1st integral. Moreover, since x is a cyclic coordinate

$$W = \tilde{L}_{\dot{x}}$$

is another 1st integral. It follows that this classical variational problem is integrable by quadratures. But, by the argument in example (II.a.23) the extremals of this problem coincide with those of the classical variational problem with Lagrangian L. Q.E.D.

Discussion of Example (IV.e.51). For an airplane travelling with constant velocity in the absence of wind currents the same argument as in the 1st simpler example shows that optimal routes are great circles.

We now assume that the wind currents are given by a vector field on S^2 that is invariant under rotation about the north pole-south pole axis. The corresponding variational problem for the optimal flight path then admits an infinitesimal symmetry. The argument in the 3rd simpler example may then be directly adapted to show that the extremals of this variational problem coincide with the extremals of a classical variational problem on S^2 given by a non-degenerate, time independent Lagrangian admitting an additional infinitesimal symmetry. As before, this problem is integrable by quadratures.

FOOTNOTES TO CHAPTER IV

(1) This definition is only possible by theorem (I.e.9), which gives a bijection between the integral manifolds of (J,ω) on Y and the integral manifolds N of (I,ω) on X that also satisfy the Euler-Lagrange equations (I.d.14).

(2) We already encountered this particular W^* in our discussion of prolongation at the end of Chapter I, Section c).

(3) Again, we emphasize that the ω^i and ω_i^j are defined on the open set $U \subset X =$ unit sphere bundle of M, and *not* downstairs over an open subset of M.

(4) This is because the Euler-Lagrange system J always contains the 1-forms θ^α.

(5) Actually, all that is required is the quadratic form be positive definite along N.

(6) This is a special case of the following general fact: If $f: X \to X'$ is a surjective map and (I',ω') an exterior differential system on X', then the Cauchy characteristic system of $(\pi^* I', \pi^* \omega')$ on X is the Pfaffian system generated by $(\pi_*^{-1}(S'))^\perp$ where $S' \subset T(X')$ is the distribution corresponding to the Cauchy characteristic system of (I',ω').

(7) We have not used equation (IV.c.19) yet. It will play an essential role in our discussion of Hilbert's invariant integral in the next section.

(8) $\delta^2\Phi$ is defined by equation (IV.c.21), and it will be proved below that it is a symmetric bilinear form on the vector space $T_N(V(I,\omega;[A,B]))$.

(9) Here we must be more precise. The neighborhood of N is an ε-neighborhood U of $N \in V(I,\omega;[A,B])$ in the C^2-*topology*. If $N' \in U$ then we may join N and N' by an arc $\{N_t\} \subset V(I,\omega;[A,B])$ with infinitesimal variation $v_t \in T_{N_t}(V(I,\omega;[A,B]))$ such that $N_0 = N$, $N_1 = N$, and

$$\frac{d^2}{dt^2}\left(\int_{N_t} \varphi\right)_{t=0} = \delta^2\Phi(N)(v,v) > 0 \quad .$$

Thus

$$\frac{d}{dt}\left(\int_{N_t} \varphi\right)' > 0 \qquad 0 \leq t \leq 1$$

and so $\Phi(N) < \Phi(N')$.

(10) When we compare with (I.b.16) we see that the variational equations of (J,ω) are given by a 1st order operator

$$L: C^\infty(\Gamma,E) \to C^\infty(\Gamma, W \otimes T^*(\Gamma))$$

where E is the normal bundle of Γ and $W^* \subset T^*(Y)$ is the sub-bundle generating the Pfaffian system J. The point is that in the present situation

$$E^* = W^* .$$

In fact, this is true whenever W^* generates a Frobenius system.

(11) Here it may be helpful to recall the symplectic linear algebra discussion in Chapter 0, Section d).

(12) From these equations we see that the integral manifolds of $(\pi^* I,\omega)$ depend on s arbitrary functions of 1-variable, plus a certain number of constants. Of the s functions, s_1 are the $V^\nu(x)$ and the remaining $s-s_1$ are the Cauchy characteristics $\lambda_\rho(x)$.

(13) Intuitively the theorem states: If N is a solution of the Euler-Lagrange equations along which the quadratic form $\|A_{\mu\nu}\|$ is positive and (the essential hypothesis) *no two points of* N *are mutually conjugate*, then N gives a local minimum. Certainly this result is fundamental in the classical theory of unconstrained variational problems (cf. [13]). In the case of the Lagrange problem there are also results along the lines of theorem (IV.d.1) given, e.g., in [5] and in the last chapter of [13]. But without employing the structure theory of exterior differential systems (in particular the derived system), it seems difficult to properly formulate the general result (recall that $\|A_{\mu\nu}\|$ is a quadratic form on $(W^*/W_1^*)^*$).

(14) In brief, given a 2-form Ψ of rank m on a vector space $V \cong \mathbb{R}^{2m+1}$, the maximal isotropic planes for Ψ have dimension $m+1$ and all contain the characteristic direction $\Psi^\perp$.

(15) This is a slightly different use of the word than in Chapter 0, Section d). Here we have the data of a vector space $T \cong \mathbb{R}^6$ (a typical tangent space to U) and both a 1-form $\theta \in T^*$ and 2-form $\Theta \in \Lambda^2 T^*$. On the hyperplane $H = \{\theta = 0\}$ the 2-form Θ_H is assumed to have maximal rank 2, and $V \in H$ spans the characteristic direction of Θ_H in the sense of Chapter 0, Section d).

(16) These are by definition the integral curves of the characteristic vector field.

(17) This will be very relevant later--in fact, it is the main point beyond theorem (IV.d.1) in the classical unconstrained case.

(18) This is the invariance of Ψ_Y under the flow $\exp(tV)$; it allows us to verify any statement about Ψ_Y on S by looking along T.

(19) Here we are using the obvious decomposition

$$T_{\tilde{y}}(S) \cong T_{\tilde{y}}(T) \oplus (\mathbb{R}V) .$$

(20) Since π is surjective it follows that $(\pi^* I)_1 = \pi^*(I_1)$ (i.e., for any Pfaffian system the pullback of the 1st derived system is the 1st derived system of the pullback under a submersion).

(21) Since $\det\|L_{\dot{y}^\alpha \dot{y}^\beta}\| \neq 0$, $(x;y^\alpha;\dot{y}^\alpha)$ are perfectly good *local* coordinates on Y. Although not so natural for the Hamiltonian formalism as $(x;y^\alpha;\lambda_\alpha)$, they are better suited to the present computation.

(22) This explains what we meant by reducing (IV.d.1) to a computation along N: The quantity Δ vanishes to 2nd order when $N_1 = N$, and we shall determine the sign of the quadratic term.

(23) We have not attempted to develop the theory of Chapter IV, Sections b)-d) (Jacobi equations, conjugate points, 2nd variation, fields, etc.) for well-posed mixed variational problems; this should be interesting to carry out.

(24) In example (I.e.39) we discussed non-degeneracy for the classical Lagrange problem *in case the additional conditions* (I.e.28) *were satisfied*. Since these conditions were not in very intrinsic form, it is not clear how the non-degeneracy condition there relates to (IV.e.27).

APPENDIX

MISCELLANEOUS REMARKS AND EXAMPLES

a) Problems With Integral Constraints; Examples.

We consider a variational problem $(I,\omega;\varphi)$ on a manifold X. Thus we are given a differential system (I,ω) on X with $V(I,\omega)$ denoting the set of integral manifolds, and we consider the functional

$$\Phi: V(I,\omega) \to \mathbb{R} \tag{A.a.1}$$

defined by

$$\Phi(N) = \int_N \varphi$$

where φ is a given differential form on X. The problem frequently arises of minimizing the functional (A.a.1) subject to a set of *integral constraints*

$$\Xi_\rho(N) = \int_N \xi_\rho = \text{constant} \tag{A.a.2}$$

where the ξ_ρ are differential forms on X. Of course we must be precise here about the boundary or endpoint conditions on N, but as in Chapter I we will ignore this point for a while and simply argue formally. Also, as usual we will assume that the number of independent variables (= degree of ω) is one.

Recall that the Euler-Lagrange system associated to (A.a.1) was obtained as follows (cf. Chapter I, Section e)): We assume that (I,ω) is a Pfaffian differential system

$$\theta^1 = \cdots = \theta^s = 0, \ \omega \neq 0 ,$$

and on $Z = X \times \mathbb{R}^s$ we consider the differential forms

$$\begin{cases} \psi = \varphi + \lambda_\alpha \theta^\alpha \\ \Psi = d\psi \end{cases} .$$

The Euler-Lagrange system (J,ω) was defined to be the restriction to the momentum space $Y \subset Z$ of the Cartan system $(C(\Psi),\omega)$. The main result was that, at least in a formal sense, the integral manifolds $N \in V(I,\omega)$ satisfying

$$\delta\Phi(N) = 0 \tag{A.a.3}$$

were in one-to-one correspondence with the integral manifolds of (J,ω) on Y. Finally, in Chapter IV, Section a) we discussed endpoint conditions and explained how (A.a.3) could be interpreted precisely.

Now suppose we denote by $V(I,\omega;\xi_\rho) \subset V(I,\omega)$ the integral manifolds of (I,ω) satisfying the constraints (A.a.2). We may view $V(I,\omega;\xi_\rho)$ as a finite-codimensional submanifold of the infinite-dimensional manifold $V(I,\omega)$. The differential of $\Phi|V(I,\omega;\xi_\rho)$ is then the restriction of $\delta\Phi$ to the subspace $\delta\Xi_\rho = 0$ of $T_N(V(I,\omega))$. Therefore:

The condition on $N \in V(I,\omega;\xi_\rho)$ *that* $\delta\Phi(N) = 0$ *is then expressed by*

$$(\delta\Phi + \mu^\rho \delta\Xi_\rho)(N) = 0 \tag{A.a.4}$$

where the μ^ρ *are constants.*
Accordingly we set $\tilde{X} = X \times \mathbb{R}^t$ where $\mathbb{R}^t$ has coordinates $(\mu^1,..,\mu^t)$ and on $\tilde{X}$ we consider the Pfaffian differential system $(\tilde{I},\omega)$ given by

$$\theta^1 = \cdots = \theta^s = d\mu^1 = \cdots = d\mu^t = 0, \quad \omega \neq 0 .$$

On $\tilde{X}$ we consider the variational problem $(\tilde{I},\omega;\tilde{\varphi})$ where

$$\tilde{\varphi} = \varphi + \mu^\rho \xi_\rho \quad , \tag{A.a.5}$$

and from this we construct the Euler-Lagrange system $(\tilde{J},\omega)$ on $\tilde{Y}$ as in Chapter II, Section e). It is then clear that (with due attention being paid to endpoint conditions as in Chapter IV, Section a)):

(A.a.6) *The integral manifolds* $N \in V(I,\omega;\xi_\rho)$ *satisfying the Euler-Lagrange equations (A.a.4) are in one-to-one correspondence with the integral manifolds of* $(\tilde{J},\omega)$.

(A.a.7) <u>Example</u>. Perhaps the oldest problem in the calculus of variations is the *isoperimetric problem of Pappus*:
Among all curves N joining two points A,B on the x-axis and enclosing with the line segment AB a region of fixed area, determine the curve of shortest length.

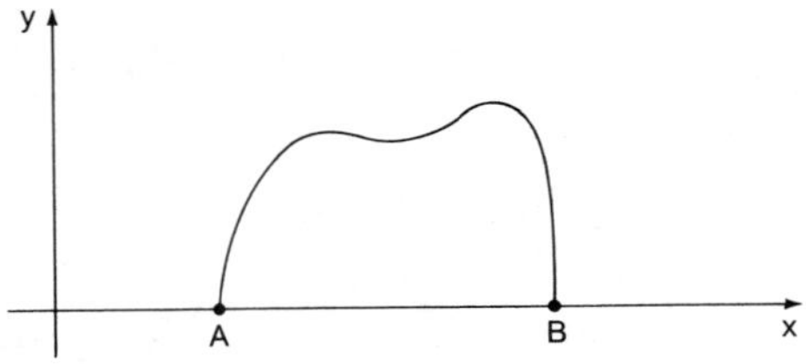

To discuss this problem we shall for simplicity consider only those curves that are graphs $x \to (x, y(x))$ of a function $y(x)$, remarking at the end what to do in the general case. On $X = J^1(\mathbb{R}, \mathbb{R})$ with coordinates $(x, y, \dot{y})$ we thus have the situation

$$\begin{cases} \theta = dy - \dot{y}dx = 0 \\ \omega = dx \neq 0 \\ \varphi = \sqrt{1 + \dot{y}^2}\, dx \\ \xi = ydx \end{cases} .$$

This is because clearly

$$\begin{cases} \text{length}(N) = \Phi(N) = \int_A^B \sqrt{1 + \left(\frac{dy(x)}{dx}\right)^2}\, dx \\ \text{area bounded by } NAB = \Xi(N) = \int_A^B y(x)\, dx \end{cases} .$$

Following our prescription, on $\tilde{X}$ with coordinates $(x, y, \dot{y}, \mu)$ we consider the variational problem given by the data

$$\begin{cases} \theta^1 = dy - \dot{y}dx = 0 \\ \theta^2 = d\mu = 0 \\ \omega = dx \neq 0 \\ \tilde{\varphi} = \varphi + \mu\xi \end{cases} .$$

For

$$\tilde{\psi} = \tilde{\varphi} + \lambda_1\theta^1 + \lambda_2\theta^2$$

the 2-form $\tilde{\Psi} = d\tilde{\psi}$ is given by

$$\tilde{\Psi} = \frac{\dot{y}}{\sqrt{1+\dot{y}^2}}\, d\dot{y} \wedge dx + \mu\theta^1 \wedge dx + d\lambda_1 \wedge \theta^1 + d\lambda_2 \wedge \theta^2 - \lambda_1 d\dot{y} \wedge dx .$$

The Cartan system $\mathcal{C}(\tilde{\Psi})$ is generated by the Pfaffian equations

$$\left\{\begin{array}{lll} \text{(i)} & \partial/\partial\dot{y} \;\lrcorner\; \tilde{\Psi} = \left(\dfrac{\dot{y}}{\sqrt{1+\dot{y}^2}} - \lambda_1\right) dx = 0 & \\ \text{(ii)} & \partial/\partial\theta^1 \;\lrcorner\; \tilde{\Psi} = -d\lambda_1 + \mu dx = 0 & \\ \text{(iii)} & \partial/\partial\theta^2 \;\lrcorner\; \tilde{\Psi} = -d\lambda_2 = 0 & \\ \text{(iv)} & \partial/\partial\lambda_1 \;\lrcorner\; \tilde{\Psi} = \theta^1 = 0 & \\ \text{(v)} & \partial/\partial\lambda_2 \;\lrcorner\; \tilde{\Psi} = \theta^2 = 0 & . \end{array}\right. \tag{A.a.8}$$

It follows from the 1st equation that $\tilde{Z}_1 \subset \tilde{Z}$ has the defining equation

$$\lambda_1 = \frac{\dot{y}}{\sqrt{1+\dot{y}^2}} \quad .$$

This equation may be inverted to yield

$$\dot{y} = \frac{\lambda_1}{\sqrt{1-\lambda_1^2}} \quad .$$

We may take $\{dx,\theta^1,\theta^2,d\lambda_1,d\lambda_2\}$ as a coframe on $\tilde{Z}_1$, and at points $p \in \tilde{Z}_1$ the Cartan system is (these are linear equations in the *ambient tangent space* $T_p(\tilde{Z})$)

$$\left\{\begin{array}{l} \eta_1 = d\lambda_1 - \mu dx = 0 \\ \eta_2 = d\lambda_2 = 0 \\ \theta^1 = 0 \\ \theta^2 = 0 \end{array}\right. \quad .$$

Since we may also take $\{dx,\theta^1,\theta^2,\eta_1,\eta_2\}$ as a coframe on $\tilde{Z}_1$, it follows that there are integral elements of $(C_1(\tilde{\Psi}),\omega)$ *tangent to* $\tilde{Z}_1$; i.e.,

$$\tilde{Y} = \tilde{Z}_1 \quad .$$

We claim that the variational problem $(\tilde{I},\omega)$ on $\tilde{X}$ is non-degenerate. To see this, we have

$$\tilde{\psi}_{\tilde{Y}} = \tilde{L}dx + \lambda_1\theta^1 + \lambda_2\theta^2 \qquad (\tilde{L} = \sqrt{1+\dot{y}^2} + \mu y)$$
$$\tilde{\Psi}_{\tilde{Y}} = \eta_1 \wedge \theta^1 + \eta_2 \wedge \theta^2 \quad .$$

From this it is clear that

$$\tilde{\psi}_{\tilde{Y}} \wedge (\tilde{\Psi}_{\tilde{Y}})^2 \neq 0 \ ,$$

which implies non-degeneracy.

Taking $(x,y,\lambda_1,\lambda_2,\mu)$ as coordinates on $\tilde{Y}$ the Euler-Lagrange system is (we take the + in the equation for $\dot{y}$)

$$\begin{cases} \text{(i)} & \eta_1 = d\lambda_1 - \mu dx = 0 \\ \text{(ii)} & \eta_2 = d\lambda_2 = 0 \\ \text{(iii)} & \theta^1 = dy - \dfrac{\lambda_1}{\sqrt{1-\lambda_1^2}} dx = 0 \\ \text{(iv)} & \theta^2 = d\mu = 0 \end{cases} .$$

To integrate this system we have

$$\begin{cases} dy = \dfrac{1}{\mu} \dfrac{\lambda_1 d\lambda_1}{\sqrt{1-\lambda_1^2}} & \text{(from (i) and (iii))} \\ y = \sqrt{\dfrac{1}{\mu^2} - \left(\dfrac{\lambda_1}{\mu}\right)^2} - \alpha & (\alpha = \text{constant}) \\ \dfrac{\lambda_1}{\mu} = x + \beta & (\beta = \text{constant}) \end{cases}$$

$$\Rightarrow y(x) = \sqrt{\frac{1}{\mu^2} - (x+\beta)^2} - \alpha \ . \tag{A.a.9}$$

To determine the constants we assume that $A = -1$, $B = +1$. It follows that $\beta = 0$ and $\alpha = \frac{1}{\mu}\sqrt{1-\mu^2}$. Then from (A.a.9) we have

$$(y+\alpha)^2 + x^2 = \frac{1}{\mu^2} \quad ,$$

which is a family of circles with centers on the y-axis. We then determine μ so as to have the required area.

Remarks. i) Without assuming N to be a graph we must set up the problem in parametric form on $J^1(\mathbb{R},\mathbb{R}^2)$ with coordinates $(t,x,y,\dot{x},\dot{y})$ and where

$$\theta^1 = dx - \dot{x}dt = 0$$

$$\theta^2 = dy - \dot{y}dt = 0$$

$$\omega = dt \neq 0$$

$$\varphi = \sqrt{\dot{x}^2 + \dot{y}^2}\, dt$$

$$\xi = (x\dot{y} - y\dot{x})\, dt \quad .$$

ii) In general we may ask if a variational problem $(I,\omega;\varphi)$ has arisen by imposing integral constraints on another problem. A necessary condition is that the derived system I_* be non-zero (cf. Chapter I, Section c)). By the Frobenius theorem, locally I_* is generated by $d\mu^1 = \cdots = d\mu^t = 0$ for functions $\mu^1,\ldots,\mu^t$, and these are candidates for the constants μ^ρ in (A.a.4).

It is interesting to see how the Pappus problem fits into the general theory developed in Chapter IV. We will i) show that it is well-posed; ii) show that it is strongly non-degenerate; and iii) investigate conjugate point behavior.

The momentum space $\tilde{Y}$ has coordinates $(x,y,\mu,\lambda_1,\lambda_2)$ with

$$\begin{cases} \theta^1 = dy - \dfrac{\lambda_1\, dx}{\sqrt{1-\lambda_1^2}} \\ \theta^2 = d\mu \\ \omega = dx \quad . \end{cases}$$

Thus $K^* = \mathrm{span}\{\omega,\theta^1,\theta^2\} = \mathrm{span}\{dx,dy,d\mu\}$ is completely integrable with reduced momentum space

$$\tilde{Q} = \mathbb{R}^2 \times \mathbb{R}$$

having coordinates (x,y,μ). The Pfaffian differential system $(\tilde{G},\omega)$ on $\tilde{Q}$ that pulls back to the 1st derived system of $(\pi^*\tilde{I},\omega)$ on $\tilde{Y}$ is

$$d\mu = 0, \quad dx \neq 0 \quad .$$

The integral manifolds $\tilde{N} \subset \tilde{Y}$ of $(\pi^*\tilde{I},\omega)$ project in $\tilde{Q}$ to integral curves $\tilde{\gamma}$ of $(\tilde{G},\omega)$. We may picture such a curve as

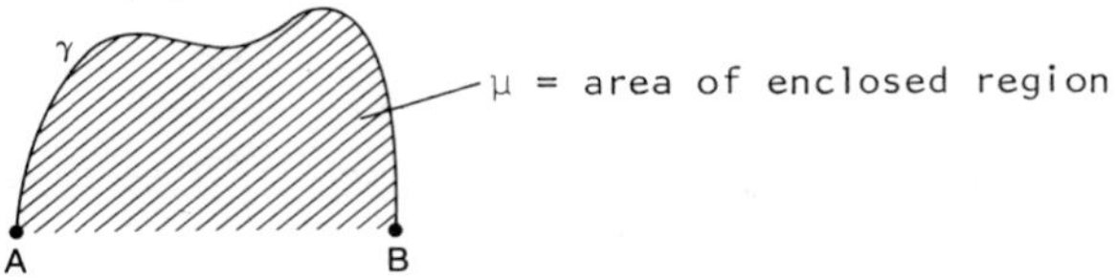

It is given parametrically by

$$x \to (x, y(x), \mu = \text{constant})$$

where $(x,y(x))$ is the curve γ. The endpoint conditions (IV.a.9) for a family $\tilde{N}_t$ of integral manifolds $x \to (x, y(x,t), \mu(t))$ are

$$dy(A,t) = dy(B,t) = d\mu(t) = 0 \ .$$

This means exactly that the endpoints of γ and enclosed area remain fixed. Thus the variational problem is well-posed with the correct endpoint conditions.

To check that it is strongly non-degenerate we consider the original variational problem $(\tilde{I},\omega;\tilde{\varphi})$ in the space of variables $(x,y,\dot{y},\mu)$. Thus $(\tilde{I},\omega)$ is the Pfaffian differential system

$$\begin{cases} \theta^1 = dy - \dot{y}dx = 0 \\ \theta^2 = d\mu = 0 \\ \omega = dx \neq 0 \end{cases}$$

and

$$\tilde{\varphi} = L\omega$$

where

$$L = \sqrt{1+\dot{y}^2} + \mu y \quad .$$

The 1st derived system is $\theta^2 = 0$, $\omega \neq 0$. Thus the structure equations (I.e.28) are satisfied where

$$\pi = \pi^1 = d\dot{y} \ .$$

Setting $K^* = \text{span}\{\omega, \theta^1, \theta^2\} = \text{span}\{dx, dy, d\mu\}$ we have

$$dL \equiv \frac{\dot{y}}{\sqrt{1+\dot{y}^2}} \pi \mod K^*$$

$$d\left(\frac{\dot{y}}{\sqrt{1+\dot{y}^2}}\right) \equiv \frac{\pi}{(1+\dot{y}^2)^{3/2}} \mod K^* \ .$$

In the notation of Chapter I, Section e) we thus obtain

$$\begin{cases} A_\mu = \dfrac{\dot{y}}{\sqrt{1+\dot{y}^2}} & \text{(vector with one component)} \\ \| A_{\mu\nu} \| = \dfrac{1}{(1+\dot{y}^2)^{3/2}} & (1 \times 1 \quad \text{matrix}) \quad , \end{cases}$$

and consequently the variational problem $(\tilde{I},\omega;\tilde{\varphi})$ is strongly non-degenerate.

Finally we turn to the Jacobi equations. These are the variational equations of the Euler-Lagrange system

$$\begin{cases} \theta^1 = dy - \dfrac{\lambda_1 \, dy}{\sqrt{1-\lambda_1^2}} = 0 \\ \theta^2 = d\mu = 0 \\ \eta_1 = d\lambda_1 - \mu dx = 0 \\ \eta_2 = d\lambda_2 = 0 \end{cases}$$

on the momentum space $\tilde{Y}$. Writing

$$V = \alpha \, \partial/\partial\theta^1 + \beta \, \partial/\partial\theta^2 + \gamma \, \partial/\partial\eta_1 + \delta \, \partial/\partial\eta_2 \quad ,$$

the Jacobi equations for $V(x)$ along an integral curve $\tilde{\Gamma}$ of $(\tilde{J},\omega)$ where $A=-1$, $B=1$ and $\lambda_1 = \mu x$ are

$$\begin{cases} d\alpha - \dfrac{\gamma \, dx}{(1-\mu^2 x^2)^{3/2}} \equiv 0 \bmod \tilde{\Gamma} \\ d\beta \equiv 0 \bmod \tilde{\Gamma} \\ d\gamma - \beta dx \equiv 0 \bmod \tilde{\Gamma} \\ d\delta \equiv 0 \bmod \tilde{\Gamma} \quad . \end{cases}$$

To determine if there are any conjugate points we must find solutions to these equations defined for $-1 \leqq x \leqq +1$ and with

$$\alpha(-1) = \alpha(1) = \beta(-1) = \beta(1) = 0 \ .$$

This gives

$$\begin{cases} \beta = 0, \quad \gamma = c = \text{constant} \\ \alpha'(x) = \dfrac{c}{(1-\mu^2x^2)^{3/2}} \end{cases} .$$

Integration of the 2nd equation gives

$$\alpha(x) = \frac{cx}{\sqrt{1-\mu^2x^2}} + c' .$$

The endpoint conditions

$$\alpha(-1) = \alpha(1) = 0$$

give

$$c = c' = 0 .$$

Thus $V \equiv 0$ and there are no conjugate points.

On the basis of (IV.d.1) we may conclude that figure

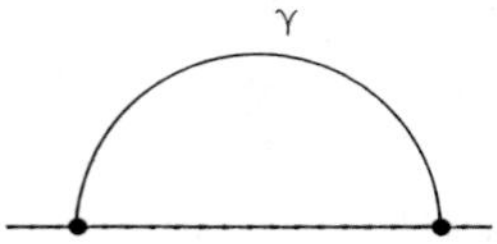

gives a local minimum to Pappus' problem. Actually, using the Weierstrass E-function plus the absence of conjugate points it may be shown that the above figure gives a *global* minimum to Pappus' problem (cf. [5], [13]).

(A.a.10) Example. Referring to example (I.d.27) we consider the variational problem corresponding to the functional

$$\Phi(\gamma) = \frac{1}{2} \int_\gamma \kappa^2(s)\, ds , \qquad \gamma \subset \mathbb{E}^2 , \qquad \text{(A.a.11)}$$

with the integral constraint

$$\Xi(\gamma) = \int_\gamma ds = \ell \qquad \text{(A.a.12)}$$

that the curve γ have constant length (cf. the 2nd figure in that example). Following the procedure in (A.a.6) and using the notation of example (I.d.27) we have

$$\begin{cases} \theta^1 = \omega^2 = 0 \\ \theta^2 = \omega_1^2 - \kappa\omega = 0 \\ \theta^3 = d\mu = 0 \\ \omega = \omega^1 \neq 0 \\ \tilde{\varphi} = \left(\frac{\kappa^2}{2} + \mu\right)\omega \end{cases} .$$

Computing the Euler-Lagrange equations as in that example leads to the O.D.E.

$$\frac{d^2\kappa}{ds^2} + \left(\frac{\kappa^3}{2} - \kappa\mu\right) = 0 \quad . \tag{A.a.13}$$

Comparing this equation with the discussion there we draw the following interesting conclusion:

The Euler-Lagrange equations for the functional (A.a.11) defined on curves $\gamma \subset \mathbb{E}^2$ *and with the constraint (A.a.12) are the same as the Euler-Lagrange equations for the functional (A.a.11) defined on curves in a surface of constant curvature* μ.

To solve these equations we represent κ by an elliptic function $\kappa(s,\mu)$ with variable modulus depending on μ. Then we determine μ so that the curve determined by $\kappa(s,\mu)$ and drawn in the 2nd figure of example (I.d.27) has the given length ℓ between α,α' and β,β'. The picture may be something like

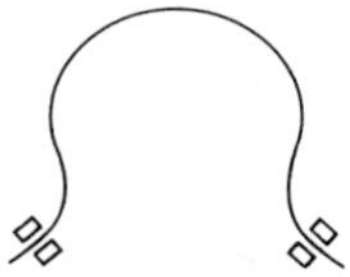

As in example (I.d.27) it is instructive to draw the phase portrait associated to (A.a.13). Following the notations used in (I.d.27) we consider in the phase plane with coordinates (x,y) the vector field

$$v = y\,\partial/\partial x + (\mu x - x^3/2)\,\partial/\partial y$$

with critical points at $(0,0)$, $(0, \pm\sqrt{2\mu})$. The integral curves of v are given by the real points on the algebraic curves

$$y^2 + \frac{x^4}{4} - \mu x^2 - c = 0 \; . \tag{A.a.14}$$

To plot these curves we consider the functions

$$f(x,c) = -\frac{x^4}{4} + \mu x^2 + c$$

and plot their graphs together with the curves (A.a.14):

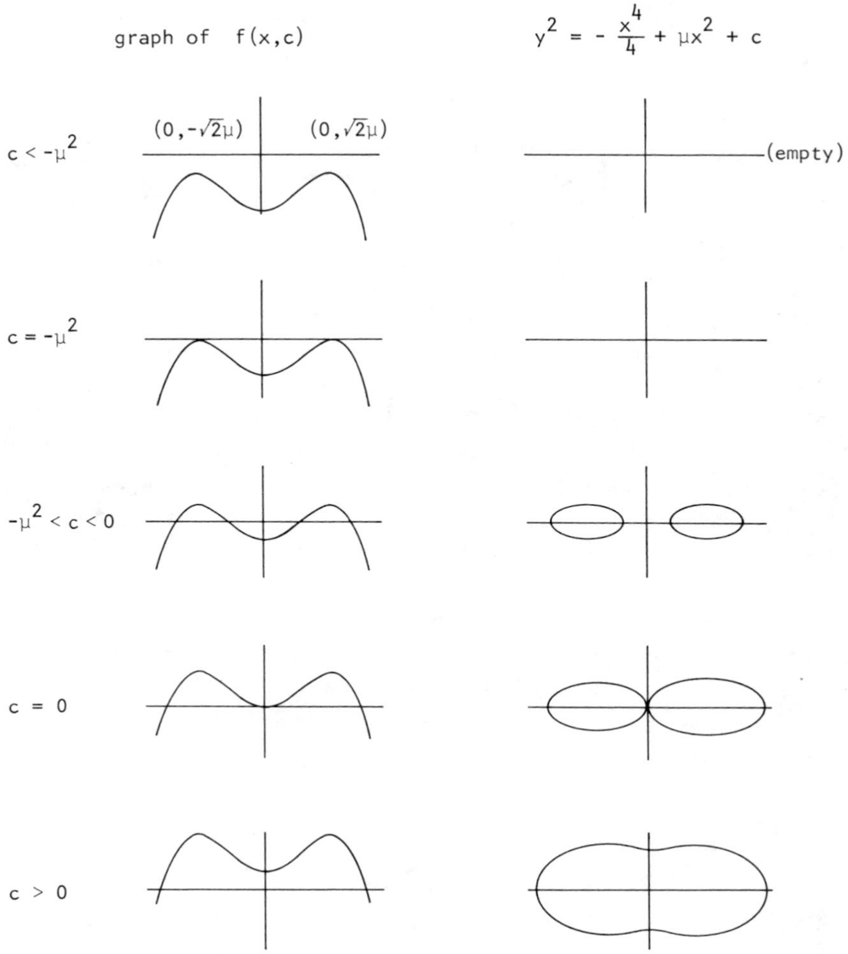

Thus the family of curves (A.a.14) looks like

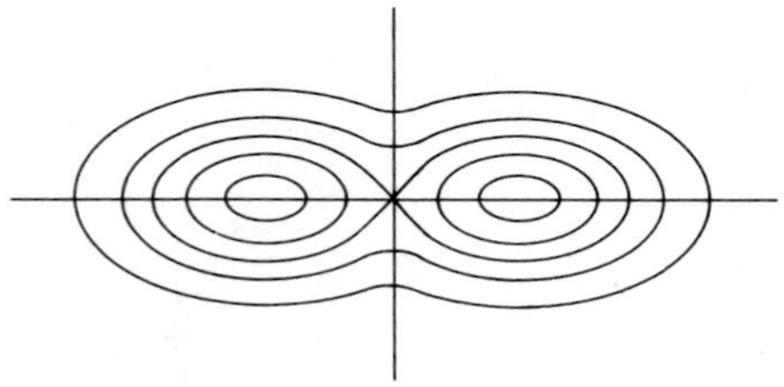

In particular for $c=-\mu^2$ there are two equilibrium solutions to (A.a.13) given by

$$\kappa = \pm\sqrt{2\mu} = \text{constant} \quad .$$

These are arcs on a circle, so we may conclude that: *with appropriate endpoint conditions the extremals of the constrained functional (A.a.11), (A.a.12) are circles.* Referring to the discussion in example (I.d.27), this was exactly the case where (A.a.13) could be integrated by elementary (or circular) functions.

b) Some Classical Problems in the Setting of Moving Frames.

Among the classical problems in the calculus of variations are Newtonian motion (using the least action principle), the brachistochrone, and the minimal surface of revolution. We have already commented on the second and third of these (cf. examples (IV.e.17) and (IV.b.41)), but it may be amusing to see how they each appear in our formalism using moving frames. What turns out is this: *the invariantly expressed consequences of the Euler-Lagrange equations (i.e., those involving the curvature and torsion) appear quite readily and naturally. But to actually solve the equations one must resort to the traditional methods of calculus* (separation of variables, etc.). In fact this conclusion is more or less clear, but it may be interesting to actually carry out the details.

i) Newtonian Motion. In $\mathbb{R}^3$ with coordinates $x=(x^1,x^2,x^3)$ we set $\rho=\|x\|$ and assume given a potential function $U(\rho)$. We want to use moving frames to discuss the mechanical system with $T=\frac{1}{2}\|\dot{x}\|^2$ as kinetic energy and U as potential energy. For this we follow the notations and prescription of Chapter III, Section b). On $X=F(\mathbb{R}^3)\times\mathbb{R}^6$ we consider the exterior differential system generated by the Pfaffian equations (corresponding to Frenet frames of curves in $\mathbb{R}^3$)

$$\left\{\begin{array}{l} \theta^1 = \omega^1 - p\,dt = 0 \\ \theta^2 = \omega^2 = 0 \\ \theta^3 = \omega^3 = 0 \\ \theta^4 = \omega_1^2 - q\,dt = 0 \\ \theta^5 = \omega_1^3 = 0 \\ \theta^6 = \omega_2^3 - r\,dt = 0 \\ \omega = dt \neq 0 \end{array}\right. \qquad \text{(A.b.1)}$$

The structure equations are readily computed to be

$$\begin{cases} d\theta^1 \equiv -q\theta^2 \wedge dt - dp \wedge dt \\ d\theta^2 \equiv \ \ q\theta^1 \wedge dt - p\theta^4 \wedge dt - r\theta^3 \wedge dt \\ d\theta^3 \equiv -p\theta^5 \wedge dt + r\theta^2 \wedge dt \\ d\theta^4 \equiv -r\theta^5 \wedge dt - dq \wedge dt \\ d\theta^5 \equiv \ \ r\theta^4 \wedge dt - q\theta^6 \wedge dt \\ d\theta^6 \equiv \ \ q\theta^5 \wedge dt - dr \wedge dt \end{cases} \quad . \tag{A.b.2}$$

On X we set

$$\varphi = \left(\frac{p^2}{2} - U(\rho)\right) dt \tag{A.b.3}$$

and consider the variational problem $(I,\omega;\varphi)$. The solutions to the Euler-Lagrange equations of this problem have position vectors $x(t) \in \mathbb{R}^3$ whose motion minimizes the action functional

$$\Phi = \int \left(\frac{\|\dot{x}(t)\|^2}{2} - U(\|x(t)\|)\right) dt \quad . \tag{A.b.4}$$

As discussed in almost any text (e.g., [1], [2]) and reviewed several times above (cf. (I.b.18)), the extremals of (A.b.4) are the solution curves to Newton's equations

$$\ddot{x}(t) - F(x(t)) = 0 \ , \qquad F = -\nabla U \quad . \tag{A.b.5}$$

Following our usual algorithm (Chapter I, Section e)) we set $Z = X \times \mathbb{R}^6$ and consider on Z the differential forms

$$\begin{cases} \text{(i)} \qquad \psi = \varphi + \lambda_\alpha \theta^\alpha = \left(\frac{p^2}{2} - U(\rho)\right) dt + \lambda_\alpha \theta^\alpha \\ \text{(ii)} \qquad \Psi = d\psi = p\,dp \wedge dt - \frac{U'(\rho)}{\rho}\,(dx,x) \wedge dt + d\lambda_\alpha \wedge \theta^\alpha + \lambda_\alpha d\theta^\alpha \end{cases} \quad . \tag{A.b.6}$$

The Cartan system $C(\Psi)$ is generated by the Pfaffian equations (using "$\equiv$" to denote $\bmod\{\theta^\alpha\}$)

$$
\begin{cases}
\partial/\partial\lambda_\alpha \lrcorner \Psi = \theta^\alpha = 0 & \\
\text{(i)} \quad \partial/\partial r \lrcorner \Psi \equiv -\lambda_6 dt = 0 & \\
\text{(ii)} \quad \partial/\partial q \lrcorner \Psi \equiv -\lambda_4 dt = 0 & \\
\text{(iii)} \quad \partial/\partial p \lrcorner \Psi \equiv (p-\lambda_1)dt = 0 & \\
\text{(iv)} \quad \partial/\partial\theta^6 \lrcorner \Psi \equiv -d\lambda_6 - q\lambda_5 dt = 0 & \\
\text{(v)} \quad \partial/\partial\theta^5 \lrcorner \Psi \equiv -d\lambda_5 + (q\lambda_6 - r\lambda_4 - p\lambda_3)dt = 0 & \\
\text{(vi)} \quad \partial/\partial\theta^4 \lrcorner \Psi \equiv -d\lambda_4 + (r\lambda_5 - p\lambda_2)dt = 0 & \\
\text{(vii)} \quad \partial/\partial\theta^3 \lrcorner \Psi \equiv -d\lambda_3 + \left(-r\lambda_2 - \frac{U'(\rho)}{\rho}(x,e_3)\right)dt = 0 & \\
\text{(viii)} \quad \partial/\partial\theta^2 \lrcorner \Psi \equiv -d\lambda_2 + \left(r\lambda_3 - q\lambda_1 - \frac{U'(\rho)}{\rho}(x,e_2)\right)dt = 0 & \\
\text{(ix)} \quad \partial/\partial\theta^1 \lrcorner \Psi \equiv -d\lambda_1 + \left(q\lambda_2 - \frac{U'(\rho)}{\rho}(x,e_1)\right)dt = 0 & .
\end{cases}
\qquad \text{(A.b.7)}
$$

These equations are an immediate consequence of (A.b.2), where in (vii)-(ix) we have used

$$(dx,x) \equiv (x,e_1)\theta^1 + (x,e_2)\theta^2 + (x,e_3)\theta^3 \mod dt \quad .$$

Although (A.b.7) appears somewhat lengthy we wish to emphasize the *algorithmic nature* of the procedure, which conceivably could be useful in other contexts.

Since the extremals are not lines (unless $U = \text{constant}$) we have that the curvature

$$\kappa = q/p^2 \neq 0 \quad . \qquad \text{(A.b.8)}$$

Then (i), (iv), (ii), (v), (vi) give respectively

$$\lambda_6 = \lambda_5 = \lambda_4 = \lambda_3 = \lambda_2 = 0 \quad .$$

More interestingly (vii) then gives

$$\frac{U'(\rho)}{\rho}(x,e_3) = 0 \ .$$

Ruling out the case $U = \text{constant}$ this implies that

$$(x,e_3) = 0 \quad .$$

It is well-known that any curve in $\mathbb{R}^3$ satisfying this equation must lie in a plane (recall that e_3 is the binormal). (Proof. Letting $'$ denote the derivative with respect to arclength and τ the torsion, the Frenet equations give for the derivative of $(x,e_3)=0$

$$\begin{aligned} 0 &= (x',e_3) - \tau(x,e_2) \\ &= -\tau(x,e_2) \qquad \text{(since } x'=e_1) . \end{aligned}$$

If $\tau \neq 0$ then $(x,e_2)=0$ and we have

$$\begin{aligned} 0 &= (x',e_2) + (x,-\kappa e_1 + \tau e_3) \\ &= -\kappa(x,e_1) . \end{aligned}$$

Thus $(x,e_1)=0$ and this gives

$$\begin{aligned} 0 &= (x',e_1) + (x,\kappa e_2) \\ &= (e_1,e_1), \end{aligned}$$

which is a contradiction.) *Thus the solution curves to (A.b.5) are planar.* From (iii) and (viii) we obtain for the curvature

$$\kappa = \frac{1}{p^2}(F,e_2), \qquad p = \|\dot{x}\| = \text{velocity}, \tag{A.b.9}$$

which is a well-known (but not so commonly stated) consequence of central force motion (cf. [66]). Finally (xi) gives

$$\dot{p} = (F,e_1) \quad . \tag{A.b.10}$$

Equations (A.b.9), (A.b.10) thus interpret the normal and tangential components of the force field along solution curves to (A.b.5).

As is also well-known, Noether's theorem gives two more 1st integrals (in addition to $\tau=0$). The vector field $v=\partial/\partial t$ gives an obvious infinitesimal symmetry of $(I,\omega;\varphi)$ with 1st integral

$$\begin{aligned} H &= v \lrcorner \psi \\ &= -\left(\frac{p^2}{2} + U(\rho)\right) \end{aligned}$$

using $\lambda_6 = \cdots = \lambda_2 = 0$, $\lambda_1 = p$. Thus the total energy

$$E = -H$$

is constant on solution curves to (A.b.6). Secondly, rotation about the origin in $\mathbb{R}^2$ gives conservation of angular momentum. At this stage we have enough 1st integrals to conclude that *the Euler-Lagrange equations associated to (A.b.4) are algebraically integrable by quadratures.* The question of finding explicit "formulas" is then whether the genus g of the algebraic curves giving the phase portrait is zero or not. (Actually, there is not agreement on terminology here; it may be argued that the elliptic or even hyperelliptic cases also give "formulas").

In any case, to complete the discussion we use polar coordinates

$$x = (\rho\cos\varphi,\ \rho\sin\varphi)$$

and assume that

$$U(\rho) = \frac{k}{\rho^m}, \qquad 0 \neq m \in \mathbb{Z} . \tag{A.b.11}$$

Then conservation of angular momentum and total energy are

$$\begin{cases} \rho^2\dot{\varphi} = c_1 \\ \frac{1}{2}(\dot{\rho}^2+\rho^2\dot{\varphi}^2) + \frac{k}{\rho^m} = \frac{c_2}{2} . \end{cases}$$

These combine to give

$$\dot{\rho}^2 + \left(\frac{c_1^2}{\rho^2} + \frac{2k}{\rho^m} - c_2\right) = 0 .$$

For $m = 1,2$ and c_1, c_2 general constants the algebraic curve

$$y^2 + \frac{c_1^2}{x^2} + \frac{2k}{x^m} - c_2 = 0 \tag{A.b.12}$$

has genus zero and the Euler-Lagrange equations may be integrated by elementary functions (cf. the discussion at the end of example (IV.d.22)). When $m = 1$ we have *Keplerian motion* and the solution curves to (A.b.5) are conics. Of course this traditionally comes out by a computation. From the viewpoint of the present text a droll way to establish this is to show that: *When* $m = 1$ *in (A.b.11) the solution curves to the Euler-Lagrange equations associated to (A.b.4) are affine geodesics* (cf. (II.b.8)). Of course this also comes out of a computation, which as far as we can see is not particularly enlightening.

When $m = 3, -1, -2$ the curve (A.b.12) is elliptic (the case $m = -1$ is special, cf. [66]), while in the remaining cases it is hyperelliptic.

ii) <u>The Brachistochrone</u>. We recall that this is the following problem: Given two points A,B in the plane, determine the curve γ joining A to B such that a point mass sliding down γ under the influence of gravity minimizes the time of travel?

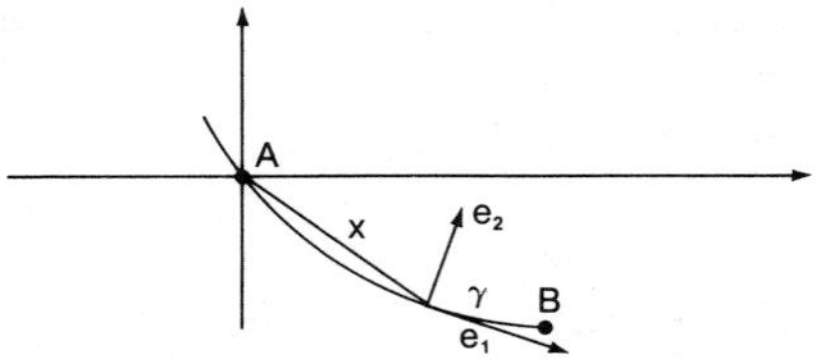

In the figure, F is the vector representing the force of gravity; thus the potential energy is $-(x,F)$.

To set up, as in ii) of Chapter II, Section b) we describe curves $\gamma \subset \mathbb{E}^2$ by their Frenet frames viewed as integral manifolds N of the differential system

$$\begin{cases} \theta^1 = \omega^2 = 0 \\ \theta^2 = \omega_1^2 - \kappa\omega \\ \omega = \omega^1 \neq 0 \end{cases} \tag{A.b.13}$$

on $F(\mathbb{E}^2) \times \mathbb{R}$. If s is the arclength parameter on γ and $s(t)$ the distance the particle has traveled during time t, then the velocity is

$$p(t) = ds(t)/dt \ .$$

Conservation of total energy gives

$$\frac{p(t)^2}{2} - (x(t),F) = 0 \quad .$$

Thus we want to minimize the functional

$$\Phi = \int_N \varphi$$

where (omitting the $1/\sqrt{2}$ factor)

$$\varphi = dt = ds/p$$

$$= \frac{\omega}{\sqrt{(x,F)}} \quad .$$

For the Euler-Lagrange equations we consider the differential forms

$$\begin{cases} \psi = \varphi + \lambda_1\theta^1 + \lambda_2\theta^2 \\ \Psi = d\psi \end{cases} \quad .$$

Then setting $L = 1/\sqrt{(x,F)}$ we have

$$\begin{cases} \varphi = L\omega \\ d\varphi = -\frac{L^3}{2}((e_1,F)\omega + (e_2,F)\theta^1)\wedge\omega - \kappa L\theta^1\wedge\omega \\ \Psi = d\varphi + d\lambda_1\wedge\theta^1 + d\lambda_2\wedge\theta^2 + \lambda_1 d\theta^1 + \lambda_2 d\theta^2 \end{cases}$$

where $d\theta^1$, $d\theta^2$ are given by the formulas

$$\begin{cases} d\theta^1 \equiv -\theta^2\wedge\omega \\ d\theta^2 \equiv -\pi\wedge\omega \qquad\qquad \pi = -d\kappa - \kappa^2\theta^1 \end{cases}$$

In Part ii) of Chapter II, Section b). Following our usual procedure, the Cartan system is generated by the Pfaffian equations

$$\begin{cases} \partial/\partial\lambda_\alpha \lrcorner \Psi = \theta^\alpha = 0 \\ \partial/\partial\pi \lrcorner \Psi \equiv -\lambda_2\omega = 0 \\ \partial/\partial\theta^2 \lrcorner \Psi \equiv -d\lambda_2 - \lambda_1\omega = 0 \\ \partial/\partial\theta^1 \lrcorner \Psi \equiv -d\lambda_1 - \left(\frac{L^3}{2}(e_2,F) + \kappa L\right)\omega = 0 \end{cases}$$

It follows that $\lambda_2 = \lambda_1 = 0$, and then

$$\kappa = -\frac{1}{2}\frac{(e_2,F)}{(x,F)} \quad . \tag{A.b.14}$$

If we describe γ by $(x_1, x_2(x_1))$ then setting $x_2' = dx_2/dx_1$, $x_2'' = d^2x_2/dx_1^2$, and $Y = \sqrt{1+(x_2')^2}$

$$\begin{cases} e_1 = \frac{(1,x_2')}{Y} \\ e_2 = \frac{(-x_2',1)}{Y} \\ \kappa = \frac{x_2''}{Y^3} \end{cases},$$

and (A.b.14) is

$$\frac{x_2''}{1+(x_2')^2} = \frac{1}{2}\frac{x_2'}{x_2} \quad .$$

This is an O.D.E. in which the variables have been separated. If we integrate both sides, separate variables again, and then integrate both sides once more we end up with the equation of a cycloid (cf. [13], [29]). Equation (A.b.14) therefore gives a curious differential-geometric characterization of cycloids.

iii) The Minimal Surface of Revolution. We use the differential system (A.b.13) on $F(\mathbb{E}^2) \times \mathbb{R}$ in the preceding example. Letting V be the vector $(0,2\pi)$ we set

$$\begin{cases} L = (x,V) & (= \text{vertical coordinate of } x) \\ \varphi = L\omega & , \end{cases}$$

so that the minimal surface of revolution corresponds to the variational problem $(I,\omega;\varphi)$

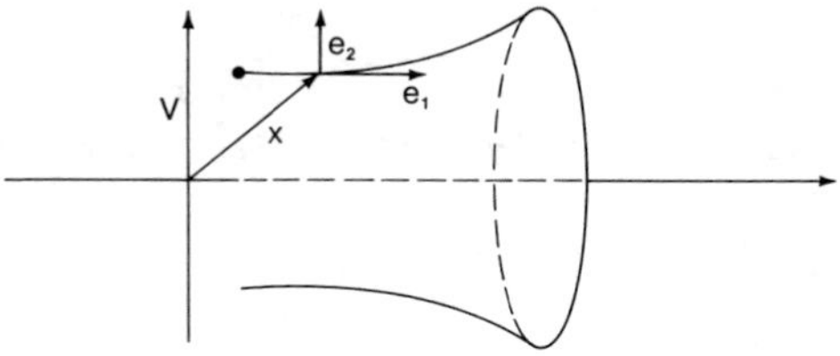

Computing as before we find that

$$\Psi \equiv (e_2,V)\theta^1 \wedge \omega - \kappa(x,V)\theta^1 \wedge \omega + d\lambda_1 \wedge \theta^1 + d\lambda_2 \wedge \theta^2 - \lambda_1\theta^2 \wedge \omega - \lambda_2\pi \wedge \omega.$$

The Euler-Lagrange system is generated by

$$\begin{cases} \partial/\partial\lambda_\alpha \lrcorner \Psi = \theta^\alpha = 0 \\ \partial/\partial\pi \lrcorner \Psi \equiv -\lambda_2\omega = 0 \\ \partial/\partial\theta^2 \lrcorner \Psi \equiv -d\lambda_2 - \lambda_1\omega = 0 \\ \partial/\partial\theta^1 \lrcorner \Psi \equiv -d\lambda_1 + ((e_2,V) - \kappa(x,V))\omega = 0 . \end{cases}$$

It follows that $\lambda_2 = \lambda_1 = 0$, and then

$$\kappa = \frac{(e_2,V)}{(x,V)} \quad . \tag{A.b.15}$$

As has already been noted, the solution curves to this equation are the catenaries (cf. (IV.b.41)). Once again these curves are characterized by a curious curvature property.

INDEX

BIBLIOGRAPHY

1. R. Abraham and J. Marsden, *Foundations of Mechanics*, Benjamin (1978).

2. V.I. Arnold, *Mathematical Methods of Classical Mechanics*, Springer Verlag (1978).

3. W. Blaschke, *Vorlesungen über Differentialgeometrie*, Vol. I, Springer (1929).

4. G. Bliss, "The Problem of Lagrange in the Calculus of Variations," *Amer. J. Math.*, Vol. 52 (1930), 673-744.

5. O. Bolza, *Vorlesungen über Variationsrechnung*, Kohler and Amelang, Leipzig (1909).

6. W. Boothby, *An Introduction to Differentiable Manifolds and Riemannian Geometry*, Academic Press (1975).

7. R.W. Brockett, "Lie Theory and Control Systems Defined on Spheres," *SIAM Jour. of Applied Math.*, Vol. 25 (1973), 213-225.

8. R.W. Brockett, "Control Theory and Singular Riemannian Geometry," *New Directions in Applied Math.* (P. Hilton and G. Young, eds.), Springer Verlag (1981), 11-27.

9. R.W. Brockett, "Lie Theory and Lie Groups in Control Theory," *Geometric Methods in Systems Theory* (D. Mayne and R. Brockett, eds.), Reidel (1973), 43-82.

10. R. Brockett, *Finite Dimensional Linear Systems*, Wiley (1970).

11. R. Bryant, S.S. Chern, and P. Griffiths, "Exterior Differential Systems," Proc. Beijing Symposium on Differential Geometry and Differential Equations (1980).

12. R. Bryant, S.S. Chern, R. Gardner, and P. Griffiths, "Essays on Exterior Differential Systems," in preparation.

13. C. Caratheodory, *Calculus of Variations and Partial Differential Equations of the First Order*, (2 Volumes), Holden-Day (1965/67).

14. C. Caratheodory, "Die Methode der Geodätischen Äquidistanten und das Problem von Lagrange," *Acta Math.*, Vol. 47 (1926), 199-236.

15. E. Cartan, *Leçons sur la géométrie des espaces de Riemann*, Gauthier-Villars, Paris (1946).

16. E. Cartan, *Leçons sur les invariants intégraux*, Herman, Paris (1924).

17. E. Cartan, *Les systèmes différentielles extérieurs et leurs applications géométriques*, Herman, Paris (1945).

18. E. Cartan, *La theorie des groupes finis et continus et la géométrie différentielle*, Gauthiers-Villars, Paris (1937).

19. E. Cartan, *Oeuvres complètes*, Gauthier-Villars, Paris (1952).

20. L. Cesari, "Existence Theorems for Optimal Solutions in Lagrange and Pontryagin Problems," *SIAM J. Control* (1965), 475-498.

21. J. Cheeger, and D. Ebin, *Comparison Theorems in Riemannian Geometry*, North-Holland (1975).

22. S.S. Chern, *Studies in Global Geometry and Analysis*, No. 4, Prentice-Hall (1967).

23. Y. Choquet-Bruhat, *Géométrie différentielle et systèmes extérieurs*, Dunod, Paris (1968).

24. W.L. Chow, "Über Systeme von linearen partieller Differentialgleichungen erster Ordnung," *Math. Ann.*, Vol. 117 (1940/41), 98-105.

25. R. Courant, "Calculus of Variations and Supplementary Notes and Exercises," Notes from NIU.

26. P. Dedecker, "Calcul des variations, formes différentielles et champs géodésiques," *Géométrie différentielle*, Colloq. Int. de la Recherche Sci., Strassbourg (1953), 17-34.

27. Th. DeDonder, *Théorie invariantive du calcul des variations*, Gauthier-Villars, Paris (1935).

28. M. DoCarno, *Differential Geometry of Curves and Surfaces*, Prentice-Hall (1976).

29. I. Gelfand and G. Fomin, *Calculus of Variations*, Prentice-Hall (1963).

30. H. Gluck, "The Converse to the Four-Vertex Theorem," *Enseignement Math.*, Vol. 17 (1971), 295-309.

31. H. Goldschmidt and S. Sternberg, "The Hamilton-Cartan Formalism in the Calculus of Variations," *Ann. L'Inst. Fourier*, Vol. 23 (1973), 203-267.

32. E. Goursat, *Leçons sur le problème de Pfaff*, Herman, Paris (1922).

33. E. Goursat, *Leçons sur l'intégration des équations partielles du prémière ordre*, Herman, Paris (1921).

34. M. Green, "The Moving Frame, Differential Invariants and Rigidity Theorems for Curves in Homogeneous Spaces," *Duke J. Math.*, Vol. 45 (1978), 735-780.

35. P. Griffiths, "On Cartan's Method of Lie Groups and Moving Frames as Applied to Existence and Uniqueness Questions in Differential Geometry," *Duke J. Math.*, Vol. 41 (1974), 775-814.

36. P. Griffiths and J. Harris, *Principles of Algebraic Geometry*, Wiley (1978).

37. J. Hadamard, *Leçons sur le calcul des variations*, Herman, Paris (1940).

38. R. Hermann, *Differential Geometry and the Calculus of Variations*, 2nd edition, Math. Sc. Press (1977).

39. R. Hermann, "Geodesics of Singular Riemannian Metrics," *Bull. A.M.S.*, Vol. 79 (1973), 780-782.

40. R. Hermann, "Some Differential Geometric Aspects of the Lagrange Variational Problem," *Ill. J. Math.*, Vol. 6 (1962), 634-673.

41. R. Hermann and C. Martin, "Lie Theoretic Aspects of the Ricatti Equation," Proc. of the 1977 CDC Conference, New Orleans, LA.

42. R. Hermann, "E. Cartan's Geometric Theory of Partial Differential Equations," *Advances in Math.*, Vol. 1 (1965), 265-317.

43. R. Hermann, *Cartanian Geometry, Nonlinear Waves, and Control Theory*, Part A, Math. Sci. Press (1979).

44. G. Jensen, *Higher Order Contact of Submanifolds of Homogeneous Spaces*, Springer Verlag (1977).

45. F. John, *Partial Differential Equations*, Springer Verlag (1971).

46. V. Jurdjevic and I. Kupka, "Control Systems on Semi-Simple Lie Groups and their Homogeneous Spaces," *Ann. L'Institut Fourier*, Vol. 31 (1981), 151-180.

47. E. Kähler, "Einführen in die Theorie der Systeme von Differentialgleichungen," *Hamburger Math. Einz.*, Vol. 16 (1934).

48. I. Kaplansky, *An Introduction to Differential Algebra*, Hermann, Paris (1957).

49. W. Klingenberg, *Lectures on Closed Geodesics*, Springer Verlag (1978).

50. B. Kostant, Quantization and Unitary Representations, Springer Verlag, No. 170 (1970), 87-208.

51. P. Li and S.T. Yau, "A New Conformal Invariant and its Applications to the Wilmore Conjecture and the First Eigenvalue of Compact Surfaces," to appear.

52. P. Liebermann, "Pfaffian Systems and Transverse Differential Geometry," *Differential Geometry and Relativity* (Cahen and Flato, eds.), Reidel (1976), 107-126.

53. J. Milnor, *Morse Theory*, Annals of Math. Studies, No. 51, Princeton (1963).

54. M. Morse, "The Calculus of Variations in the Large," A.M.S. Colloquium Publ. No. 18 (1934).

55. E. Noether, "Invariante Variationsprobleme," *Nachr. Ges. Wiss. Göttingen* (1918), 235-257.

56. B. O'Neill, *Elementary Differential Geometry*, Academic Press (1966).

57. J. Radon, "Zum Problem Lagrange," *Hamburg Math. Einzelschriften*, Teubner, Leipzig (1928).

58. H. Rund, *The Hamilton-Jacobi Theory in the Calculus of Variations*, D. van Nostrand (1966).

59. I. Shafarevich, *Basic Algebraic Geometry*, Springer Verlag (1974).

60. I. Singer and S. Sternberg, "The Infinite Groups of Lie and Cartan," *Jour. d'Analyse Math.*, Vol. 15 (1965).

61. J.M. Souriau, *Structure des systèmes dynamiques*, Dunod, Paris (1970).

62. D.C. Spencer, "Overdetermined Systems of Linear Partial Differential Equations," *Bull. A.M.S.*, Vol. 75 (1969), 179-239.

63. M. Spivak, *Differential Geometry*, (5 volumes), Publish or Perish (1970).

64. S. Sternberg, *Lectures on Differential Geometry*, Prentice-Hall (1964).

65. H. Weyl, "Geodesic Fields in the Calculus of Variations for Multiple Integrals," *Ann. of Math.*, Vol. 36 (1935), 607-629.

66. E.T. Whittaker, *A Treatise on the Analytical Dynamics of Particles and Rigid Bodies*, (4th edition), Cambridge Univ. Press (1937).

67. T.J. Wilmore, "Note on Embedded Surfaces," *Anal. Stüntifice ale Univ.*, Isai. Sect. I a Mat. II (1965), 493-496.

Progress in Mathematics

Edited by J. Coates and S. Helgason

Progress in Physics

Edited by A. Jaffe and D. Ruelle

- A collection of research-oriented monographs, reports, notes arising from lectures or seminars
- Quickly published concurrent with research
- Easily accessible through international distribution facilities
- Reasonably priced
- Reporting research developments combining original results with an expository treatment of the particular subject area
- A contribution to the international scientific community: for colleagues and for graduate students who are seeking current information and directions in their graduate and post-graduate work.

Manuscripts

Manuscripts should be no less than 100 and preferably no more than 500 pages in length.

They are reproduced by a photographic process and therefore must be typed with extreme care. Symbols not on the typewriter should be inserted by hand in indelible black ink. Corrections to the typescript should be made by pasting in the new text or painting out errors with white correction fluid.

The typescript is reduced slightly (75%) in size during reproduction; best results will not be obtained unless the text on any one page is kept within the overall limit of 6x9½ in (16x24 cm). On request, the publisher will supply special paper with the typing area outlined.

Manuscripts should be sent to the editors or directly to: Birkhäuser Boston, Inc., P.O. Box 2007, Cambridge, Massachusetts 02139

PROGRESS IN MATHEMATICS

Already published

PM 1 Quadratic Forms in Infinite-Dimensional Vector Spaces
Herbert Gross
ISBN 3-7643-1111-8, 432 pages, paperback

PM 2 Singularités des systèmes différentiels de Gauss-Manin
Frédéric Pham
ISBN 3-7643-3002-3, 346 pages, paperback

PM 3 Vector Bundles on Complex Projective Spaces
C. Okonek, M. Schneider, H. Spindler
ISBN 3-7643-3000-7, 396 pages, paperback

PM 4 Complex Approximation, Proceedings, Quebec, Canada, July 3-8, 1978
Edited by Bernard Aupetit
ISBN 3-7643-3004-X, 128 pages, paperback

PM 5 The Radon Transform
Sigurdur Helgason
ISBN 3-7643-3006-6, 202 pages, paperback

PM 6 The Weil Representation, Maslov Index and Theta Series
Gérard Lion, Michèle Vergne
ISBN 3-7643-3007-4, 348 pages, paperback

PM 7 Vector Bundles and Differential Equations
Proceedings, Nice, France, June 12-17, 1979
Edited by André Hirschowitz
ISBN 3-7643-3022-8, 256 pages, paperback

PM 8 Dynamical Systems, C.I.M.E. Lectures, Bressanone, Italy, June 1978
John Guckenheimer, Jürgen Moser, Sheldon E. Newhouse
ISBN 3-7643-3024-4, 298 pages, paperback

PM 9 Linear Algebraic Groups
T. A. Springer
ISBN 3-7643-3029-5, 314 pages, hardcover

PM10 Ergodic Theory and Dynamical Systems I
A. Katok
ISBN 3-7643-3036-8, 346 pages, hardcover

PM11 18th Scandinavian Congress of Mathematicians, Aarhus, Denmark, 1980
Edited by Erik Balslev
ISBN 3-7643-3040-6, 526 pages, hardcover

PM12 Séminaire de Théorie des Nombres, Paris 1979-80
Edited by Marie-José Bertin
ISBN 3-7643-3035-X, 404 pages, hardcover

PM13 Topics in Harmonic Analysis on Homogeneous Spaces
Sigurdur Helgason
ISBN 3-7643-3051-1, 152 pages, hardcover

PM14 Manifolds and Lie Groups, Papers in Honor of Yozô Matsushima
Edited by J. Hano, A. Marimoto, S. Murakami, K. Okamoto, and H. Ozeki
ISBN 3-7643-3053-8, 476 pages, hardcover

PM15 Representations of Real Reductive Lie Groups
David A. Vogan, Jr.
ISBN 3-7643-3037-6, 776 pages, hardcover

PM16 Rational Homotopy Theory and Differential Forms
Phillip A. Griffiths, John W. Morgan
ISBN 3-7643-3041-4, 258 pages, hardcover

PM17 Triangular Products of Group Representations and their Applications
S.M. Vovsi
ISBN 3-7643-3062-7, 142 pages, hardcover

PM18 Géométrie Analytique Rigide et Applications
Jean Fresnel, Marius van der Put
ISBN 3-7643-3069-4, 232 pages, hardcover

PM19 Periods of Hilbert Modular Surfaces
Takayuki Oda
ISBN 3-7643-3084-8, 144 pages, hardcover

PM20 Arithmetic on Modular Curves
Glenn Stevens
ISBN 3-7643-3088-0, 236 pages, hardcover

PM21 Ergodic Theory and Dynamical Systems II
A. Katok, editor
ISBN 3-7643-3096-1, 226 pages, hardcover

PM22 Séminaire de Théorie des Nombres, Paris 1980-81
Marie-José Bertin, editor
ISBN 3-7643-3066-X, 374 pages, hardcover

PM23 Adeles and Algebraic Groups
A. Weil
ISBN 3-7643-3092-9, 138 pages, hardcover

PM24 Ennumerative Geometry and Classical Algebraic Geometry
Patrick Le Barz, Yves Hervier, editors
ISBN 3-7643-3106-2, hardcover

PROGRESS IN PHYSICS

Already published

PPh1 Iterated Maps on the Interval as Dynamical Systems
Pierre Collet and Jean-Pierre Eckmann
ISBN 3-7643-3026-0, 256 pages, hardcover

PPh2 Vortices and Monopoles, Structure of Static Gauge Theories
Arthur Jaffe and Clifford Taubes
ISBN 3-7643-3025-2, 294 pages, hardcover

PPh3 Mathematics and Physics
Yu. I. Manin
ISBN 3-7643-3027-9, 112 pages, hardcover

PPh4 Lectures on Lepton Nucleon Scattering and Quantum Chromodynamics
W.B. Atwood, J.D. Bjorken, S.J. Brodsky, and R. Stroynowski
ISBN 3-7643-3079-1, 574 pages, hardcover

PPh5 Gauge Theories: Fundamental Interactions and Rigorous Results
P. Dita, V. Georgescu, R. Purice, editors
ISBN 3-7643-3095-3, 406 pages, hardcover